T0327501

Resilient Control Architectures and Power Systems

Resilient Control Architectures and Power Systems

Edited by Craig Rieger (Lead), Ronald Boring, Brian Johnson, and Timothy McJunkin

IEEE Press Series on Power and Energy Systems
Ganesh Kumar Venayagamoorthy, Series Editor

WILEY

Published by John Wiley & Sons, Inc., Hoboken, New Jersey.
Published simultaneously in Canada.

For general information on our other products and services or for technical support, please contact our Customer Care Department within the United States at (800) 762-2974, outside the United States at (317) 572-3993 or fax (317) 572-4002.

Wiley also publishes its books in a variety of electronic formats. Some content that appears in print may not be available in electronic formats. For more information about Wiley products, visit our web site at www.wiley.com.

Library of Congress Cataloging-in-Publication Data

Names: Rieger, Craig G., editor. | Boring, Ronald, editor. | Johnson, Brian (Engineering professor), editor. | McJunkin, Timothy, editor.
Title: Resilient control architectures and power systems / edited by Craig Rieger, Ronald Boring, Brian Johnson, Timothy McJunkin.
Description: Hoboken, New Jersey : Wiley-IEEE Press, [2022] | Series: IEEE press series on power and energy systems | Includes bibliographical references and index.
Identifiers: LCCN 2021035609 (print) | LCCN 2021035610 (ebook) | ISBN 9781119660415 (cloth) | ISBN 9781119660224 (adobe pdf) | ISBN 9781119660422 (epub)
Subjects: LCSH: Electric power distribution–Automation. | Electric power system stability. | Electric power failures.
Classification: LCC TK3091 .R47 2022 (print) | LCC TK3091 (ebook) | DDC 621.319–dc23
LC record available at https://lccn.loc.gov/2021035609
LC ebook record available at https://lccn.loc.gov/2021035610

Cover Design: Wiley
Cover Images: © Lisa-S/Shutterstock, JordiDelgado/iStockphoto, landbysea/Getty Images, William Kiestler

Set in 9.5/12pt STIXTwoText by Straive, Chennai, India

10 9 8 7 6 5 4 3 2 1

Contents

Foreword

With the growing dependence on control system technologies and concerns over stresses on existing energy infrastructure, specifically the automated operation of the power grid, the resilience of control systems to malicious and/or unexpected threats has received greater focus by the government. This attention includes the implementation of a smart grid, where existing loads can be more readily monitored and controlled, allowing the existing power generation to be operated more efficiently. However, the complexity generated with such extended monitoring requires a clear understanding of those system interactions, human and automated, which are necessary to bring resilience to the overall design. In addition, the cyber vulnerability of these systems has raised specific concerns, documented in many recent articles on state-sponsored attacks to electric power systems and similar infrastructure. It is therefore critical in the next generation of control systems that resilience plays a large role in its design and development. As a necessary contributor, the paradigm of education should reflect this need, and while other electrical engineering and computer science programs in the nation have included a cybersecurity perspective, few if any have focused on the unique control system aspects.

To address this need, this text is intended to provide a primer for universities to cultivate interdisciplinary teaming considerations for resilient control systems. While each undergraduate or graduate student will have in-depth knowledge in one discipline, each chapter is written at an entry skill level to enable greater comprehension and appreciation of other disciplines. The infrastructure domain considered is the power system, but the disciplinary aspects provide a basis for other infrastructure applications. Additional chapters were added at the end of the text to provide additional resilience metrics and design considerations as special studies.

Preface

There are seven Parts to this Book consisting of 19 chapters:

Part I, "Introduction," provides background on the definition of a resilient control system and its application to the power system use case.

Chapter 1: This chapter outlines a course designed to introduce students from multiple science and engineering disciplines to the challenges of automation in the power system. As more automatic control systems are applied, the resulting complexity and vulnerabilities increase the need for resilient control systems. A resilient control system "maintains state awareness and an acceptable level of operational normalcy in response to disturbances" [1]. The chapter also discusses the expected outcomes of the course.

Chapter 2: The electric power system is a fundamental infrastructure that is critical to everyday life. Resilient design and resilient control of the power grid are essential. This chapter will introduce the power system as a use case to demonstrate the concepts of resilient control system design. The use case will illustrate how power systems measurements and control are implemented using autonomous control devices, both in normal operation and during larger disturbances. Later chapters will show how modern control approaches can improve system resilience. The use case also considers the human system operator interface and the importance of applying human factors to allow automation to support the human operators to ensure the human-in-the-loop can concentrate on what humans do best. The use case also allows exploration of cybersecurity and cyber-defense concerns.

Part II, "Infrastructure Fundamentals," provides a background on the design of current power system designs, including the integrated control and communications systems.

Chapter 3: Power system architectures evolved over a period of more than 100 years. The power grid from the 1960s to the late 1990s will be referred to as a "traditional architecture." This architecture was the result of 60 years of gradual evolution. Recent decades have seen accelerating changes. Emerging trends, especially those driven by the significant increase of renewable power generation sources, the evolution of power markets, and the advent of microgrids, will be described. The chapter will discuss power systems operations and control, including the roles of human operators. In addition, the power system planning process will be introduced in this chapter, followed by a discussion of measures of operational performance used for transmission and distribution operations.

Chapter 4: The most important control of electric power generation is the inherent detection of load demand changes. The first response mechanism is to keep the production and the consumption operating in balance. The second response mechanism is to maintain the voltage level within tolerance for the operation of loads. The power and voltage control at generation units is a primary problem in power-system design. The control of individual generators has evolved into a hierarchical control for the management of large interconnections. Modern energy control centers command the generation levels and supervise the flow of power across the grid.

The control of alternative current (AC) power systems is benefited by the inherent ability of electric generation to detect the load demand changes without any communication and control infrastructure. The basic response mechanism to keep the balance between electric power production and consumption comes from the turbine-generator response to the conservation of energy. The control of generator units is the primary control problem of power systems. The methods developed for control of individual generators and of large interconnections play a vital role in energy control centers.

Chapter 5: The electric power utility system has, over the past several decades, become highly dependent upon high-speed, reliable communications systems. This evolution has gone from simple human-to-human communication for the manual operation of the system to a variety of systems and subsystems. These include systems such as Supervisory Control and Data Acquisition (SCADA), Distribution Automation (DA), system protection including specialized systems dependent upon communication, and more modern systems for security and surveillance, condition monitoring, asset management, and customer billing.

Part III, "Disciplinary Fundamentals," provides background on the unique disciplinary foundations that are brought to bear in this text.

Chapter 6: This chapter argues that an interdisciplinary education is critical to addressing the complex problems of today. Engineering curricula traditionally provide students with a broad education, but additional work must be done to help students appreciate the unique contributions of members of an interdisciplinary team. Because resilient solutions are not found in any one system, interdisciplinary teams are critical to success. Initiatives such as the Resilient Control Systems for the Power Grid course and the GridGame promote have been developed to help students understand multiple roles and perspectives within the resilience community.

Chapter 7: Cyber–Physical Systems (CPS) or Industrial Control Systems (ICS), such as the power grid and manufacturing plants, are systems that are comprised of an array of interconnected physical, control, computing, and networking devices. Often, such systems bear vulnerabilities in either their physical or digital components, which in turn may expose them to threats and render them susceptible not only to physical but also cyberattacks. In this chapter, we will examine the main elements of security within the context of ICS/CPS and focus on its cybersecurity aspects. We will analyze the main properties of cybersecurity, namely confidentiality, integrity, and availability, and study the most important technical mechanisms that exist to ensure these properties, including cryptography, authentication, authorization, accountability, access control, and redundancy. We will describe the common types of vulnerabilities in ICS/CPS and inspect the main stages of a cyberattack. We will also provide pointers of system design principles that must be followed during the various stages of the ICS/CPS lifecycle to increase their security. Finally, the most important approaches for threat and risk mitigation will also be outlined.

Chapter 8: Control Theory addresses the feedback principles of any dynamical system where the output is fed back via a controller for comparison with the desired input to make any necessary changes to satisfy the customer specifications. Dynamical systems exist in various forms such as linear or nonlinear, continuous or discrete, deterministic or stochastic, etc. The field of control systems has a long history dating back to 300 BCE when the Greeks invented a water clock and with a formal work on governors by James Clerk Maxwell in 1868, leading to classical control era (Routh-Hurwitz, Bode, Nyquist) and modern control era (Lyapunov, Pontryagin, Kalman, etc.). This chapter presents an overview of the theory and techniques arising in modern control systems such as optimal control, and briefly touch upon nonlinear control, adaptive control, intelligent control, etc. Any engineering system to be controlled needs to have three components of modeling, analysis or performance, and synthesis or design. Optimization is a very desirable feature in day-to-day life. We like to work and use our time in an optimum manner, use resources optimally, and so on. The main objective of optimal control is to determine control signals that will cause a process (plant) to satisfy some physical constraints and at the same time extremize (maximize or minimize) a chosen performance criterion (performance index or cost function). Thus, we address optimal control systems where the theory is rooted in the field of calculus of variations developed during sixteenth and seventeenth centuries over 300 years ago [2] and flourished right into the twenty-first century.

Chapter 9: This chapter reviews user-centered design for human–system interfaces of control systems. The premise of user-centered design is that the designer must consider the user, in this case the operator of a control system. User-centered design also advocates for iteration, in which feedback from operator testing is used to improve the design of the system. This chapter walks through the importance of keeping humans in the loop in control systems design and then outlines approaches for design planning, prototyping, and evaluation. It concludes with a checklist to help the control systems engineer follow a user-centered process in the design of human–system interfaces.

Part IV, "Metrics Fundamentals," establishes a basis for measuring success in the area of resilience.

Chapter 10: The improvement of resilience in electric power systems has been of growing importance in the United States for several years. Progress has been made in various areas, but much remains to be done in terms of the basic architecture of the power grid. A limiting factor has been the lack of a connection between foundational grid architecture principles and methods on the one hand and clearly defined relationships between resilience improvement objectives and actual means for assessing, planning, and implementing resilience measures on the other. At the core of this limitation is the need for principled definition, quantification, and valuation of the resilience impacts of grid architectures and architecture changes. The use of structural concepts provides a framework for these issues and provides a new means to obtain insight into how resilience may be analyzed and improved.

Chapter 11: To improve anything, there must be a way to assess its character with respect to the definition of the desired characteristics. With this in mind, a solid definition that people or organizations with an interest in the performance of the system, the stakeholders, must be stated. From there, a method of measuring the system against that definition needs to be created. It is often useful to express a concept in a notional manner but to put it into use the metric must be made tangible. This chapter will present a definition of the word resilience in the context of critical infrastructure. For that definition, a notional representation that has become common in the description of resilience that captures performance through time as it proceeds from an event that disturbs the system through the stages of resilience. Next, we will construct a tangible form of a metric to that can be used as a design tool to determine what improvements should be made to produce a more resilient system keeping in mind that cost is always a consideration.

Part V, "Resilience Application," provides a resilient control system perspective for application of disciplinary contributions, with the intent to evolve from multidisciplinary to interdisciplinary. A system application gaming environment provides a thoughtful means for students to apply these considerations.

Chapter 12: What is with a game in a resilience class? Well, there are many reasons to use a game for education. One, it is something for you to look forward to as part of a class. An event where you compete and cooperate with your fellow students can reveal more about resilience than reading papers, doing homework, or studying for an examination. The only thing that you might learn more from is creating your own game or really using a project development to dig into the understanding of a subject. The Grid Game has evolved from a simple swing equation simulation of the real power aspects of a microgrid to a multiplayer game that enables players to experience the impacts of unexpected events. As resilience is multidisciplined as you have been learning in this textbook, a simulated game gives you a chance to think about strategy and improvements to the human interaction with a system.

Chapter 13: Modern power grids rely heavily on TCP/IP networks to monitor and control physical processes. This reliance opens the door to potentially new and powerful cyberattacks against them. In this chapter, we introduce technologies that are used to operate the power grid and security challenges facing the power grid, present previous attacks, and discuss research efforts to improve the security and resiliency of the grid.

Chapter 14: In this chapter, we introduce methods to address resiliency issues for control systems. The main challenge for control systems is its cyber–physical system nature that strongly couples the cyber systems with physical layer dynamics. Hence, the resiliency issues for control systems need to be addressed by integrating cyber resiliency with physical layer resiliency. We introduce frameworks utilizing a games-in-games paradigm that can provide a holistic view of the control system resiliency and enable an optimal cross-layer and cross-stage design at the planning, operation, and recovery stage of control systems. The control systems are often large-scale systems in industrial application and critical infrastructures. Decentralized control of such systems is indispensable. We extended the resiliency framework to address distributed and collaborative resiliency among decentralized control agents.

Chapter 15: Technological advancements have resulted in highly critical infrastructure, which has increased the infrastructure's attack surface and made them more vulnerable to cyberattacks. The constantly evolving threat landscape and sophisticated attack vectors boasts intelligent and adaptive threat actors that can surpass traditional engineered and deployed defenses. A skilled cybersecurity workforce is essential; furthermore, there is an immediate need for anticipatory defense measures that reflect the adaptive and dynamic nature of the threat actors. Developing anticipatory cyber strategies require understanding the human aspects of cyberattacks: how adversaries organize, strategize, adapt, and function effectively, and how defenders secure grids and make effective decisions in cyber defense and system operation when experiencing cyberattacks. One effective mechanism to train the future workforce in this space is by gamifying cybersecurity.

Part VI, "Additional Design Considerations," as an optional chapter, includes considerations that extend the resilience considerations to different domains, consider interdependencies among infrastructures and provide some thoughts for the future of distributed control.

Chapter 16: Critical infrastructure is ubiquitous in modern societies and its reliable and resilient operation is of paramount importance to national security, economic vitality, and public confidence [3, 4]. The nation's critical infrastructure is diverse and complex. Electricity transmission and distribution networks, telecommunication networks, and transportation systems are common representative examples. Their high degree of inter- and intra-connectedness make them vulnerable to cascading disruptions when exposed to man-made or natural hazards. These critical infrastructures must be secure and able to withstand or rapidly recover from all hazards. Safeguarding the reliability of the nation's infrastructure will require a greater understanding of the complex interdependencies of these systems, from their subtle emergent behaviors to large-scale cross-sector consequences in an all-hazard environment.

In the context of this paper, all-hazard vulnerability analysis of critical infrastructure is assumed to be a quantitative process, used to facilitate risk-informed decision-making by identifying which infrastructures are susceptible to what hazards. The resulting outcome of this process is then used to reduce the probability of adverse events and mitigate their consequences, should they occur. Modeling and simulation play an important role in identifying, understanding, and analyzing these events and their effects on the robustness and resilience of the nation's critical infrastructure. This is because, for most scenarios, it is impossible or impractical to create experimental conditions to directly measure the effects of hazards on these complex and diverse systems.

Chapter 17: The idea of distributed control has been considered for decades and became the namesake of a certain type of ICS architecture released in the 1980s. While the digital components were, in fact, distributed throughout a facility, the concept of distributed control was not in play. Not just the dependence on a centralized set of operator consoles provided this limitation but also the ability to autonomously negotiate shifts in operations. Current ICSs are still dependent upon human input down to some common feedback loops, if not direct action. Evolving from current system designs to distributed control will require tiers of recognition and response, at the top providing the management and coordination, currently based in procedures and skill of the craft, but in future extracting the management parameters for operation (e.g. production rate) and engineering parameters through coordination of resources (i.e. settings that safely and efficiently transition the operations from one state to another). At the lowest tier, the execution layer provides a true time-based dynamic but taking intelligent instruction from the settings and autonomous control action in response. To establish resilience to threats, including cyberattack or damaging storms causing physical degradation, the tiers are decomposed into agents, which maintain state awareness and adapt to maintain the overall management philosophy. Even if the communications are lost, also a possibility with threats, those elements that survive can recognize and respond to maintain an optimized state. The result is a distributed and resilient control system.

Chapter 18: Previous chapters focus on resilient architectures for the electrical power grid. Emphasis on this "uniquely critical" infrastructure system is merited, and other critical infrastructures can also benefit from design of resilient control systems. This chapter discusses resilient design considerations that generally apply across a broad spectrum of critical infrastructures. The chapter introduces four resilient design capacities, that is, fundamental system attributes that contribute to or detract from resilient operations. The chapter also discusses design issues and system constraints that often need to be considered when balancing the capacities in resilient designs.

Part VII, "Conclusions," summarizes the book and challenges the students to consider the future and the new science of resilience.

Chapter 19: The previous chapters of this book take an interdisciplinary approach to discussing resilience in control systems, designed to encourage students from diverse disciplines to consider this critical concept. This chapter concludes that resilience is not a design layer. Instead, it is a philosophy. This chapter summarizes the challenges of designing resilient control systems and the relationship between humans and automation. Autonomy is not the final goal, but one tool to achieve a resilient system.

References

1 Rieger, C. (2010). Notional examples and benchmark aspects of a resilient control system. Resilient Control Systems (ISRCS). *3rd International Symposium* (August 2010), 64–71.

2 Sussmann, H.J. and Willems, J.C. (1997). 300 Years of optimal control: from the Brachys-Tochrone to the maximum principle. *IEEE Control Systems Magazine* 17: 32–44.

3 President's Commission on Critical Infrastructure Protection. (1997). Critical Foundations: Protecting America's Infrastructures the Report of the President's Commission on Critical Infrastructure Protection. United States. President's Commission on Critical Infrastructure Protection, Washington, DC.

4 The White House (1998). *Presidential Decision Directive 63*. Washington, DC: The White House.

Acknowledgments

This text is the culmination of a multi-university course that focused on the interdisciplinary considerations to achieve resilience. While much of the science is yet to be established for this area, we must thank the many professors that supported the course. We would also like to thank Idaho National Laboratory for its vision in the pioneering of this field of study and in the completion of the book writing effort.

Editors Biography

Craig Rieger
Chief Control System Research Engineering and Directorate Fellow

Craig Rieger, PhD, PE, is the Chief Control Systems Research Engineer and a Directorate Fellow at the Idaho National Laboratory (INL), pioneering interdisciplinary research in next generation resilient control systems. The grand challenge provided an integrated research strategy to address the cognitive, cyber–physical challenges of complex control systems into self-aware, trust-confirming, and threat-resilient architectures.

In addition, he has organized and chaired 13 co-sponsored symposia and one National Science Foundation workshop in this new research area and authored more than 70 peer-reviewed publications.

Craig received BS and MS degrees in Chemical Engineering from Montana State University in 1983 and 1985, respectively, and a PhD in Engineering and Applied Science from Idaho State University in 2008. Craig's PhD coursework and dissertation focused on measurements and control, with specific application to intelligent, supervisory ventilation controls for critical infrastructure.

Craig is a senior member of IEEE and has 20 years of software and hardware design experience for process control system upgrades and new installations. Craig has also been a supervisor and technical lead for control systems engineering groups having design, configuration management, and security responsibilities for several INL nuclear facilities and various control system architectures.

Ronald Boring
Distinguished Scientist, Human Factors and Reliability

Ronald Boring, PhD, is a Distinguished Human Factors Scientist and Department Manager at Idaho National Laboratory, where he has led research projects for the US Nuclear Regulatory Commission, NASA, the US Department of Energy, the Canadian Nuclear Safety Commission, the Department of Defense, and the Norwegian Research Council. He previously worked as a human reliability researcher at Sandia National Laboratories, a usability engineer for Microsoft Corporation and Expedia Corporation, a guest researcher in human–computer interaction at the National Research Council of Canada, and a visiting human factors scientist at OECD Halden Reactor Project.

Ronald and his research team developed the Guideline for Operational Nuclear Usability and Knowledge Elicitation (GONUKE) for conducting human factors in support of nuclear technologies, the Human Unimodel for Nuclear Technology to Enhance Reliability (HUNTER) dynamic human reliability framework, and the Advanced Nuclear Interface Modeling Environment (ANIME) for prototyping digital interfaces in nuclear power environments. Dr. Boring is the founder of the Human Systems Simulation Laboratory.

Ronald has a PhD in Cognitive Science from Carleton University, a Master's degree in Experimental Psychology from New Mexico State University, and dual Bachelor's degrees in Psychology and German from the University of Montana. He was a Fulbright Academic Scholar to the University of Heidelberg, Germany.

Ronald has published over 300 research articles in a wide variety of human reliability, human factors, and human–computer interaction forums. He is the founder and chair of the Human Error, Reliability, Resilience, and Performance conference, he was co-chair for the 2019 American Nuclear Society Nuclear Power Instrumentation, Controls and Human-Machine Interface Technology (ANS NPIC&HMIT) conference, and he is ongoing Chair for the Annual Meeting of the Human Factors and Ergonomics Society. He is a fellow of the Human Factors and Ergonomics Society.

Brian K. Johnson, PhD, PE

University Distinguished Professor, Schweitzer Engineering Laboratories Endowed Chair in Power Engineering

Brian K. Johnson, PhD, PE, is a University Distinguished Professor and the Schweitzer Engineering Laboratories Endowed Chair in Power Engineering in the University of Idaho Department of Electrical and Computer Engineering. Brian received BS, MS, and PhD degrees in electrical engineering from the University of Wisconsin-Madison in 1987, 1989, and 1992, respectively. He joined the University of Idaho shortly after completing his doctoral degree.

He was chair of the Department of Electrical and Computer Engineering from 2006 to 2012. His teaching and research interests include power system protection, power systems transients, HVDC and FACTS, and resilience controls for critical infrastructure systems. He has advised over 200 part-time and full-time Master's and doctoral students. He has published over 170 papers in journals and conferences.

Dr. Johnson was chair of the IEEE Power and Energy Education Committee from 2014 to 2015, and is currently the chair of the IEEE HVDC and FACTS subcommittee. Dr. Johnson is a registered professional engineer in the State of Idaho.

Timothy McJunkin

Distinguished Researcher, Power and Energy Systems

Timothy McJunkin is a Distinguished Research in the Power and Energy Systems Department of Idaho National Laboratory (INL). At INL since 1999, his current research and development interests include resilient control of critical infrastructure, Smart Grid for renewable energy integration, and cybersecurity. He has performed research in robotics and automation, intelligent systems, and acoustic-based nondestructive examination. Mr. McJunkin has published 20+ peer review journal articles, two book chapter and been awarded 13 patents on topics of computer systems, analytical chemistry instrument systems, industrial automation, Smart Grid, and nondestructive examination. He has served as an Adjunct Faculty member of Idaho State University Electrical Engineering Department and was a co-initiator of the multi-university class in resilient control systems, centered at the public Idaho universities. He is the architect and principal developer of the Grid Game. Prior to joining INL, he was with Compaq Computer Corporation's Industry Standard Server Group (1994–1999) leading board level motherboard design of multiple server products. He holds a Master of Science in electrical and computer engineering from Utah State University and is pursuing his PhD in the University of Idaho's Electrical Engineering Department. At Utah State he was awarded a Rocky Mountain NASA Space Grant Consortium fellowship for his work on autonomous planetary vehicles.

List of Contributors

Rômulo Bainy
Electrical and Computer Engineering
University of Idaho
Moscow, ID
USA

Thomas Baldwin
Electrical and Computer Engineering, Idaho
State University
Pocatello, ID
USA

Ronald Boring
Center for Advanced Energy Studies, Idaho
National Laboratory
Idaho Falls, ID
USA

Alvaro A. Cárdenas
Computer Science and Engineering, University
of California
Santa Cruz, CA
USA

Chris Dyer
SCADA and Analytical Services, Power
Engineers
Meridian, ID
USA

Ryan Hruska
National and Homeland Security, Idaho
National Laboratory
Idaho Falls, ID
USA

Brian K. Johnson
Electrical and Computer Engineering
University of Idaho
Moscow, ID
USA

Constantinos Kolias
Computer Science, University of Idaho
Idaho Falls, ID
USA

Keerthi Koneru
Computer Science and Engineering, University
of California
Santa Cruz, CA
USA

Daniel Conte de Leon
Computer Science, University of Idaho
Moscow, ID
USA

Kelvin Mai
Computer Science, University of Texas-Dallas
Dallas, TX
USA

Georgios M. Makrakis
Computer Science, University of Idaho
Idaho Falls, ID
USA

Timothy McJunkin
Energy and Environment Science and
Technology, Idaho National Laboratory
Idaho Falls, ID
USA

Desineni S. Naidu
Electrical Engineering, University of
Minnesota-Duluth
Duluth, MN
USA

Neil Ortiz
Computer Science and Engineering, University
of California
Santa Cruz, CA
USA

Xi Qin
Computer Science and Engineering, University
of California
Santa Cruz, CA
USA

Aunshul Rege
Criminal Justice, Temple University
Philadelphia, PA
USA

Craig Rieger
National and Homeland Security, Idaho
National Laboratory
Idaho Falls, ID
USA

Meghan G. Sahakian
National Security Programs, Sandia National
Laboratories
Albuquerque, NM
USA

Jeff Taft
Energy and Environment, Pacific Northwest
National Laboratory
Richland, WA
USA

Eric D. Vugrin
National Security Programs, Sandia National
Laboratories
Albuquerque, NM
USA

Quanyan Zhu
Electrical and Computer Engineering, New
York University
Brooklyn, NY
USA

Part I

Introduction

Resilient Control Architectures and Power Systems, First Edition.
Edited by Craig Rieger, Ronald Boring, Brian Johnson, and Timothy McJunkin.
© 2022 The Institute of Electrical and Electronics Engineers, Inc. Published 2022 by John Wiley & Sons, Inc.

1

Basis, Definition, and Application

Craig Rieger

National and Homeland Security, Idaho National Laboratory, Idaho Falls, ID, USA

1.1 Introduction

As energy companies and governments attempt to get more from the existing power grid and other critical infrastructures, more automatic control systems are being applied [1, 2]. With this greater reliance on network-based, digital automation, and the stresses of pressing the existing infrastructure for greater performance, the power grid and underlying systems have become more susceptible to both malicious attacks and unexpected, natural threats. Governments and other stakeholders have chosen to address infrastructure issues by the implementation of a smarter grid. In the smart grid, operators and control systems supervise power generation, distribution, transmission, and loads to utilize these assets most efficiently [3]. Such extensive monitoring and control over a distributed system cause complexity that challenges system designers and human operators in new ways. In addition, cyber vulnerability of these systems has been illustrated in many recent articles on state-sponsored attacks to electric power systems and other similar infrastructure for natural gas, water, and communications [4]. Therefore, it is critical in the next generation of control systems that resilience plays a large and critical role in the grid design and development. Resilient control systems are a field of research that seeks solutions to complexity through a holistic approach that combines cognitive science, computer security, communications, and control systems. To enable future researchers and practitioners to assist with designing more resilient systems, science, technology, engineering, and mathematics education needs to incorporate interdisciplinary topics. While electrical engineering and computer science programs in the nation include a cybersecurity perspective, few if any have focused on the unique control system aspects. Human cognitive aspects are most definitely not addressed in technology education discourse. To this end, a class and education tools in resilient controls systems have been created.

1.2 Definition and Application

Benefiting from an already ongoing interdisciplinary field of study [5], a course was created to establish a perspective for college students on the unique challenges of automation in our society. The course was broadcast to the participating universities through interactive web-based lectures and provides perspectives based upon the definition of a resilient control system, which is one that "maintains state awareness and an acceptable level of operational normalcy in response to

Resilient Control Architectures and Power Systems, First Edition.
Edited by Craig Rieger, Ronald Boring, Brian Johnson, and Timothy McJunkin.
© 2022 The Institute of Electrical and Electronics Engineers, Inc. Published 2022 by John Wiley & Sons, Inc.

disturbances, including threats of an unexpected and malicious nature" [5]. The course was first organized in the fall 2013 as a series of lectures in resilient controls, without a central application theme. The course was refined for fall 2014 to include institutions outside of Idaho and incorporate a focus on the application of power systems. Lecturers and faculty from Idaho National Laboratory and participating university institutions covered the disciplinary contexts, with a mentor at each institution to facilitate the student questions and grading as part of a special topics or catalogue course.

Resilient control systems architecture, as shown in Figure 1.1, offers additional perspective on a subset of interdisciplinary topics that impact real-world critical infrastructure. The course addressed how systems fail due to threats from cybersecurity, human error, and complex inter-dependencies, and how the application of resilient control system technologies addresses these challenges. The broad range of topics in resilient control systems would typically be addressed in different courses and in different departments or colleges. When taught together, a course becomes relevant to multiple engineering and science disciplines, drawing students into the sometimes challenging but equally rewarding interdisciplinary conversation. The course has the potential to lead to the desired academic and social outcome of more broadly developed engineers and scientists with the ability to connect the "languages" of the distinct disciplines to tackle increasingly coupled problems in complex systems.

The power grid was chosen due to its importance to the support of modern society, the distributed and complex nature of the control systems, and the current and planned efforts to modernize through smart grid initiatives. The goal of the course is for students from multiple disciplines, ranging from college juniors to graduate students, to arrive at an intuitive perspective on the control, human, and cybersecurity aspects of the electric grid. Understanding of the multiple challenges and failure modes in critical infrastructure is performed through weekly sessions in a semester-long course. The weekly sessions cover a survey of resilient control topics as well as sufficient background discussion on the electric power grid to prime students from a variety of levels in engineering studies for the discussions.

This book is organized in the sequence that discussions would be expected to occur, with the exception of Part VI, which provides additional special topics that could be addressed as desired. The focus of each book section is provided below:

Part I. The introduction and use case for reference in the remainder of the book.

Part II. Overviews of the power system infrastructures that would be recognized in practice in the community, including the power grid, control system, and communications architectures.

Part III. Disciplinary fundamentals for the student for each of the primary disciplinary considerations considered on resilient control.

Part IV. For relevance, metrics are required for measurement of benefit and success, not unlike those know for reliability. This section will provide a differentiation of how resilience is quantified and valued.

Part V. Building upon the fundamentals and the means to measure, this section provides the interdisciplinary challenges with examples of applications that can be addressed to achieve resilience.

Part VI. Additional design considerations provide a basis for other factors that influence the resilience of control systems, specifically in addressing the current complexity and the future of systems that are designed to engender resilience and prevent brittle failures.

Part VII. Concluding the book will be a summary and a brief overview of interdisciplinary research challenges, borne out in the current understanding and addressed as this foundational area matures.

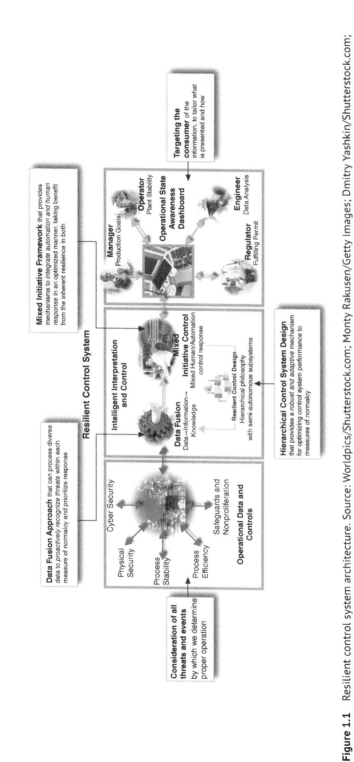

Figure 1.1 Resilient control system architecture. Source: Worldpics/Shutterstock.com; Monty Rakusen/Getty Images; Dmitry Yashkin/Shutterstock.com; ritanan/Getty Images; Marko Rupena/Getty Images; Alexey Stiop/Shutterstock.com; pryzmat/123RF.

Each chapter will provide objectives and overview of the particular topic. Designed to be readable across disciplines, the chapters are written at a high level with additional references provided for future reading. Questions are provided to aid the student in testing comprehension of the main points of the chapter.

References

1 Cecati, C., Mokryani, G., Piccolo, A. et al. (2010). An overview on the smart grid concept. IECON 2010 – 36th Annual Conference on IEEE Industrial Electronics Society (November 2010), 3322–3327.

2 Shladover, S. (2007). PATH at 20 – history and major milestones intelligent transportation systems. *IEEE Transactions on Intelligent Transportation Systems* 8 (4): 584–592.

3 Sridhar, S., Hahn, A., and Govindarasu, M. (2012). Cyber-physical system security for the electric power grid. *Proceedings of the IEEE* 100 (1): 210–224.

4 Bradley, T. (2010). Critical Infrastructure under Siege from Cyber Attacks, PC World. http://www.pcworld.com/article/188095/Critical_Infrastructure_under_Siege_from_Cyber_Attacks.html (accessed 30 August 2021).

5 Rieger, C. (2010). Notional examples and benchmark aspects of a resilient control system. Resilient Control Systems (ISRCS), 2010 3rd International Symposium, 64–71.

2

General Use Case Introduction

Brian Johnson

Electrical and Computer Engineering, University of Idaho, Moscow, ID, USA

2.1 Introduction

Resilient control approaches can be applied in any cyber-physical system that utilizes distributed communication and control architectures that provide automation support to enable human operators to better run systems. Such applications can range from individual industrial facilities to power grids that span nations.

The authors in this book use the power grid as an use case example to illustrate the key aspects of modern resilient control design. Resilient control systems combine communication-enhanced distributed control, improved operator support through human factors engineering, and applied cybersecurity concepts. All these are aspects are critical to modern power systems operations. Improving grid resilience is of national significance and is an active area of research.

2.2 Importance of Resilient Controls for Power Systems

The power grid was chosen as the use case for this book since most people have some understanding of the pervasiveness of electrification in modern society and the importance of power system resilience.

Most aspects of daily life are affected by the power infrastructure, including heavy industries, commercial sectors, health systems, and residential sectors. The interconnected power system of the United States and Canada is one or the largest and most critical infrastructures in the world [1].

The power infrastructures of most nations are large, complex systems that are of critical importance to a nation's financial well-being. They are large, interconnected networks that can span continental distances, and local events can have far-reaching impacts. The Northridge earthquake in 1994 caused power outages over 1000 miles away. A little over a year later, an incorrect protective device response to a short circuit in Idaho caused millions of customers in southern California to lose power.

2.3 Power Systems Operations and Control

Power systems operations utilize a decentralized control scheme with human operators playing a critical role. Generation sources and end use load points are interconnected using transmission and

Resilient Control Architectures and Power Systems, First Edition.
Edited by Craig Rieger, Ronald Boring, Brian Johnson, and Timothy McJunkin.
© 2022 The Institute of Electrical and Electronics Engineers, Inc. Published 2022 by John Wiley & Sons, Inc.

distribution lines, as will be discussed in Chapter 3. Measurements from devices in the substations are communicated to control centers owned by utilities. Most utilities have developed an extensive communications infrastructure over the past 50 years. The measurements are processed in a control center to correct errors and focus what is presented to human operators. Operator responses are communicated back to components in the system to take actions such as changing generator set points or changing switch states [2].

The control of the power system takes place over a range of time scales. Operators largely respond to slowly changing conditions with time scales ranging from tens of minutes to hours, with support from computer simulation and automation tools. Faster disturbances are dealt with by autonomous controls. The fastest controls use local measurements in a substation, possibly enhanced with measurements from neighboring substations. Examples include protections systems that respond to faults with time responses on the order of tens of milliseconds. Other autonomous control schemes respond over time periods of seconds to minutes, which often encompass larger areas.

Most of these measurement, protection, and control systems are implemented in intelligent electronic devices (IEDs). These devices are special purpose computing platforms. IEDs can communicate measurements to the control center or receive commands from operators or control devices. In many cases, these devices have communication interfaces that allow remote access to modify settings.

Many utilities maintain multiple communication systems which overlay the power system, as shown in Figure 2.1. The system is often referred to as the operational system. Historically, the operational technology communication networks were isolated from the outside Internet. However, the utility control center often forms several bridges between the enterprise system and the utility uses for financial operations, such as retrieving information for billing from the operational network. Other communication links allow engineering access for engineers in the headquarters to read event logs from protection and control devices. These bridges open possible cyber-vulnerabilities, which can compromise the resilience of the power grid. Cybersecurity

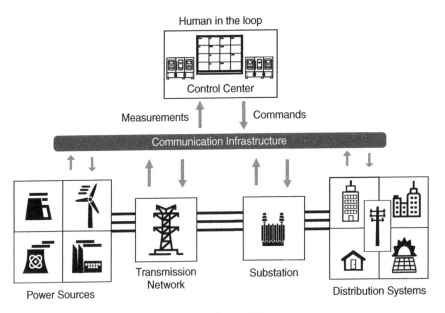

Figure 2.1 System power and communication architecture.

is becoming an increasing concern for power systems operations. This book introduces basic concepts of power grid cybersecurity.

Chapter 12 will introduce the Grid Game, a simulation environment that allows students to learn to operate a small electric power system and apply the concepts presented in this book. Teams of students control their small power systems through an interactive human machine interface and compete with their classmates to see who can operate their system in the most cost-effective manner [3]. The power system operator teams have opportunities to invest in cyber defenses. Meanwhile, a red team made up of other students in the class or a course instructor launch cyberattacks against the different power system operator teams. The operators respond to the attacks, in some cases needing to reset their control infrastructure to restore operation. The Grid Game enables students from multiple disciplines to apply what they learn in the class [4].

2.4 Summary

The goal of this book is for students, ranging from college juniors to graduate students from multiple disciplines to develop an understanding of the resilient operation and control of the power grid from the perspective of combined autonomous and human-based distributed control, human factors, and cybersecurity aspects.

References

1 Marston, T.U. (2018). *The US Electric Power System and Its Vulnerabilities*, vol. 2, 48. The Bridge, National Academy of Engineers.

2 Thomas, M.S. and McDonald, J.D. (2015). *Power Systems SCADA and Smart Grid*. Boca Raton, Florida: CRC Press.

3 McJunkin, T.R., Rieger, C.G., Johnson, B.K. et al. (2015). Interdisciplinary education through "Edu-tainment" electric grid resilient control systems course. 2015 ASEE Annual Conference and Exposition, Seattle, Washington, ASEE Conferences (June 2015). https://peer.asee.org/24349 (accessed 31 August 2021).

4 Rege, A., Parker, E., and McJunkin, T. (2017). Using a critical infrastructure game to provide realistic observation of the human in the loop by criminal justice students. 2017 Resilience Week (RWS) (September 2017), 154–160. https://doi.org/10.1109/RWEEK.2017.8088665.

Part II

Infrastructure Fundamentals

Resilient Control Architectures and Power Systems, First Edition.
Edited by Craig Rieger, Ronald Boring, Brian Johnson, and Timothy McJunkin.
© 2022 The Institute of Electrical and Electronics Engineers, Inc. Published 2022 by John Wiley & Sons, Inc.

3

Power Grid Architecture

Brian Johnson and Rômulo Bainy

Electrical and Computer Engineering, University of Idaho, Moscow, ID, USA

Objectives

This chapter describes the basic architecture of the power grid and differentiates the predominant power architectures of previous decades from emerging ones, which are broadly classified as smart grids. This chapter provides overviews of power system configurations, power systems operations, and of the power system planning process. Readers will also learn measures of power system performance.

3.1 Introduction

Electricity is of critical importance to modern society. The development of the electric power system began in the late 1800s where early applications using electric power included electric lighting and street cars. The number of applications utilizing electricity increased significantly over time, supporting the development and growth of many everyday applications taken for granted in modern society, such as electric lighting, home appliances, television, the phone system, and the Internet to name a few. As a result, electrification was listed as the most significant engineering development of the twentieth century [1]. Electrification impacts all parts of modern society, and interconnected electrical systems cover large parts of most continents.

Large and interconnected power systems are among the most complex systems humanity has built. Modern power systems are controlled by human operators with critical support from computer systems, autonomous intelligent control devices, and communication networks. The decision-making reaches high complexity because of the continental scale of the power grid, the large number of nonlinear devices, and the occurrence of unpredictable events. The system operational plans also account for the impacts of seasonal variations. For example, water resources for hydroelectric generation are limited during dry seasons, and summer's high temperatures may limit the power flow in transformers, transmission lines, and other devices.

Power systems have experienced a drastic revolution in the last 50 years. Computer simulation tools enabled power system planners and operators to optimize the behavior of large interconnected systems. The availability of fast and reliable communication networks allowed systems to be operated closer to their limits while also increasing efficiency and system reliability. The introduction of intelligent electronic devices (IEDs) enabled smarter protection functions, which can detect and

Resilient Control Architectures and Power Systems, First Edition.
Edited by Craig Rieger, Ronald Boring, Brian Johnson, and Timothy McJunkin.
© 2022 The Institute of Electrical and Electronics Engineers, Inc. Published 2022 by John Wiley & Sons, Inc.

isolate faults faster; therefore, increasing the ability to maintain stable operation at higher levels of system loading. Grid applications of power electronics became more common, resulting in more flexibility and faster control for the system operator. The system topological architecture evolved to meshed systems spanning thousands of miles.

Communication systems help the operator make informed decisions that improve efficiency of operation and allow stable operation of systems with increasing levels of renewable generation. Supervisory control and data acquisition (SCADA) systems continuously monitor conditions at substations throughout the system and send commands back to the remote stations. Communication systems also enable faster detection and response to short circuit faults. Recent trends in communication systems with higher data rates and time-aligned wide-area measurements enable decentralized autonomous control systems to further improve reliability and resilience of the power system. The development of power semiconductor switching devices for high-voltage high-power applications substantially enhanced the control of transmission systems and enabled efficient integration of wind generation and photovoltaic (PV) generation. High-voltage direct current (HVDC) transmission systems enable transmission of high levels of power over longer distances than would be possible for alternating current (AC) lines. Flexible alternating current transmission system (FACTS) devices can provide dynamic reactive power support, improving the margin to maneuver to improve system performance and resolving transmission bottlenecks.

This chapter will provide an overview of classical power system architectures of the past 50 years and discuss recent trends toward smarter power grids. Power systems operations will be described, along with the typical planning process. This chapter will set the stage for later developments in this book.

3.2 Classical Power System Architectures

The first power distribution systems used direct current (DC), with DC generators supplying DC loads through DC distribution lines. Power generation, distribution, and end-use were all at the same low voltage since there were no efficient ways to step voltage up or down. The voltage was determined based on the requirements for lighting, which formed a significant fraction of the load in the earliest systems. Distributing larger current levels as load increased or distributing power over longer distances resulted in unacceptable voltage drops, limiting power rating, and geographical dimension of early systems. The modern power infrastructure would be very different if DC systems had become the norm, with many widely distributed small systems, probably with limited interconnections between systems.

The development of AC systems shortly after the introduction of the DC systems presented an alternative approach for system design and operation. In an AC system, the generator voltage varies sinusoidally with time and the load current will exhibit the same frequency. The use of AC systems facilitated several key innovations that led to the development of the modern power system. The first was the transformer, which provided the ability to step up the voltage to a higher level by using a fixed ratio while simultaneously stepping down the current by the same ratio. The use of transformers led to a second innovation: the development of transmission lines, which operated at higher voltage levels with lower currents levels. High-voltage transmission lines allowed more-efficient long-distance power transfers because of the reduced voltage magnitude drops and reduced energy losses. The use of high-voltage transmission lines allowed larger generators to be built at remote locations. The first AC power plant built at Niagara Falls demonstrated the economic benefits of AC systems leading to the power systems of today [2].

The early power system operators developed a business model where the same organization owned power generation, transmission, and distribution. This model subsequently spread across the United States (US) and around the world. Local utilities gradually expanded and grew to cover larger territories. A financial model based on customers paying for energy usage was implemented. The inefficiencies of competing companies trying to build transmission and distribution facilities in the same neighborhoods led to the development of state-regulated monopoly franchises with each franchise having a specific service territory. The power providers were a mix of private investor-owned utilities, municipal government-owned companies, and rural electrical cooperatives.

Each utility owned a mix of generation sources to meet their local loads. Several early systems used hydroelectric generation. Steam boilers burning coal and oil added more flexibility for locating power generation. Expanding transmission systems allowed utilities to expand to spread generation and loads over larger territories. Utilities added connections with their neighbors to increase reliability when generation was disconnected due to faults or unscheduled maintenance. This web of interconnections led to the emergence of large interconnection transmission systems that spanned much of the United States and southern Canada.

Interconnecting systems created vulnerabilities for cascading outages. The November 1965 blackout in the northeast showed the risks of creating large interconnected systems. The Federal Power Commission (predecessor to the Federal Energy Regulatory Commission [FERC]) recommended the formation of a power coordination council, leading to the establishment of the North American Electric Reliability Corporation (NERC). Some of NERC's roles are to develop planning standards, recommend operating policies, and compliance requirements. Regional coordinating councils under NERC work with utilities to ensure reliable operation of the interconnected systems.

The power system infrastructure is primarily associated with power generation, transmission, and distribution to enable power delivery from generation to end users. Conventional power plants produce large amounts of energy through hydro and steam turbines. The former generates electricity by producing energy from water spinning a rotating shaft. The latter uses steam created by burning different types of fuels, such as natural gas, oil, coal, or through nuclear fission.

To supply the distant consumers, substations located at power plants use transformers to raise the voltage levels of hundreds of kilovolts (kV) so the energy can travel over long distances through high-voltage transmission lines. Transformers at substations installed close to a consumption center reduce voltage levels to tens of kilovolts for delivery through distribution lines. Large consumers, such as industrial facilities, may have dedicated substations. Some install backup generators. The classical power grid is divided into four main parts: power generation, transmission, distribution, and end users, as shown in Figure 3.1.

The power system can also be defined in terms of its architecture or topology, which depends on financial limitations and technical aspects [3]. Radial and meshed topologies are the two main types of architectures, as shown in Figure 3.2. Energy is generated at lower voltage levels, ranging from a few thousand volts to 24 kV. Transformers step voltage up to hundreds of kilovolts for transmission. Increasing the voltage decreases the current, which decreases transmission losses for long distance transmission. Closer to its end-user load centers (e.g. cities), the voltage is reduced using transformers until it reaches final users such as homes and businesses. Distribution systems typically have a radial layout as shown in Figure 3.2a with power flowing from a substation out to individual loads such as homes. A short circuit fault on a distribution feeder is interrupted by circuit protection equipment that opens a circuit breaker (switch). But this action also results in a power interruption for everything downstream of the open switch. Most distribution systems have

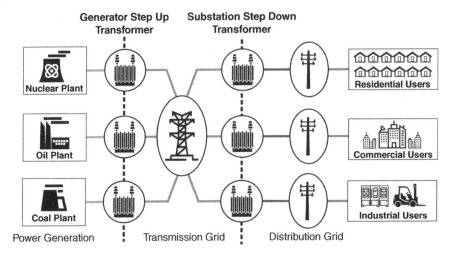

Figure 3.1 Classical power system.

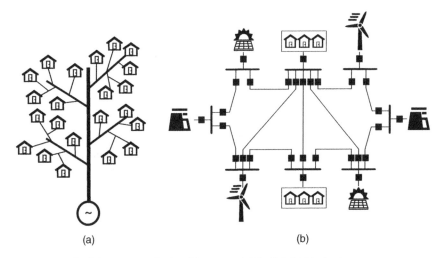

Figure 3.2 Radial versus meshed architectures. (a) Radial, (b) Meshed.

ability for manual reconfiguration to supply loads from an alternate feeder following an outage, typically with response times between 30 minutes and 1 hour for crews to reach the location.

To achieve better reliability, the transmission grid uses a meshed topology, as shown in Figure 3.2b. This approach results in redundancy, thus allowing the isolation of the failed part of the system while maintaining supply to most or all the consumers. However, a meshed topology is more complex to protect and operate since it requires circuit breakers at both ends of each line. The higher reliability of meshed connections is cost-effective for transmission systems, but in most cases, it is not cost-effective for distribution systems. However, meshed distribution architectures are used in some dense urban areas.

3.3   Emerging Architecture Trends

Grid architecture [4, 5] is one of the main tools to describe, define, and understand an electrical power grid. It consists of the application of network theory, system architecture, and control theory to the power system and is the highest-level description of the electrical power infrastructure. The abstracted model obtained through grid architecture is used to understand the behaviors and predict the characteristics of complex power systems. Two trends of interest for the remainder of this book are the emergence of smarter grids and growing interest in microgrids.

3.3.1   Smart Grids

An emerging trend among power systems architecture is commonly referred to as the smart grid [6–8]. While the term smart grid was coined as a term for government policymakers, it has also caught on in the power engineering research literature. The present-day concept of a smart grid is the outcome of 50 years of accelerating integration of digital communication and control technologies in power systems operations. It represents a considerable change in how engineers deal with energy production and delivery, coupled with the growth of financial energy market resulting from changes to the regulatory environment. A smart grid consists of a deep integration of different types of technologies that result in savings for many aspects such as operation costs, maintenance costs, and capital expenditures, while providing environmental and customer benefits [9]. Figure 3.3 shows a key point differentiating a smart grid from the classical power grid; a smart grid consists of advanced integration between communication and electrical systems supporting increased use of information technologies in control and automation [10]. The greater collaboration of those systems results in the modernization of all parts of the power grid, including distribution, transmission, and generation. The initial installations of automated communication and control systems started 50 years ago, with the advent of SCADA systems with digital protection and control technologies appearing in the following decade. Decreasing capital costs for information technologies combined with significant improvements in processing and communication speeds have led to a rapid increase in the deployment of new technologies in the last 20 years. A smarter grid enables optimized operation and better use of assets, such as water and fuel [11]. Improved communication and control allow the grid to support high levels of renewable generation such as wind and PV. While hydro generation is typically a renewable generation resource, there is little growth in hydro generation in much of the world. The discussion to follow will concentrate on wind and PV

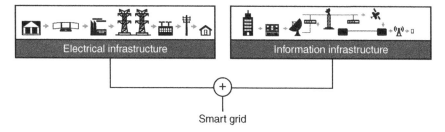

Figure 3.3   Communication and electrical Infrastructure integration to create smarter grids.

when referring to renewable generation. The connection of distributed energy storage installations enables optimal use of renewable generation that cannot regulate their output in the same fashion as primary generation sources.

The transmission system of a smart grid is more complex and more interconnected than the traditional system; it supports a greater proportion of renewable energy generation. The smart transmission grid can be divided into substations, transmission network, and control centers. The smart substations are increasingly digitalized and equipped with at least one local area network (LAN). In a digital substation, current and voltage measurements are converted into digital packets of information by merging units (MUs). Modern families of standards, such as IEC-61850, present a philosophy for achieving a smart substation by defining interoperability frameworks for equipment from different vendors to work together. The installation FACTS devices and HVDC transmission substantially increase reliability and flexibility. The modern control center is equipped with high-end state estimator modules that process the data collected via SCADA and remote terminal units (RTUs) to provide the operator with state awareness. These data are supplemented with increased measurements from smart meters at end user sites and time-synchronized measurements in substations. The utility control centers send their state measurements down to distribution control centers and up to regional independent system operators. Distribution operators use data visualizations and identification of patterns supported by frameworks like the geographical information system (GIS) data to enable faster, more-accurate dispatch of repair crews [12]. In some regions of the United States, independent system operators coordinate transmission system operation to facilitate energy markets.

The distribution system of a smart grid is substantially different from the classical model where all the consumers were loads. With the growth of renewable energy sources such as rooftop PV generation, consumers can become producers as well. Smart meters at end user points track this bidirectional flow of energy and can facilitate time-of-use billing for energy use or for sales of energy to the grid. In addition, smart meters at the end-user sites provide the distribution system operator with immediate information about faults, enabling fast autonomous reconfiguration and reducing outage times from tens of minutes to seconds or less. The modern smart grid has two-way communication throughout the system. An example of smart grid is shown in Figure 3.4.

The quality of energy supply from emerging renewable sources such as wind or PV generation is less controlled than classical thermal and hydro plants, mainly because of the desire to extract the maximum energy from the renewable resource coupled with unpredictable fluctuation due to weather conditions [6]. In the classical grid, generation is controlled, and load is an uncontrolled variable where generation is adjusted to maintain a balance between generation and load. With increased use of renewable generation, a growing portion of the generation is also uncontrolled, creating challenges for power system control. Frequency regulation is a challenge for operation since there are unpredictable variations in the output large-scale renewable generation and with demand for loads that have their own PV generation. This has led to an increased use of autonomous response schemes for scenarios that unfold too quickly for an operator to respond.

Renewable energy portfolio standards (RPSs) are created at government levels to motivate integration of renewable sources into the power grid. They include initiatives, such as tax credits and feed-in tariffs, to encourage higher penetration of renewables. One of the main concerns with connecting high proportions of wind and solar energy are their intermittent generation patterns. Integration of large levels of renewable energy requires increased use of smart grid technologies

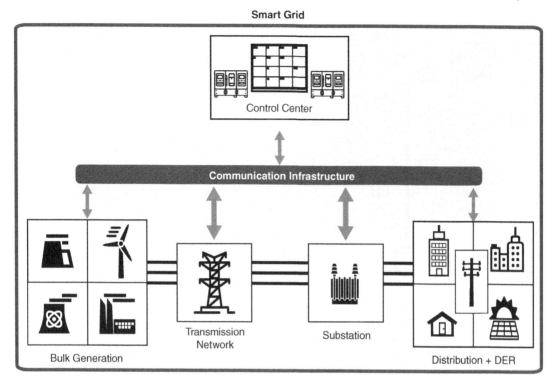

Figure 3.4 Example of a smart grid.

and has led to deep changes in planning, operation, and protection to maintain reliable operation of the grid. Some of the effects of high penetration of wind and PV generation include:

- Decreased short circuit current levels complicating fault detection and isolation.
- Decreased voltage quality.
- Reduced stability margins.
- Requirements to ensure renewable generation stays online following disturbances (fault ride through).

The classical power grid has a relatively small number of power plants owned by utilities that are largely located relatively far from load centers. Regulatory changes coupled with decreased capital and operating costs for smaller generators, such as rooftop PV, enabled moving to a mix of centralized generation with an increasing amount of third party owned distributed generators (DGs). This trend started with legislation to encourage industrial facilities to self-generate as the construction of large power plants and long transmission lines became less desirable as environmental concerns increased [6]. Therefore, distributed energy resources (DERs), such as DGs or distributed storage, located closer to the consumers became more popular, either utility owned DER or DER owned by independent power producers (IPPs). The DGs are often connected to the distribution system as shown in Figure 3.5. The differences between centralized generation and DER generation are that

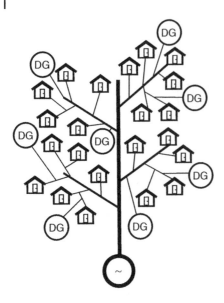

Figure 3.5 Distribution grid with high penetration of DERs.

the latter have smaller capacity, are closer to loads, do not contribute to frequency regulation and have limited ability to provide voltage regulation. Another challenge is the fact that the distribution network was designed to operate as a radial system with energy coming from the connection to the transmission system. Adding DER to distribution systems impacts protection and operation philosophies, requiring installation of more capable protective relays and additional circuit breakers [8, 12]. The utilization of the added smart grid communication technologies helps provide operators with situational awareness that can facilitate the integration of intermittent power generators.

As shown in Figure 3.4, communication systems are present in all parts of the power system, from generation to the consumers. The smart grid has an increased reliance on the communication system when compared to the classical power system; new techniques and software must be developed to secure the electrical grid from cyberattacks [6], as will be discussed in detail in later chapters. The design of the communication network requires end-to-end reliable and secure communication with adequate bandwidth and low latency.

The power system can be defined in terms of reliability and resilience. According to the US Department of Energy (DOE), reliability is the ability of the system to face unanticipated loss of system components or uncontrolled events without cascading failures or loss of stability [13]. Reliability is a binary way to measure the performance of a system where components are either functional or failed. Resilience is defined by the DOE as the ability of the system to adapt to changing conditions and to withstand and rapidly recover from disruptions. A given system can be considered more resilient than another if, during an extreme event, it can adapt while maintaining a higher level system performance than a less resilient system and then recovering to near normal operation more quickly, as shown in Figure 3.6.

3.3.2 Microgrids

A microgrid is a made up of a group of interconnected DERs and interconnected loads with clearly defined boundaries that can be connected to a larger power grid and can also cleanly separate to operate as an isolated islanded system [14, 15]. The connection point between a microgrid and

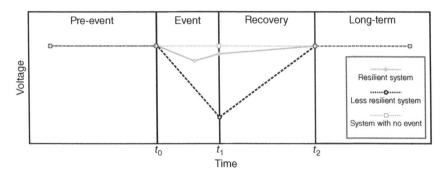

Figure 3.6 Comparison of the resilience of two systems.

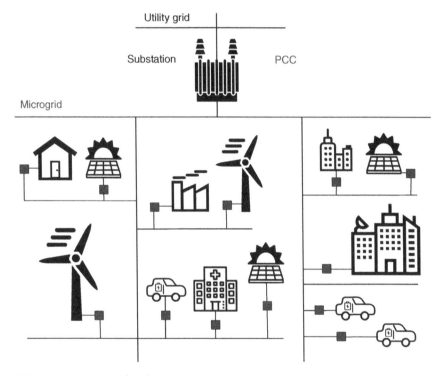

Figure 3.7 Example of a microgrid.

the larger power system is a point of common coupling (PCC). When the microgrid is connected to the main power grid, it can appear either as uncoordinated energy sources and loads or as a coordinated entity which manages its supply or demand facilitated by a microgrid controller. The microgrid controller enables the system to isolate from the main grid and continue to supply loads with local generation following a major disturbance on the main grid. Many researchers define microgrids as operating at distribution voltage levels, but some existing microgrids include portions of a transmission system as well. Microgrids are under quick development because they present environmental incentives and enhanced efficiency, and in many situations are more reliable. An example of a microgrid is shown in Figure 3.7.

One of the main challenges faced by the microgrid is maintaining the balance between load and generation when operating in isolated mode. Unintentional islanding needs to be detected to ensure protection and safety of personnel during emergency maintenance. A controlled transition to islanded operation may require changing control settings for generation, shedding lower priority loads, and changing protection settings. A key component for some microgrids is energy storage units because they compensate for load variations and the fluctuations of intermittent renewable sources [16]. The location and capacity of those energy storage units can be designed according to a combination of reliability/resilience indices, the location of renewable generation sources, and locations with available space.

The control systems of a microgrid [9, 17] can be divided into different levels. The first level is local controllers for individual components, such as energy storage controllers, protective relays, and generation controllers. They respond to control signals sent by the microgrid controllers and are designed to react to real-time events, such as change of frequency and detection of faults. The second level is the microgrid controller, which monitors and controls all components of a microgrid. The maintenance of reliability and power quality are the main objective of the microgrid controller [16]. Some microgrids have a third level of control in the form of a master energy management system (EMS), which deals mainly with the optimization and financial aspects of operating the system when connected to the main grid. This controller adjusts the setpoints for controllable load and generation to optimize efficiency based on energy prices and generation cost.

Types of microgrids include institutional microgrids (hospitals, governmental buildings), commercial/industrial (malls, factories), and community grids (neighborhoods, apartments). The customers can be segmented as [17]:

- Blue Ocean: regions not connected to the main grid of a country.
- Network relief: areas that demand increased stability but are connected to grids that are saturated.
- Energy security: segments where stable and high-quality energy are strategic or can save lives.

Controllable load is also a key part of smarter grids. Classical power grid control includes undervoltage or under frequency load shedding that trips predefined blocks of loads to maintain stability of the system. Some customers have contracts that allow them to be interrupted for limited periods of time in exchange for paying lower rates. Modern demand response schemes create a partnership between the end user and the utility where building EMSs or residential smart homes can be configured to vary demand in components, such as water heaters or air conditioners in response to price signals or requests from a utility control center.

3.4 Power Systems Operations and Control

Larger power grids are divided into smaller control areas. The operation of the US power system is divided into federal entities, state entities, and independent system operators (ISOs). FERC is responsible for defining mandatory reliability standards for interstate transfer of energy. NERC delegates authority to regional entities that assure that the standards defined by the FERC are being followed. Regional transmission organizations (RTOs) monitor systems within a single state or spanning multiple states, and have the exclusive responsibility for stable grid operation, short-term reliability, and optimized transmission service.

The operation of the power system mainly relies on the control of active and reactive power. Frequency control depends on the balance of active power generation and demand while reactive power control is related to voltage control. Changes in the balance of the active power between sources and loads will affect the frequency of the system; one basic control strategy is the inclusion of a speed governor on each generation complex. The speed governor provides primary speed control, which immediately tries to balance the relationship between the turbine speed and electrical power by either speeding up or slowing down the turbine, using frequency as a system-wide control signal [18].

The control of the reactive power is responsible for maximizing the use of the transmission grid while maintaining satisfactory operation within voltage stability limits. There are different devices that can be used for reactive power control, including shunt/series reactors and capacitors, generator excitation systems, synchronous condensers, and tap changing transformers. FACTS devices provide fast dynamic reactive power support. Reactive power needs are most efficiently met by local sources rather than transferring reactive power over long distances. The system operator can change setpoints for devices that automatically control the amount of reactive power supplied or absorbed to regulate the voltage at specific buses.

The human role in power system operation and control is supplemented by communication networks and decentralized controllers. A large proportion of the control systems is automatic since an increased delay may compromise reliability and system stability. However, the system operators work in conjunction with the automatic control system and can be considered "in-the-loop" of the control elements [19, 20]. Their short-term planning and operation decisions directly impact the performance and resilience of the power system. Therefore, smart grid data and advanced visualization of the collected data can positively impact operation and reduce human errors. Dispatching generation setpoints is also one of the roles of the system operator, which is performed with the risk of uncertainties (e.g. weather) or insufficient situation awareness due to misinterpretation of the system state.

The exchange of data between components of the power system follows proper guidelines that are defined by communication protocols. The data can be the processed values of measured voltages, currents, and power flows. Switches and breaker status are also monitored to provide the real-time topology of the system.

SCADA systems are responsible for automation of the power system and provide information to the system operator about any particular device of the system [7, 21]. The SCADA system monitors and collects data from components of the system to send to the operator over a scheduled polling period, typically on the order of seconds. The master station at the control center in Figure 3.4 receives all the field data and performs state estimation to determine the whether the system is operating in an acceptable condition. The state variables are voltage magnitude and angle at each bus in the system. Combining the state of the system with the network topology, the EMS determines if there are overloaded components, unacceptable voltage magnitudes, or other operational problems. In addition, the EMS performs contingency analysis to determine if there will be violations if any single point in the system fails $(n-1)$ or a combination of two devices are out of service at the same time $(n-2)$. Based on the results of these studies, the system operator will change generator dispatch, adjust power curves, change topology, and adjust reactive power devices to ensure secure operation should those failures occur. The SCADA system is composed of four main parts:

- RTU: collects the data and transmit the control signals to all field devices.
- Communication system: channels and bandwidth employed between the master station and field equipment.
- Master station: devices that enable monitoring of the state of the power system by the operator.
- Human–machine interface (HMI): interface that informs the operator about the state of the components and allows for interaction.

Table 3.1 OSI seven-layer communication architecture.

Layer		Functions
07	Application	Services FTP, MMS, Telnet, etc.
06	Presentation	Encrypt, compress, translate data
05	Session	Establish, manage, terminate sessions
04	Transport	End to end delivery
03	Network	Routing of data
02	Data link	MAC, errorcheck, sequencing, media access
01	Physical	Physical medium, data in bits

Source: Thomas and McDonald [22].

Another key aspect enabled by the HMI is the reduction of human error since different data visualizations can help the decision-making process of the system operator and their supporting real-time engineers [20, 21]. Utilities also benefit with savings on maintenance costs since SCADA can provide data from self-diagnostic programs and predictive maintenance warnings, resulting in better-informed decisions.

Network communication can be defined by the open system interconnection (OSI) seven-layer architecture hierarchy and functions shown in Table 3.1. Whenever one device initiates communication to another device, a command starts at the application layer and works down to the physical layer that connects the devices. When the message reaches the next device, the information works its way up through the layers. Some power system communication protocols may bypass some of these layers.

Modern IEDs simultaneously perform multiple functions to protect and control power systems. A phasor measurement unit (PMU) is utilized to determine voltage and current magnitudes and angles. A PMU consists of an IED that is time synchronized using a global positioning system (GPS) source and estimates phasors in real time. The phasor data is communicated to a phasor data concentrator (PDC) over an independent channel with a higher update rate than that of the SCADA system. The addition of wide area monitoring systems (WAMSs) using data from PMUs can result in substantial improvement of the operator's decision-making process. Other advantages of PMUs are model validation to improve accuracy of planning studies and identification of system oscillatory modes to improve system stability.

Control and protection schemes can be configured to operate autonomously in situations where humans cannot respond fast enough. Some schemes use only local measurements, while others are supplemented with communication schemes. Protective relays can detect transmission level faults and open circuit breakers within 50 ms. Remedial action schemes operating on timescales of seconds to a few minutes can shed loads, disconnect transmission lines, or even trip generators to prevent cascading blackouts.

3.5 Power Systems Planning

Power systems' owners perform planning studies on multiple time scales, with different objectives for each time scale. Short-term planning is primarily performed a day ahead or an hour ahead providing support for power system operation. Mid-term planning includes scheduling of system maintenance outages and upgrades to substations or construction of new substations and adjustments on the transmission grid in response to addition of load sites or integration of DER. Long-term planning encompasses large expansion of the grid and connection of new power plants or transmission lines based on forecast trends [3].

Long-term system planning is commonly divided into integrated resource (or generation) planning, transmission planning, and distribution planning [23].

Integrated resource planning starts with forecasting the long-term increase on demand, which is followed by evaluation of reliability impacts of this growth to determine the best technical location of additional generation if needed. Proposals for third party generation such as large PV or wind generation installations are considered in the process as well. The process ends with the optimization of the location and type of the utility-owned generator according to economic considerations. Load forecasting utilizes the historic measurement data from past years and is performed over a time horizon of 1–10 years. The outputs of the planning process are the predicted annual energy consumption (kWh) and the annual peak demand (kW). Generation planning also includes reliability considerations because some generation stations are not available year-round. Some stations can be out of service due to unforeseen equipment failures or for scheduled maintenance. The energy production cost is combined with the capital investment cost to decide which type of generation station is optimal for the planning study (hydro, coal, nuclear, solar, etc.).

Transmission system planning aims for the optimal use of generation by providing the most efficient path to reach the supply. It ensures that the energy will be delivered while maintaining reasonable operation limits even under system contingencies, such as unexpected failures. The analysis of stability limits is part of transmission system planning. Rotor-angle stability is studied in terms of steady-state analysis (small signal) and transient stability (large disturbances). Transmission planning also takes into consideration heavily loaded transmission lines, which may present voltage stability issues under contingency situations. Finally, frequency stability is also considered for transmission line planning [24].

3.5.1 Modeling and Simulation

Power system planning requires modeling and simulating large and complex networks. Some of these studies include power-flow, short-circuit, and transient stability analyses [18, 25, 26].

The power-flow analysis considers the power system in steady-state and the goal is to calculate the voltage, active power, and reactive power at every bus. Transmission level analysis treats the system as balanced; therefore, only one phase is studied [27]. The buses can be represented in four different ways:

- Real power and voltage controlled (PV) bus where the voltage magnitude and active power operating points are set through dispatch. The modeled source varies reactive power to regulate voltage. This type of bus usually represents generators or devices that provide reactive power, such as a synchronous condenser.
- Load (PQ) bus represents a load that has constant active and reactive power.
- A device bus considers special conditions, such as a load that varies active and reactive power according to the bus voltage magnitude.
- A slack bus is a generator bus that serves as the phase angle reference for all other buses. Voltage magnitude and angle are specified at this bus, and the simulation varies the real and reactive power to supply system losses. The setpoints for the generators in the system are determined assuming the slack bus is dispatched at a set level and supplies additional power to meet losses supplied on top of that level.

The transmission and distribution lines are represented as lumped equivalent circuit models. The network is represented in terms of its node admittance matrix Y_{bus} [3]. The matrix for a large interconnected system could include over 50 000 buses. Boundary conditions based on real and reactive power injections turn the problem nonlinear, thus requiring iterative techniques to solve. Popular methods are the Gauss–Seidel (GS) and the Newton–Raphson (NR), which are incorporated in commercial software tools.

Fault analysis enables the calculation of ratings for switches, circuit breakers, and protective relay setting thresholds [28]. The main goal of fault analysis is to calculate the fault current for symmetrical (e.g. three-phase) or asymmetrical faults (e.g. single-phase, or two-phase). A common approach combines impedance matrices, Z_{bus}, for three different sequence networks (i.e. positive phase rotation, negative phase rotation, and zero phase rotation components) to simplify analysis by using a mathematical transformation combined with superposition [23]. The Z_{bus} can be calculated by utilizing the matrix inverse of the Y_{bus} or other direct procedures. Symmetrical faults are very rare but can have the most severe consequences. Single-phase to ground faults caused by vegetation or weather are the most common types of faults in systems with overhead lines. Faults can also occur due to human errors, insulation failures, or equipment failures [23].

Short circuit faults cause currents 10–20 times normal and can also create high frequency transient voltage or current conditions [29]. The power system is predominantly inductive, typically behaving like an RL circuit in response to a fault; however, high frequency resonances can occur when parasitic capacitance interact with the resistances and inductances.

The ability to maintain synchronism in response to a major unexpected event like a three-phase fault on a critical line is measured using transient stability analysis [18]. The power-angle relationship of a synchronous generator is used to define the stability limits of the system. After the disturbance occurs, the rotor angle tends to increase, as shown Figure 3.8. The initial disturbance causes a deficit of energy, which can be described as area A_1. The recovery area, A_2, represents the effect of the electromechanical response of the generators to the disturbance, and the two areas should be equal. The area to the right of e, below the post fault curve and above P_m in the left plot in Figure 3.8 is the stability margin. The graph on the right in Figure 3.8 has a longer clearing time, causing an increase in the clearing angle, δ_{c2}, leaving no stability margin. If the clearing time of the fault is too long, the area A_1 is greater than the available area A_2, resulting in loss of synchronism in response to the disturbance. Heavily loaded generators have smaller available deceleration areas, A_2, which make it easier to lose synchronism. Decreased system inertia due to the increase in power electronic coupled generation such as PV and wind generation can increase the speed at which phase angle δ increases, decreasing the critical clearing times needed to maintain stability.

Voltage stability analysis studies the operational limits of the system to avoid voltage collapse. Factors such as heavily loaded transmission grids, limited reactive power provided by compensation devices, and insufficient control over the reactive power and voltage output of generators influence the voltage stability limits. The analysis utilizes the characteristics of a voltage–power (V–P) curve, as shown in Figure 3.9. The critical value of power delivered to the load is indicated by the dot (knee of the curve). If the load demand is higher than the maximum power, then any variation of the load results in voltage instability.

3.6 Measures of Performance

The power system performance is typically defined in terms of operational limits, reliability indices, and power system stability margins.

Figure 3.8 Rotor-angle stability curves.

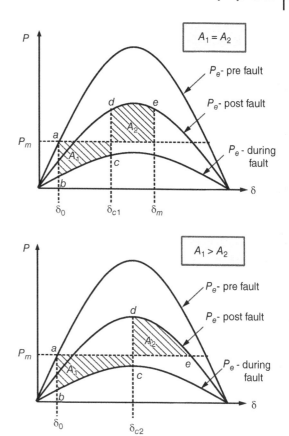

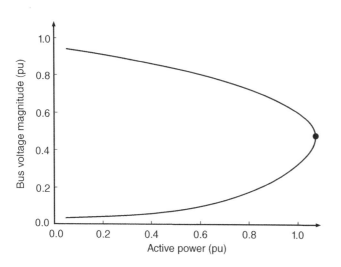

Figure 3.9 V–P curve and critical point.

The power grid is composed of many different devices that are connected to produce and deliver energy to all customers. However, these devices have operational limits related to current or power flows that must be followed to ensure personnel safety and equipment longevity and avoid catastrophic damage. These limits are determined in the planning process and must be considered when designing the power system. In addition, voltage magnitude limits are proscribed by standards to avoid insulation failures from overvoltages or equipment mis-operation from low voltages. The system operator uses these operational limits to identify if any violations are present in real-time operations. One example of a type of limit is thermal limits, which are related to the heating due the internal resistance and current flowing through the device. For example, transmission lines are designed to operate continuously with a maximum rated power. They can operate overloaded for a limited time. If they operate overloaded for too long, the integrity of the conductors may be compromised. These limits are weather dependent, a line will have a higher limit during winter than in summer. Another example of thermal limit impact is with transformers; their internal insulation will age faster and will fail well before its design life if operation limits are exceeded, even for brief periods of time. Breakers have maximum fault currents they can interrupt without risk of causing a damaging flashover. Generators present mechanical and thermal limits. The former results in non-instantaneous changes on the generated power. The latter affects the internal voltage of the generator and maximum active and reactive power flows.

Power system operational reliability depends mainly on predefined limits of electrical quantities, such as frequency, voltage, and power flow. Failure to develop power to end use customers is also an important measure of reliability of performance. It is defined as the probability that the power system will perform adequately, with acceptable quality, and on a continuous basis [30]. Power system reliability and risk assessment can be numerically evaluated by using probabilistic methods (e.g. Monte Carlo simulation). The overall system reliability is estimated by analyzing a combination of the reliability of each component of the power system through different system operating scenarios. Scheduled maintenance frequency, predicted equipment failure rates, and mean repair times are utilized to calculate the reliability of each component. The main reliability indices are presented below [31]:

- Loss of load expectation (LOLE) or loss of load probability (LOLP) [days per *year*]: Considers forced or unexpected outages of generators for different load profiles to determine the expected number of days in the year where the generated power cannot meet the load demand. In turn, the generation capacity and location are determined based on LOLE/LOLP and a reliability target.
- Expected frequency of load curtailment (EFLC) [occurrences per year]: Load curtailment is an emergency resource that the operator uses when they cannot meet voltage, current, power, and frequency constraints. Basically, it consists of disconnection of loads to alleviate problems on the power grid. The EFLC is estimated for different load profiles and considering forced or unexpected outages of generators.
- Expected energy not supplied (EENS) [MWh per year]: Combines the index LOLE/LOLP with the expected energy that the power system will be unable to supply.
- System average interruption frequency index (SAIFI) [interruptions per year]: This index estimates how often a customer experiences interruption of energy supply. SAIFI is the total number of customer interruptions divided by the total number of customers served.

$$\text{SAIFI} = \frac{\sum_i \lambda_i \cdot N_i}{\sum_i N_i} \tag{3.1}$$

- System average interruption duration index (SAIDI) [hours of interruption per year]: This index corresponds to the average outage duration for each customer per year. SAIDI is the ratio of the sum of the customer interruption durations divided by the total number of customers served.

$$\text{SAIDI} = \frac{\sum_i U_i \cdot N_i}{\sum_i N_i} \qquad (3.2)$$

- Customer average interruption duration index (CAIDI): This index corresponds to the duration of the average interruption. CAIDI is the sum of the durations of all customer interruptions divided by the total number of customer interruptions.

$$\text{CAIDI} = \frac{\sum_i U_i \cdot N_i}{\sum_i \lambda_i \cdot N_i} \qquad (3.3)$$

CAIDI can also be defined in terms of SAIDI and SAIFI.

$$\text{CAIDI} = \frac{\text{SAIDI}}{\text{SAIFI}} \qquad (3.4)$$

However, these metrics do not adequately measure resilience. The development of resilience metrics is an active research area. Chapter 11 will describe steps to produce resilience metrics that can be applied to different types of systems.

3.7 Summary

This chapter has provided a brief history of power systems, followed by an overview of classical power system architectures. Recent developments leading toward smarter power grids were discussed. Power systems operations were described, along with the typical planning process. Measures of power system performance were discussed. This chapter provides a basic understanding of modern power systems as background for discussions of control theory, cybersecurity, and human factors to create resilient control systems in the chapters to follow.

Thoughtful Questions to Ensure Comprehension

1 What elements of a power system (and quantities) are controlled in power system operations with classical generation resources, and what quantities are not controlled (and are instead reacted to)?

2 How does your answer to Problem 1 change if the system has significant percentages of the total generation coming from PV generation?

3 What are the advantages of allowing a microgrid to operate in islanded mode if the larger grid loses power? Are there disadvantages?

4 How does a grid operator determine voltage magnitude and current flows at a remote substation?

5 How does a smart grid differ from the classical grid?

6 The Table 3.2 gives 2020 annual outage data for sustained interruptions (longer than 5 minutes) from a utility's database for a distribution system that serves 4500 customers with a total load of 9 MW. Note that a major event occurred on 4 November due to an ice storm. Calculate SAIDI, SAIFI, and CAIDI for this feeder (a) with all events and (b) without the 4 November ice storm event. Comment on the differences.

Table 3.2 Outage data for Problems 6–8.

Outage date	Outage duration (min)	Number of customers interrupted
01/15/2020	15	950
04/04/2020	160	442
07/08/2020	90	125
09/10/2020	655	15
10/11/2020	33	2500
11/04/2020	10 053	4000
12/01/2020	40	370

7 Suppose a significant portion of the feeder in Problem 6 can be supported as a microgrid supplied by PV inverters. As a result, 3500 customers are not interrupted for 6 hours out of every 24 during the 4 November event. Update your results from Problem 6a.

8 Repeat Problem 7 if the microgrid also has energy storage so those 3500 customers can be supported 24 hours/day for the duration of the event?

Further Reading

In addition to the references cited in this chapter, the following resources listed below would be beneficial for further reading.

Power Systems Analysis and Design

Glover, J.D., Sarma, M.S., and Overbye, T. (2016). *Power System Analysis & Design*, 6e. Cengage Learning.

Smart Grids

Cornwall C.E. and Hurd C.L. (2013). Smart Grids Infrastructure, *Technology, and* Solutions, 73.

Microgrids

Hatziargyriou, N. (2014). *Microgrids: Architectures and Control*. In: Wiley.

Power Systems Planning and Operations

Bebic, J. (2008). Power System Planning: Emerging Practices Suitable for Evaluating the Impact of High-Penetration Photovoltaics. NREL/SR-581-42297, US Department of Energy.

Khuntia, S.R., Tuinema, B.W., Rueda, J.L., and van der Meijden, M.A.M.M. (2016). Time-horizons in the planning and operation of transmission networks: an overview. *IET Generation Transmission and Distribution* 10 (4): 841–848. https://doi.org/10.1049/iet-gtd.2015.0791.

Wood, A.J., Wollenberg, B.F., and Sheble, G.B. (2013). *Power Generation, Operation, and Control*, 3e. Wiley.

Power System Reliability

IEEE Std 1366-2012 (2012). IEEE guide for electric power distribution reliability indices (Revision IEEE Std 1366-2003).

Karki, R., Billinton, R., and Verma, A.K. (2014). *Reliability Modeling and Analysis of Smart Power Systems*. Springer.

References

1 Constable, G. (2003). *A Century of Innovation*. Washington, DC: Joseph Henry Press.

2 Tesla Memorial Society of New York. https://www.teslasociety.com/adams.htm (accessed 05 December 2020).

3 Glover, J.D., Sarma, M.S., and Overbye, T. (2016). *Power System Analysis & Design*, 6e. Cengage Learning.

4 PNNL (2021). Grid architecture. https://gridarchitecture.pnnl.gov (accessed 29 May 2021).

5 Taft, J.D. (2019). Grid architecture. *IEEE Power and Energy Magazine* 17: 104.

6 Kulkarni, S. N. and Shingare, P. (2016). A review on smart grid architecture and implementation challenges. International Conference on Electrical, Electronics, and Optimization Techniques, ICEEOT 2016, 3285–3290, doi: https://doi.org/10.1109/ICEEOT.2016.7755313.

7 Budka, K., Deshpande, J., and Thottan, M. (2014). *Communication Networks for Smart Grids: Making Smart Grid Real*. Springer.

8 Cunjiang, Y., Huaxun, Z., and Lei, Z. (2012). Architecture design for smart grid. 2012 International Conference on Future Electrical Power and Energy Systems, 1524–1528, doi: https://doi.org/10.1016/j.egypro.2012.02.276.

9 Borlase, S. (2016). *Smart Grids: Infrastructure, Technology, and Solutions*. CRC Press.

10 Jolfaei, A. and Kant, K. (2017). A lightweight integrity protection scheme for fast communications in smart grid. Proceedings of the 14th International Joint Conference on e-Business and Telecommunications, vol. 4, (January), 31–42, doi: https://doi.org/10.5220/0006394200310042.

11 Uslar, M., Specht, M., Dänekas, C. et al. (2013). *Standardization in Smargrids*. Springer.

12 Tomar, A. and Kandari, R. (2020). *Advances in Smart Grid Power System*. Elsevier.

13 Momoh, J. (2012). *Smart Grid: Fundamental of Design and Developements*. Wiley.

14 Hatziargyriou, N. (2014). *Microgrids: Architectures and Control*. Wiley.

15 IEEE Std 1547.2-2008 (2008). IEEE application guide for IEEE Std 1547. IEEE Standard for Interconnecting Distributed Resources with Electric Power Systems.

16 PES-TR71 (2019). *Microgrid Protection Systems*. IEEE Power and Energy Society.

17 IEEE Std 2030.7-2017 (2018). IEEE standard for the specification of microgrid controllers.

18 Kundur, P. (1994). *Power System Stability and Control*, 2e. McGraw-Hill Professional.

19 Hu, W.L., Rivetta, C., MacDonald, E., and Chassin, D.P. (2019). Modeling of operator performance for human-in-the-loop power systems. *Lecture Notes in Computer Science* (including Subser. Lect. Notes Artif. Intell. Lect. Notes Bioinformatics) LNAI 11571: 39–54. https://doi.org/10.1007/978-3-030-22507-0_4.

20 Bao, Y., Guo, C., Zhang, J. et al. (2018). Impact analysis of human factors on power system operation reliability. *Journal of Modern Power Systems and Clean Energy* 6, 27 (1): –39. https://doi.org/10.1007/s40565-016-0231-6.

21 Thomas, M.S. and McDonald, J.D. (2012). *Power System SCADA and Smart Grids*. In: CRC Press.

22 Thomas, M.S. and McDonald, J.D. (2012). *Power System SCADA and Smart Grids*. In: *CRC Press*.

23 Bebic, J. (2008). *Power System Planning: Emerging Practices Suitable for Evaluating the Impact of High-Penetration Photovoltaics* NREL/SR-581-42297. US Department of Energy.

24 Khuntia, S.R., Tuinema, B.W., Rueda, J.L., and van der Meijden, M.A.M.M. (2016). Time-horizons in the planning and operation of transmission networks: an overview. *IET Generation Transmission and Distribution* 10 (4): 841–848. https://doi.org/10.1049/iet-gtd.2015.0791.

25 Grainger, J.J., Stevenson, W.D., and Stevenson, W.D. (1994). *Power System Analysis*. McGraw-Hill.

26 Anderson, P.M. and Fouad, A.A. (2008). *Power System Control and Stability*. Wiley.

27 Saadat, H. (1999). *Power System Analysis*, vol. 2. McGraw-Hill.

28 Kasikci, I. (2018). *Short Circuits in Power Systems: A Practical Guide to IEC 60909-0*. Weinheim: Wiley-VCH Verlag GmbH & Co. KGaA.

29 Greenwood, A. (1991). *Electrical Transients in Power Systems*, 2e. Wiley-Interscience.

30 IEEE Std 493-2007 (2007). IEEE recommended practice for the design of reliable industrial and commercial power systems. (Revision IEEE Std 493-1997), 1–383, doi: https://doi.org/10.1109/IEEESTD.2007.380668.

31 IEEE Std 1366-2012 (2012). IEEE guide for electric power distribution reliability indices. (Revision IEEE Std 1366-2003).

4

Control System Architecture

Thomas Baldwin

Electrical and Computer Engineering, Idaho State University, Pocatello, ID, USA

Objectives

The objective of this chapter is to provide an overview of control method applications to the power grid. As such, this chapter will discuss feedback controls that are applied to individual generators, transmission systems, and associated regulation of power quality. These systems are responsible for the maintenance of stable power flow across wide areas.

4.1 Introduction

4.1.1 Background

Generators turn mechanical energy into electrical energy by moving electrical conductors in a magnetic field. An excitation system creates a constant electromagnetic field on the generator's rotor. The spinning rotor with the magnetic field induces an alternating current (AC) voltage in the electrical coils and windings of the generator's armature. The prime mover is the source of mechanical power that spins the rotor. The armature voltage energizes the generator's terminals and connected loads. The loads draw a current from the generator. This current produces its own magnetic field in the airgap between the rotor and armature iron. The interaction between the rotor and armature magnetic fields produces a counter-torque that resists the prime mover in spinning the generator. In this manner, mechanical power in the form of torque and angular velocity converts to electrical power in the form of voltage and current.

Various mechanical systems may serve as prime movers. The most common are steam turbines, which are powered by the burning of fossil fuels or nuclear reactions. Hydro and wind turbines are frequently used where resources are readily available. To a lesser extent, diesel reciprocating engines serve as prime movers. Turbine-generator sets make up virtually all the power sources associated with the electrical power grid.

The behaviors of turbine-generators dominate the operation of power systems. Transmission lines and transformers passively carry the power and match voltage levels from power plants to loads. Substations house busbars and circuit breakers that connect lines and transformers together and provide switches to isolate failed components. Power system protection is similar to controls but addresses abnormal operating conditions that occur infrequently and can potentially damage the system.

Resilient Control Architectures and Power Systems, First Edition.
Edited by Craig Rieger, Ronald Boring, Brian Johnson, and Timothy McJunkin.
© 2022 The Institute of Electrical and Electronics Engineers, Inc. Published 2022 by John Wiley & Sons, Inc.

As such, the primary control of the power system is the control of generators. Generators deliver power and support the magnitude and frequency of the voltages. Fortunately, within the power system, there is a natural decoupling between the flow of power and the voltage magnitude. This decoupling permits separate control loops to manage the power balance between generation and consumption and maintain the voltage magnitude required by the loads. Small changes in active power affect the frequency and sinusoidal phase shifts of the voltages. Small changes in voltage magnitude cause change in reactive power flow and the voltage profile across the grid.

Generators use two important control loops to drive the prime movers and regulate the voltage excitation. The load frequency control (LFC) loop manages the active power produced and regulates the system frequency. The automatic voltage regulator (AVR) loop regulates the voltage magnitude and balances the reactive power. Utilities employ energy management systems (EMSs) that perform supervisory control of large multi-generator systems. The EMS consists of online computers that gather telemetered measurement data from outlining substations and perform error detection and signal processing to provide snapshots of the current state of the power grid. The automatic generation control (AGC) provides control signals to the LFC and AVR. The AGC ensures that the power system operates within normal voltage magnitude and system frequency tolerances while maintaining contractual exchanges of power over tie lines connecting utilities. The supervisory control and data acquisition (SCADA) system manages the gathering and transfer of instrumentation data and the command signals between the EMS and remote substations. The whole control system aims to operate the power grid economically within voltage and frequency tolerances while providing stability margins to withstand a couple of major contingencies. Only an introductory level of power system control is given here.

4.1.2 Basic Generator Control Loops

Each grid-connected generator has an LFC and AVR. Figure 4.1 gives the schematic diagram with the primary control loops. The AGC establishes the set points or reference values for the two controllers. The LFC and AVR are designed to take care of small changes in load demand to maintain the AC frequency and voltage magnitude within the specified limits. The turbine-generator set responds to small perturbations in active power with changes in shaft angle δ and speed ω. The time constant of the prime mover's throttle and rotational inertia of the system is on the order

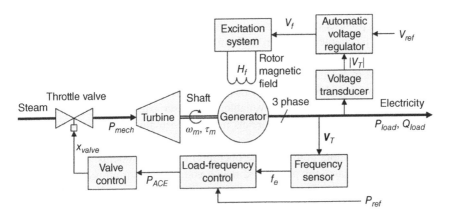

Figure 4.1 Schematic diagram of a synchronous generator with the load frequency control (LFC) and automatic voltage regulator (AVR).

of 0.1 s. The generator and excitation respond to changes in reactive power demand with changes in the voltage magnitude. The time constant of the excitation system is on the order of milliseconds – much smaller than the prime mover time constant. Voltage magnitude transients decay much faster and do not affect the LFC dynamic. Cross-coupling between the LFC loop and the AVR loop is negligible. This permits the analysis, design, and operation of the load frequency and excitation voltage controls independently.

Power system controls are challenging at best to analyze and design. There are many nonlinear behaviors among the power system components. Modeling the modern power system requires us to linearize the mathematical functions at a given operating point. We must make proper assumptions and approximations as a part of this process. In the following discussions, we introduce several components and describe them with first-order models. There are numerous resources that specialize and give in-depth details of higher order models, nonlinearities, and limits. The objective here is to show the generalities of power system controls and what is accomplished without overwhelming with details.

4.1.3 Load Frequency Control

The control objectives of the LFC are to maintain a uniform AC frequency and balance the power between generation and load. When combined with the AGC, the overall control objective includes dividing the load between the generators and keeping the power flow schedules on the tie lines. Because of the physical distances and handling cost of multiple feedback signals, the LFC only senses the change in the turbine-generator shaft speed (or equivalently, the AC frequency) and the aggregate signal representing the active power flowing across the tie lines.

4.1.4 The Generator

A prime mover, such as a steam turbine or waterwheel, provides rotational power at a constant rate. Generally, the power available is controllable by a throttle valve or penstock. The rotating shaft connects to the generator, which uses a magnetic field to convert the rotational energy into electrical power by the principle of induction. Drawing electrical power from the generator creates a retarding mechanical torque that counteracts the driving mechanical torque of the prime mover. When kept in balance with a net torque of zero, the speed of the shaft remains constant. Any imbalance causes the shaft speed to accelerate or decelerate.

The prime mover provides the mechanical torque, T_{mech}, and acts to increase the rotational speed, ω. The electrical load provides the counter-torque, T_{elec}, and acts to decrease the speed. The electrical frequency and the mechanical speed are related by a constant, which depends on the number of magnetic poles designed into the generator. Most steam systems use a two-pole generator, which gives a one-to-one relationship between the frequency and speed. If the electrical load is increased so that T_{elec} is larger than T_{mech}, the entire rotating system will begin to slow down, and the AC frequency droops. This inherent balancing of power and energy enables electrical systems to operate without prior knowledge of load increases or decreases.

The rotating masses of the turbine and generator's rotor form the system's inertia. This rotating inertia provides energy storage capability. It also maintains the security of the system as loading changes. The energy storage allows sufficient time for corrective control actions to take place to adjust the input power to the prime mover to match the electrical power demands. To keep the rotational speed at its reference value, the LFC controls the input power supplied to the prime mover.

The torque-acceleration equation describes the torque balance and energy storage as:

$$I\frac{d^2\delta_m}{dt^2} = T_{mech} - T_{elec} \tag{4.1}$$

where I, moment of inertia for the turbine-generator set; T_{mech}, mechanical torque exerted on the machine by the turbine (Newton-meters); T_{elec}, electrical torque exerted on the machine by the generator (Newton-meters); δ_m, mechanical shaft offset angle (radians).

For power systems, we rewrite the torque-acceleration equation in the power form by multiplying each term by the synchronous speed ω_s. This equation, known as the Swing Equation, is valid for both large system disturbances as well as small perturbation in load power demands. With disturbances like short-circuit faults, nonlinear terms in the electrical power flow make the mathematical analysis more challenging. However, for normal operating conditions, we linearize the equation around a given operating point and use percent changes in the mechanical and electrical powers and the shaft position and speed. The swing equation simplifies to:

$$\frac{2H}{\omega_s}\frac{d^2\Delta\delta}{dt^2} = \Delta P_{mech} - \Delta P_{elec} \tag{4.2}$$

where H, angular momentum of the turbine-generator set (seconds); ω_s, the synchronous (nominal) shaft speed (radians/second); $\Delta\delta$, percent change of the shaft offset angle (% change); ΔP_{mech}, percent change of the mechanical power input (% change); ΔP_{elec}, percent change of the electrical power output (% change).

In the implementation of generator control systems, we model the turbine-generator set in terms of the shaft angular velocity instead of the shaft displacement. We simplify the math effort once more to get:

$$\frac{d}{dt}\left(\frac{\Delta\omega}{\omega_s}\right) = \frac{1}{2H}(\Delta P_{mech} - \Delta P_{elec}) \tag{4.3}$$

where $\Delta\omega/\omega_s$, percent change of the shaft speed or AC frequency (% change).

Taking the Laplace transform of Eq. (4.3) gives the generator's transfer function, and Figure 4.2 shows the equivalent block diagram for:

$$\Delta\Omega(s) = \frac{1}{2H}[\Delta P_{mech}(s) - \Delta P_{elec}(s)] \tag{4.4}$$

4.1.5 The Load

The load on a power system consists of a variety of end-user devices found in industry, businesses, and homes. For resistive-type and electronic-based loads, such as lighting, heating, and computer loads, the power demand is independent of the AC frequency. However, motor loads, such as refrigerators and fans, are sensitive to changes in frequency. This level of sensitivity depends on the

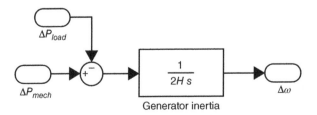

Figure 4.2 Block diagram of a synchronous generator.

Figure 4.3 Expanded and reduced block diagrams of the generator and frequency-dependent loads.

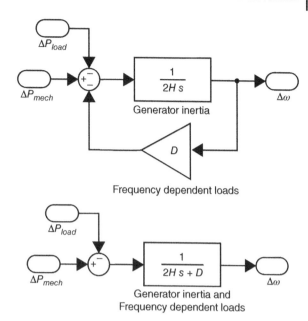

composite behaviors of the speed–load characteristics of all the connected motors. We approximate the speed–load characteristic of the aggregate motor load using:

$$\Delta P_{elec}(s) = \Delta P_{load}(s) + D\, \Delta\omega(s) \tag{4.5}$$

where ΔP_{load} is the change of load demand and $D\, \Delta\omega$ is the variation in load demand due to changes in the AC frequency. D is expressed as the percent change in load power divided by the percent change in frequency. For example, a 1% decrease in the frequency causes the total load demand to drop by 1.6%. The value of D then becomes 1.6. Figure 4.3 illustrates the combination of the generator model and the frequency-dependent loading. The first block diagram shows the details, while the second diagram presents the commonly used reduced form.

4.1.6 The Turbine-Based Prime Mover

The most common sources of mechanical power for the prime mover are turbine-based, such as hydroelectric turbines, steam turbines (from burning coal, gas, and nuclear fuel), and gas turbines. The transfer function for the turbine captures the time response for the power transferring fluid from the valve, P_{valve}, to reach the whole turbine and convert into mechanical shaft power, ΔP_{mech}. There are many types of turbines with multiple stages for high-pressure, medium-pressure, and low-pressure fluids. The IEEE standards provide detailed transfer function models for the various turbine constructions. We will use the simplest prime mover model of a single-stage, non-reheat steam or hydro turbine. We approximate the response with a single time constant τ_T. The resulting transfer function is:

$$\Delta P_{mech}(s) = \frac{1}{1 + \tau_T s} \Delta P_{valve}(s) \tag{4.6}$$

The value of the time constant τ_T ranges from 0.2 to 2.0 seconds. From the controls perspective, there are a few minor differences among the various energy sources, and the IEEE standards give more details. Figure 4.4 shows the block diagram for the simple turbine.

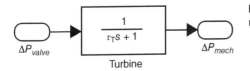

Figure 4.4 Block diagram for a simple hydro or non-reheat steam turbine.

4.1.7 The Speed Governor

Whenever the electrical load output power differs from the mechanical input power at the turbine-generator set, the shaft speed and AC frequency will change. A speed tachometer or an electrical frequency meter provide the feedback signal indicating the generator's power mismatch response. The governor compares the difference between the speed set point/frequency reference and the feedback signal. The output of the governor controls the power input valve (steam throttle valve or water penstock gate). Governors generally use electronic interfaces to drive a servomotor that adjusts the turbine's throttle valve or penstock. However, one can find pneumatic and hydraulic versions in older control systems.

Like many mechanical power devices, prime movers exhibit a speed-droop characteristic. An increase in mechanical output power is associated with a small decrease in shaft speed. The speed-droop characteristic is valuable and ensures that the turbine-generator operates at one point. The governor can also provide the speed-droop characteristic as part of the feedback loop. The slope of the curve represents the speed regulation R. We define the regulation as the negative change in shaft speed for a change in output power or:

$$R = -\frac{\Delta\omega}{\Delta P} \tag{4.7}$$

Hence, the regulation is a positive value for a stable turbine-generator set. Figure 4.5 presents the steady-state characteristics of a governor system.

Governors typically have a speed regulation of 5–6% from zero to full-load power. We express the transfer function of the governor with the speed regulation as:

$$\Delta P_{err}(s) = \Delta P_{ref}(s) - \frac{1}{R}\Delta W(s) \tag{4.8}$$

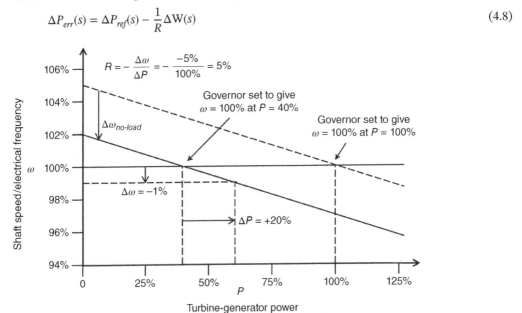

Figure 4.5 Governor steady-state speed regulation characteristic.

Figure 4.6 Block diagram of speed governing system for steam turbines.

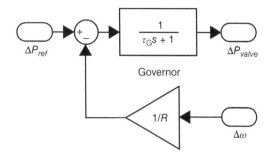

where ΔP_{err}, percent change of the power error or mismatch (% change); ΔP_{ref}, percent change of the reference power (% change); $\Delta\Omega$, percent change of the shaft speed or AC frequency (% change).

By having the speed regulation in the feedback loop, the set point or reference represents the desired output power at rated shaft speed or the nominal AC frequency. Operators can schedule or dispatch the power output level for each generator by adjusting the reference value.

The change in the error or mismatch power, ΔP_{err}, is forward from the governor to the throttle valve or penstock gate as a position command. For convenience, we express the change in valve position in terms of the change in power passing through the valve, ΔP_{valve}. Repositioning the valve takes time, and a first-order transfer function models the delayed response with a simple time constant τ_G. By using percentage change values for the power, we get as the transfer function:

$$\Delta P_{valve}(s) = \frac{1}{1 + \tau_G s} \Delta P_{err}(s) \tag{4.9}$$

Figure 4.6 shows the block diagram where the governor and valve components are combined.

4.1.8 The Load Frequency Control Loop

Combining the control elements together gives us the block diagram of Figure 4.7. This simple control loop is sufficient for an isolated generator power station. The dynamic input is the change in the electrical load ΔP_{load} and the controlled output is the frequency deviation $\Delta\omega$. The reference power input is the set point value that power station operators can control.

The closed-loop transfer function for the LFC loop is:

$$\frac{\Delta\Omega(s)}{\Delta P_L(s)} = -\frac{R(1 + \tau_{gen}s)(1 + \tau_T s)}{R(Ms + D)(1 + \tau_{gen}s)(1 + \tau_T s) + 1} \tag{4.10}$$

To gain some understanding of the LFC behavior, we look at the steady-state response for a step input change in the load. The final value theorem allows us to see the steady-state error value of

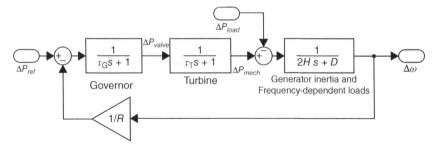

Figure 4.7 Block diagram of the load frequency control of an isolated power system.

Table 4.1 Turbine-generator parameters for the power station example.

Parameter	Value
Turbine rated output power, P_{mech}	250 MW
Nominal shaft speed, n_{mech}	3600 rpm
Nominal electrical frequency, f	60 Hz
Moment of inertia, J	10.554×10^3 kg m^2

Table 4.2 LFC loop parameters for the power station example.

Parameter	Gain values	Time constants
Governor	$1/R = 18$ %rad/s/%MW	$\tau_G = 0.22$ s
Turbine	$K_T = 1.0$ %MW/%MW	$\tau_T = 0.47$ s
Angular momentum	$H = 3$ s	
Frequency sensitivity	$D = 1.5$ %MW/%rad/s	

$\Delta\omega$, which as a goal should approach zero to meet the utility regulations. We apply a step change for the input (i.e. $\Delta P_L(s) = \Delta P_L/s$), then the steady-state value of $\Delta\omega$ is:

$$\Delta\omega(t \rightarrow \infty) = \lim_{s \rightarrow 0} s\Delta\Omega(s) = -\Delta P_L(s)\frac{R}{RD+1} \tag{4.11}$$

For the simplified case without the frequency-sensitive load component (i.e. $D = 0$), the steady-state deviation in frequency is determined by the governor speed regulation R, and becomes:

$$\Delta\omega(t \rightarrow \infty) = -R\,\Delta P_L \tag{4.12}$$

We can apply these concepts to an example of an isolated power station and connected load. Table 4.1 lists the power station parameters, and Table 4.2 lists the control loop parameters.

A sudden load increase of 25 MW ($\Delta P_{load} = 10\%$) occurs. Figure 4.8 shows the step response of the frequency deviation. Using Eq. (4.11), we verify that the steady-state deviation in the frequency is -0.513%. This is equivalent to a final shaft speed of 3581.5 rpm and an AC frequency of 59.692 Hz.

4.1.9 Multiple Generators Operating with LFC

As power systems expand, utilities connect several generators to the system. The increased generation can supply larger load demands while providing greater reliability in electricity delivery. Each generator has its own prime mover and governor. The impact of n generators on the system frequency is a function of all the speed regulation values, $R_1, R_2, ..., R_n$. The steady-state deviation in the electrical frequency is given by:

$$\Delta\omega_{SS} = -\Delta P_L(s)\frac{1}{D + \dfrac{1}{R_1} + \dfrac{1}{R_2} + \cdots + \dfrac{1}{R_n}} \tag{4.13}$$

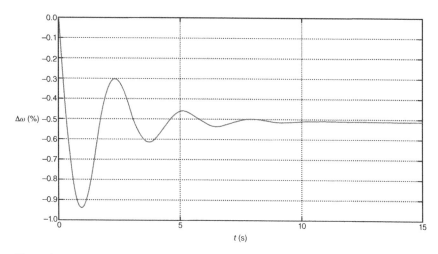

Figure 4.8 The speed/frequency response for a 10% load increase in the power station example.

Table 4.3 Generation parameters for three-unit power system example.

Generator	Initial output power, P_{elec} (MW)	Governor speed regulation, R
Unit 1	450	0.08
Unit 2	500	0.06
Unit 3	550	0.05

Table 4.4 Final power output values for three-unit power system example.

Generator	Final output power, P_{elec} (MW)
Unit 1	487.0
Unit 2	549.3
Unit 3	609.2
Total Gen.	1645.6

Now consider a power system, having three generating units with the following characteristics (Table 4.3).

The three generating units are operating in parallel and share 1500 MW at the nominal frequency of 60 Hz. The motor load dampening coefficient D is 1.5. The load demand increases by 150 MW or 10%. The resulting percentage change in frequency is −0.1974%, resulting in a final value of the frequency being 59.882 Hz and the shaft speed being 3592.9 rpm. The final value of the load power becomes 1645.6 MW. The drop in the final frequency causes a 4.4 MW loss in the total load. Using Eq. (4.12), we find the output powers for the three generators shown in Table 4.4.

4.2 Automatic Generation Control

4.2.1 Background

When the electrical load changes, the prime mover cannot respond instantaneously. Energy storage in the rotating masses of the turbine-generator permits the balance of power at the cost of changing the shaft speed and the system frequency. For a load increase, the initial system response draws power from the rotating mass of the turbine-generator set, causing a decrease in the speed and frequency. The governor senses the drop in speed and adjusts the throttle valve to add power to the turbine.

The governor forms a simple proportional-gain control loop to regulate the speed and frequency. This loop lacks an integrator element, and this closed-loop control cannot completely erase the speed/frequency error to zero. The AGC adds an integrating loop in conjunction with the governor loop. The integrator restores or resets the speed and frequency to the reference value. The AGC also provides additional functions for systems having more than one generator. The most important function is the division of the power demand among the generators.

In the interconnected system, consisting of several utilities and many generators, the AGC works with the hieratical control structure of the EMS to achieve several control objectives.

(1) Achieving maximum economic efficiency in fuel consumption and cost.
(2) Maintaining scheduled power exchanges between utility areas. Tie lines interconnect the areas and meter the power flows.
(3) Maintaining a uniform and scheduled electrical frequency.

The EMS and AGC controls perform their functions during normal operating condition of a stable power system. These controls can achieve a steady-state response for small and smooth load variations over the course of a day. On the other hand, specialized control loops override these functions during emergency conditions. The loss of use of a generator or a transmission line circuit due to a fault may cause the system to enter the emergency state. The topics of power system protection and contingency analysis address the emergency state. The goal of the specialized systems is to restore the power system to a state that is close to normal operating conditions.

4.2.2 The AGC in Single Area Systems

For a single turbine-generator, the AGC consists of the primary LFC loop and the reset integrator as shown in the block diagram of Figure 4.9. The integrator changes the control loop to a Type 1,

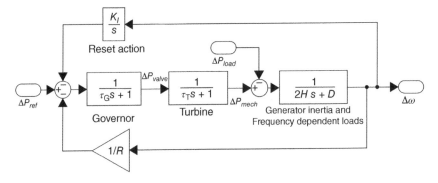

Figure 4.9 AGC for an isolated power system.

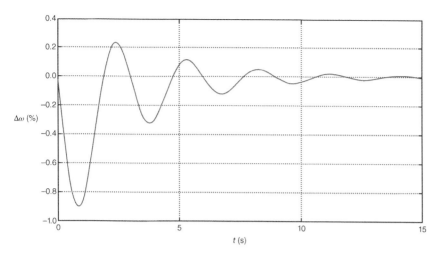

Figure 4.10 The speed/frequency response for a 10% load increase in the power station example, modified by the reset integrator action.

which permits the steady-state error to become zero. We must adjust the integral controller gain, K_I, for a satisfactory transient response. The resulting closed-loop transfer function between the change in load demand and the change in shaft speed or electrical frequency becomes:

$$\frac{\Delta\Omega(s)}{\Delta P_{load}(s)} = -\frac{sR(1 + \tau_G s)(1 + \tau_T s)}{sR(2Hs + D)(1 + \tau_G s)(1 + \tau_T s) + s + K_I R} \tag{4.14}$$

Let us return to the power station example and add the reset integrator with the gain K_I set to 8.0. All the other generation parameters are kept the same as given in Table 4.2. Figure 4.10 graphs the speed/frequency response for a 10% increase in the load demand. Note that this time, the change in frequency, $\Delta\omega$, returns to zero, representing a shaft speed of 3600 rpm and a frequency of 60 Hz.

4.2.3 The AGC in Multi-Area Systems

It is often the case at power plants that a group of generators form a closely coupled set. Together they share a common dynamic behavior and swing in unison for a perturbation. Such a group of generators are said to be coherent. It is possible to let a single LFC loop represent the coherent group. This concept can be extended up to the whole utility under favorable parameters. Each coherent grouping is referred to as a control area.

We can realize the AGC of a multi-area system by first starting with the AGC for a two-area system. The understanding gained with the first two areas can be extended repeatedly with additions of the next new area. We connect areas together through electric power transmission lines that are known as tie lines. Tie lines have no differences compared to other lines except that the power flowing in the line is metered and telemetered to the AGC. Tie lines form the boundary of the control area.

4.2.4 The Tie Line

A tie line connects two areas together so that power from one area can be exported to the other. A power meter measures the flow across each tie line and reports the flow to both area controls. The AGC for a control area sees the aggregate power total for all power imports and exports across

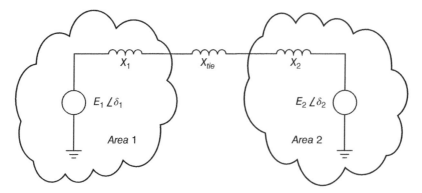

Figure 4.11 Equivalent network for a two-area system.

the adjacent tie lines. For the control, how power leaves or enters the area across any of the tie lines is not important; only the total of the power transfers is needed. In fact, for power systems with multiple control areas, power transfers may pass through neighboring areas before reaching the intended area. This operation is called wheeling. The AGC in each control area adjusts the local generation level to meet the prearranged total import or export power level.

To analyze and model the power flow across the tie line, we represent each area by an equivalent generating unit. The tie line connects the two equivalent generators. For simplicity, we represent the tie line as a lossless transmission line, having a reactance X_{tie}. Inside each area, we model the equivalent generator as a voltage source behind an equivalent reactance, as shown in Figure 4.11. The difference in the offset or phase angles between the two voltage sources determines the transferred power level. The power flow value is found from:

$$P_{12} = \frac{|E_1|\,|E_2|}{|X_1 + X_{tie} + X_2|} \sin(\delta_1 - \delta_2) \tag{4.15}$$

For inclusion in the AGC model, we must linearize Eq. (4.15) around the initial operating value, $P_{12,0}$. The linearization gives:

$$\Delta P_{12} = \left.\frac{dP_{12}}{d\delta_{12}}\right|_{\delta_{12,0}} (\Delta\delta_1 - \Delta\delta_2) = P_{S12}(\Delta\delta_1 - \Delta\delta_2) \tag{4.16}$$

where:

$$P_{S12} = \left.\frac{dP_{12}}{d\delta_{12}}\right|_{\Delta\delta_{12,0}} = \frac{|E_1|\,|E_2|}{|X_{12}|} \cos(\delta_{12,0}) \tag{4.17}$$

$$X_{12} = X_1 + X_{tie} + X_2 \text{ and } \delta_{12,0} = \delta_{1,0} - \delta_{2,0}$$

The synchronizing power coefficient P_{S12} is the slope of the power angle curve at the initial operating angle. Changes to the tie line's power flow appear as a load increase in one area and a load decrease in the other area. The direction of flow change is dictated by the sign of the phase angle difference. When $\Delta\delta_1 > \Delta\delta_2$, the incremental power flows from Area 1 to Area 2.

Figure 4.12 shows a simplified control block diagram for a two-area system. The controls have just the simple LFC loop for each generator set. The power balance for each equivalent generator

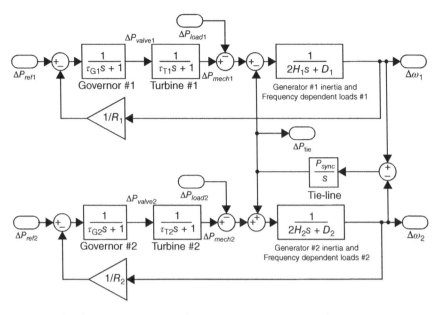

Figure 4.12 Control block diagram for a two-area system with LFC.

and its local load is given by:

$$\Delta P_{mech1} = \frac{-\Delta\omega}{R_1} = \Delta P_{load1} + \Delta\omega D_1 + \Delta P_{S12} \qquad (4.18a)$$

$$\Delta P_{mech2} = \frac{-\Delta\omega}{R_2} = \Delta P_{load2} + \Delta\omega D_2 - \Delta P_{S12} \qquad (4.18b)$$

Let us consider a two-area example with the control loops, as shown in Figure 4.12. Table 4.5 provides the parameters for the controls and the system. Area 2 experiences a load increase ΔP_{load2} of 10%.

Figure 4.13 graphs the speed/frequency response of each area. In 15 seconds, the example system settles to a new steady state, both areas sharing the same frequency deviation, $\Delta\omega$. Due to the frequency deviation, both control areas will have changes in the frequency-dependent loads, and the governor speed control will compensate with changes in mechanical power.

Table 4.5 Parameters for the two-area system with LFC example.

Parameters	Area 1	Area 2
Generator-turbine inertia, H	3.00	4.00
Frequency sensitive load, D	1.50	1.20
Speed regulation, $1/R$	18.0	15.0
Governor time constant, τ_G	0.22	0.27
Turbine time constant, τ_T	0.47	0.43

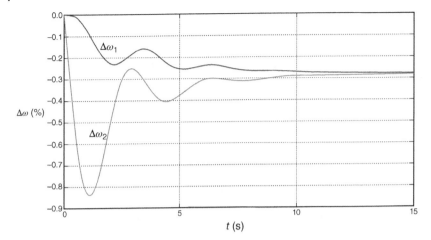

Figure 4.13 Speed/frequency response for a two-area system with LFC. Load demand in Area 2 increased by 10%.

The difference between the two area frequencies drives changes in the two phase-angles that appear across the tie line. The angle difference causes power to flow between the two areas. Figure 4.14 graphs the power of the two generators at the valve point in the control loop. The figure also plots the power flow across the tie line. Note that in control Area 1, the generator shares in the power generation increase. Because of the frequency drop in both areas, the load in Area 1 decreases slightly. The combination of increased generation and drop in load provides a larger power flow in the tie line from Area 1 to Area 2.

Generally, the response of the generator in Area 1 is unacceptable, except in emergency situations. An increase in the load of Area 2 should be addressed by an increase in the generation of Area 2 without changing the tie line power flow. This suggests that we need additional control actions.

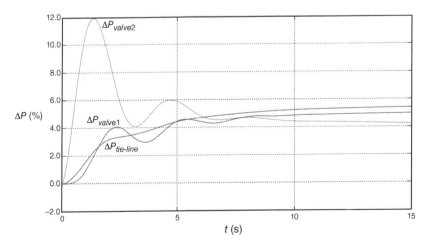

Figure 4.14 Turbine powers and tie line power flow for a two-area system with LFC. Load demand in Area 2 increased by 10%.

4.2.5   Tie Line Control

As we have just seen in the last example, a system with simple LFC loops for the generators, does not correct the shaft speeds and the AC frequency back to a zero error and some areas will produce extra power for unscheduled export to other areas when loads change. The desired objective is to have each control area address local power demand variations, reset the AC frequency back to the rated or reference value, and maintain any contractual tie line power flows.

Conventional LFC relies on the tie line bias control. Each area linearly combines the speed/frequency error signal and the tie line power flow error signal into the area control error (ACE). The area control adjusts the mechanical power to drive the ACE to zero. We can control the proportion of speed/frequency error to the tie line power error with the area bias factor, K_{bias}.

$$\text{ACE} = \sum_{i=1}^{n_{ties}} \Delta P_{tie,i} + K_{bias} \Delta \omega \tag{4.19}$$

The area bias factor K_{bias} determines the amount of interaction the area will have during a disturbance in the neighboring areas. When K_{bias} is set equal to the frequency bias factor of that area, an overall satisfactory performance can be achieved. That is:

$$K_{bias} = \frac{1}{R} + D \tag{4.20}$$

The frequency factor D is never precisely known and is load dependent. Hence, the bias factor will come from estimated values. The ACE equations for a two-area system are:

$$\text{ACE}_1 = \Delta P_{tie,12} + K_{bias1} \Delta \omega_1 \tag{4.21a}$$
$$\text{ACE}_2 = \Delta P_{tie,21} + K_{bias2} \Delta \omega_2 \tag{4.21b}$$

where $\Delta P_{tie,12}$ and $\Delta P_{tie,21}$ are the departures from the scheduled power interchange. The ACE is used as an actuating signal to activate a change in the generator's power set point. When steady state is reached, $\Delta P_{tie,12}$ and $\Delta \omega$ will be zero. The integrator gain constant K_I must be chosen small enough not to cause the area to go into a hunting or chase mode. Figure 4.15 shows the block diagram of a simple AGC for a two-area system. We can extend the tie line bias control to any size system with n control areas.

Continuing with the last example, we add the ACE and reset integrator to the two generators. Table 4.6 gives the parameters for the added elements. Area 2 again experiences a load increase ΔP_{load2} of 10%.

Figure 4.16 graphs the speed/frequency response of each area. In 20 seconds, the example system settles to a new steady state with the frequency deviation $\Delta \omega$ in both areas returning to zero.

Figure 4.17 graphs the power of the two generators at the valve point in the control loop. The figure also plots the power flow across the tie line. In control Area 1, the generator power returns to its original value ($\Delta P_{valve1} = 0\%$) after the transient response. The generator in control Area 2 takes on all the increased load. The tie line power flow returns to its pre-disturbance value. Because the frequency across both areas returns to the rated value, the frequency-dependent load component also returns to its original level; therefore, the total generation increases by 10%.

4.2.6   AGC with Generation Allocation

We see in Figure 4.15 the scheduled power input to each generator. The scheduled power command comes from the energy control center as part of the hierarchical control scheme. The scheduled

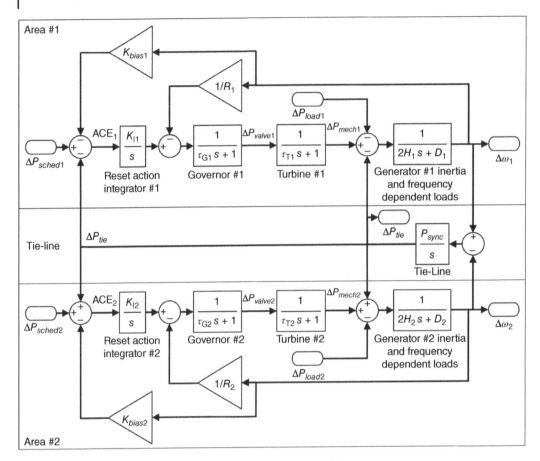

Figure 4.15 Control block diagram for a two-area system with AGC.

Table 4.6 Additional parameters for a two-area system with AGC.

Parameters	Area 1	Area 2
Reset action integral gain, K_I	0.40	0.40
Bias factor, K_{bias}	27.0	18.0

power command consists of the desired generation output to meet minimum operating cost, fuel cost, or economic dispatch and any constraints to ensure normal operation in case of the loss of a generators or transmission line asset.

The optimal dispatch of generation may be treated within the framework of LFC. In direct digital control systems, the digital computer is included in the control loop which scans the unit generation and tie line flows. These settings are compared with the optimal settings derived from the solution of the optimal dispatch program. If the actual settings are off from the optimal values, the computer generates the raise/lower commands, which are sent to the individual generator units. The allocation program will also account for the tie line power contracts between the areas.

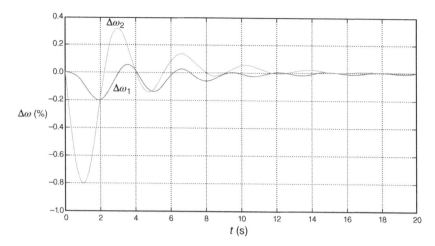

Figure 4.16 Speed/frequency response for a two-area power system with AGC. Load demand in Area 2 increased by 10%.

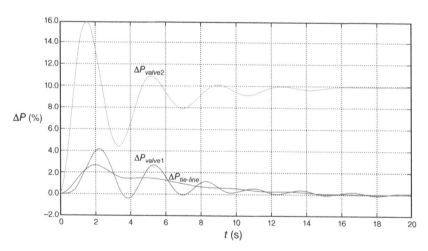

Figure 4.17 Turbine powers and tie line power flow for a two-area power system with AGC. Load demand in Area 2 increased by 10%.

The energy control center also employs other concepts of the modem control theory, such as state estimation and optimal control with linear regulator utilizing constant feedback gains. In addition to the structures that aim at the control of deterministic signals and disturbances, there are schemes that employ stochastic control concepts (i.e. minimization of some expected value of an integral quadratic error criterion).

4.3 Reactive Power and Voltage Control

4.3.1 Background

Within the turbine-generator set, the excitation system establishes the magnetic field, which induces a voltage on the winding conductors when the shaft is rotated. The diagram given in

Figure 4.1 shows a simplified voltage control with an AVR and exciter. The strength of the magnetic field on the generator's rotor determines the magnitude of the induced voltage. The generator's terminal voltage is the difference between the induced voltage and a voltage drop across the winding conductors. The magnitude of the load or armature current determines the size of the voltage drop, which is proportional to the winding inductance. The voltage control uses a feedback loop to hold the magnitude of the terminal voltage at a specified level. The AVR compares the terminal voltage with the reference or desired operating voltage and outputs an error signal to increase or decrease the size of the magnetic field. An amplifier boosts the error signal and applies appropriate limits to drive the power supply for the rotor's electromagnet.

When load demand increases, armature current grows and causes a bigger voltage drop across the generator's windings. The terminal voltage droops to a lower value. A voltage transducer measures and converts the sinusoidal AC voltage waveform into a signal that is proportional to the root mean square (RMS) value. The AVR senses the voltage drop and commands the amplifier to boost the magnetic field. The stronger magnetic field induces a larger voltage in the windings, compensating for the voltage drop in the windings. The result is a recovery of the terminal voltage.

AC power circuits have two types of power flows: active and reactive power. The active power flow represents the mechanical power from the prime mover transported to the end-users connected loads, performing real work. Reactive power is like the active power flowing through the generator, but with one major difference. Reactive power does not provide or consume energy, but merely causes energy to move between the magnetic fields surrounding conductors and the electric fields between circuit elements. The capacitance and inductance elements (both intentional and parasitic) cause the flow of reactive power. Reactive power flows across the power system according to the voltage magnitude differences between locations. Most power system components and customer loads are inductive in nature (e.g. transmission lines, transformers, and motors) and will absorb reactive power. We speak as if components supply or absorb reactive power, which is an arbitrary nomenclature.

Generators can supply or absorb reactive power depending on the magnitude of the induced voltage. To support voltages across the power system, the generator must maintain the terminal voltage and supply some reactive power to the system. Increases in the reactive power loading cause voltage drops across the system and across the generator windings. The voltage control indirectly controls the flow of reactive power from the generator by the difference between the terminal voltage and then induced voltage. Changes in the active power demand affect essentially the generator frequency, whereas a change in the reactive power affects mainly the voltage magnitude. The interaction between voltage and frequency controls is generally weak enough to justify separate control loops.

Besides generators, capacitors also supply reactive power. Utilities use switched capacitor banks in the medium voltage distribution networks to support sagging voltages on rural lines and near industrial loads. By injecting small amounts of reactive power, voltage drops along feeder lines are reduced. Autonomous slow controls sense local load conditions along with the feeder voltage and engage or disconnect stages of capacitors as needed. Capacitor controls are designed to operate at the most a couple of times during the day. Capacitor switching is a very localized control that does not impact the transmission network significantly.

Another supplementary method of improving the voltage profile on distribution networks are voltage regulators. In simple terms, the regulator is a variable transformer that can boost or buck the voltage by a few percentages. Voltage regulators are generally controlled autonomously for local improvement of the voltage. Many regulators can receive a voltage set point remotely from the EMS. We will not discuss any further the supplementary reactive power support or their controls.

4.3.2 Voltage Sensor

We connect a potential transformer with rectification or a voltage transducer to the electrical terminals of the generator. The voltage sensor converts the AC voltage magnitude V_t to an RMS signal, V_S, which serves as the feedback signal. A simple first-order transfer function models the rectification delays in the sensor. The time constant τ_R is very small, and we may assume a range of 0.01–0.06 seconds. When working with normalized control signals, the value of K_R is approximately 1. The transfer function is given as:

$$\frac{V_S(s)}{V_t(s)} = \frac{K_R}{1 + \tau_R s} = \frac{1}{1 + \tau_R s} \tag{4.22}$$

4.3.3 Amplifier

The voltage control's amplifier compares the voltage reference signal with the feedback signal and amplifies the error difference of the two signals. This amplifier has a settable gain along with upper and lower signal limits and rates of change limits. A simple first-order transfer function models the basic amplifier. Typical values of K_A are in the range of 10–400. The amplifier time constant τ_A is very small, in the range of 0.02–0.1 second, and we often neglect it in analysis. The transfer function for the amplifier is given as:

$$\frac{V_R(s)}{V_e(s)} = \frac{K_A}{1 + \tau_A s} \tag{4.23}$$

where:

$$V_e(s) = V_{ref}(s) - V_S(s) \tag{4.24}$$

4.3.4 Exciter

There are varieties of exciter types, and the IEEE Standards document them with standard models. Modem excitation systems use an AC power source through a power converter. The output voltage of the exciter is a nonlinear function of the field voltage to compensate for the saturation effects in the magnetic circuit of the rotor. There is no simple relationship between the terminal voltage and the field voltage of the exciter. A reasonable model of a modem exciter is a linearized model for a given operating point, which accounts for the major time constant and ignores the saturation or other nonlinearities. In the simplest form, the transfer function of a modem exciter may be represented by a single time constant τ_E and a gain K_E. The time constant of modem exciters is very small. When working with normalized control signals, the value of K_E is approximately 1. The transfer function is given as:

$$\frac{V_F(s)}{V_R(s)} = \frac{K_E}{1 + \tau_E s} = \frac{1}{1 + \tau_E s} \tag{4.25}$$

4.3.5 Generator

The generator's induced voltage or EMF is a function of the iron core's magnetization curve. In the linearized model, the transfer function relating the generator terminal voltage to its field voltage can be represented by a gain K_G and a time constant τ_G. These constants are load dependent; K_G may vary between 0.7 and 1, and τ_G between 1.0 and 2.0 seconds from full load to no-load.

$$\frac{V_t(s)}{V_{ref}(s)} = \frac{K_A(1 + \tau_R s)}{(1 + \tau_A s)(1 + \tau_E s)(1 + \tau_G s)(1 + \tau_R s) + K_A}$$

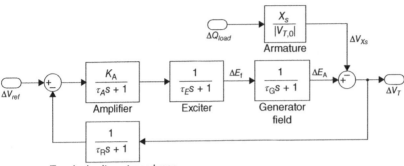

Figure 4.18 Control block diagram for a simple AVR.

When working with normalized control signals at the current operating point, the value of K_G is approximately 1. The transfer function is given as:

$$\frac{V_t(s)}{V_F(s)} = \frac{K_G}{1 + \tau_G s} = \frac{1}{1 + \tau_G s} \tag{4.26}$$

4.3.6 The Voltage Control Loop

The AVR uses the four elements to form the excitation control. A block diagram is shown in Figure 4.18. The closed-loop transfer function relating the generator terminal voltage $V_t(s)$ to the reference voltage $V_{ref}(s)$ is given as:

$$\frac{V_t(s)}{V_{ref}(s)} = \frac{K_A(1 + \tau_R s)}{(1 + \tau_A s)(1 + \tau_E s)(1 + \tau_G s)(1 + \tau_R s) + K_A} \tag{4.27}$$

Table 4.7 gives a set of parameters for an example generator with a simple AVR control loop. The signal values are percent changes from nominal values so that anyone can apply the control and its analysis to any size generator. Figure 4.19 gives the transient response of the terminal voltage for a 20% step increase in the reactive power demand. Note that the terminal voltage takes an initial drop, corrects with an underdamped response, and settles with a steady-state error of −0.8% voltage drop.

4.4 Excitation System Stabilizer

4.4.1 Rate Feedback Method

The example of a simple AVR demonstrates that for a small amplifier gain, the step response is not very satisfactory. High amplifier gain values result in an unbounded response. For better performance, we must increase the relative stability by introducing additional controller elements. One possible improvement comes by adding a zero to the open-loop transfer function. This is accomplished by adding a rate feedback loop, as shown in Figure 4.20. We can obtain a better response by

Table 4.7 Parameters for an example generator with a simple AVR.

Parameter	Gain values	Time constants (s)
Sensor	$K_R = 1.0$ %V/%V	$\tau_R = 0.40$
Amplifier	$K_A = 2.5$ %V/%V	$\tau_A = 0.25$
Exciter	$K_E = 1.0$ %V/%V	$\tau_E = 0.35$
Generator	$K_G = 1.0$ %V/%V	$\tau_G = 1.00$
Reactance	$X_S = 0.15$% of the generator base	
Terminal voltage	$|V_{T,0}| = 105\%$	

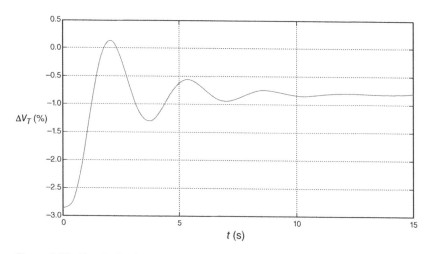

Figure 4.19 Terminal voltage response for a generator with a simple AVR. Reactive power demand increased by a 20% step.

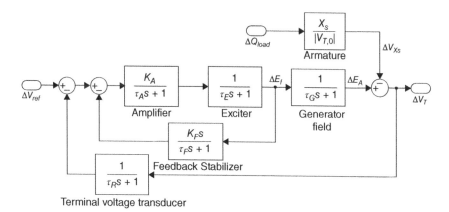

Figure 4.20 Control block diagram for a simple AVR with a feedback stabilizer.

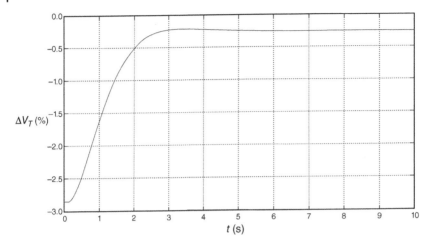

Figure 4.21 Terminal voltage response for a generator with a simple AVR and feedback stabilizer. Reactive power demand increased by 20%.

properly adjusting K_F. Continuing with the previous example, the added feedback stabilizer block has $K_F = 1.0$ %V/%V and $\tau_F = 1$ s. Figure 4.21 gives the transient response of the terminal voltage for the same 20% step increase in the reactive power demand. This time the terminal voltage is slightly underdamped, and the steady-state error is about 0.2%.

4.4.2 PID Controller

A proportional integral derivative (PID) controller can give even better performance over the rate feedback method. The PID is a common commercially available controller and can improve both the dynamic response and reduce or eliminate the steady-state error. The derivative controller adds a finite zero to the open-loop transfer function and improves the transient response. The integral controller adds a pole at origin and increases the system type by one and reduces the steady-state error due to a step function to zero. The PID controller transfer function is:

$$G_C(s) = K_P + \frac{K_I}{s} + K_D s \tag{4.28}$$

The block diagram of an AVR compensated with a PID controller is shown in Figure 4.22. Returning to the example, the PID controller has the parameters: $K_P = 1.0$, $K_I = 0.55$, and $K_D = 0.5$.

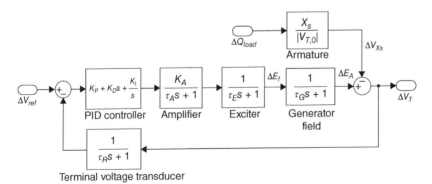

Figure 4.22 Control block diagram for a simple AVR with a feedback stabilizer.

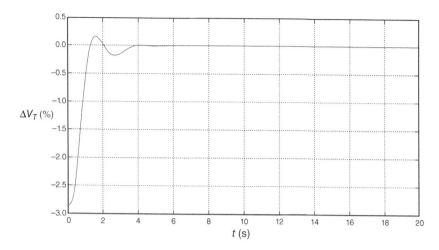

Figure 4.23 Terminal voltage response for a generator with a PID stabilizer. Reactive power demand increased by 20%.

Figure 4.23 gives the transient response of the terminal voltage for the same 20% step increase in the reactive power demand. Again, the terminal voltage is slightly underdamped, but the rise time is greatly improved, and the steady-state error is zero. This transient behavior is a more acceptable response.

4.5 Summary

The operational controls of the power system are associated with the generators and their prime movers. Other controls like network protection and relaying are generally associated with abnormal operating conditions such as short-circuit faults. The natural decoupling between real and reactive power flows permits separate control schemes for the AC frequency and the voltage magnitude. This gives the system operator two control points, the prime mover power and the rotor excitation, for each generator. The automatic generator control scheme is composed of two main control loops to keep the frequency and voltage magnitude within acceptable tolerances as load demand levels vary throughout the day. Tie lines aid a utility area by exchanging electricity for economic efficiency and during emergency conditions. The AGC regulates the generator output to ensures that power flowing across the tie lines is within the agreed amount while meeting the area's load demand.

Thoughtful Questions to Ensure Comprehension

1　Why are generators the primary control point in the electric power grid?

2　What is the function of the LFC in the AGC?

3　What is the function of the AVR in the AGC?

4　How does speed droop in prime movers help with control and stability?

5　How does voltage droop in the generator help with control and stability?

6 Why does the excitation control require a PID or a rate feedback?

7 What are control areas and what is the function of tie lines?

8 What is the difference between an AGC setup for a signal area and a multi-area power system?

9 How can AGC's operate concurrently in a multi-area power system?

Further Reading

Modeling and Analyzing the AGC and the Power Plant

Anderson, P.M. and Fouad, A.A. (2003). *Power System Control and Stability*, IEEE Press Series on Power and Energy Systems, 2e. Hoboken, NJ: Wiley. ISBN: 0-471-23862-7.

Grigsby, L.L. (ed.) (2012). *The Electric Power Engineering Handbook, Power System Stability and Control*, 3e. Boca Raton, FL: CRC Press, Taylor & Francis Group. ISBN: 13-978-1-4398-8321-1.

Ilić, M. and Zaborszky, J. *Dynamics and Control of Large Electric Power Systems*. New York, NY: Wiley. ISBN: 13-978-0471298588.

Machowski, J., Lubosny, Z., Bialek, J.W., and Bumby, J.R. (2020). *Power System Dynamics: Stability and Control*, 3e. Hoboken, NJ: Wiley. ISBN: 13-978-1119526346.

Murty, P.S.R. *Operations and Control in Power Systems*, 2e. Boca Raton, FL: CRC Press, Taylor & Francis Group. ISBN: 13-978-0415665650.

Vittal, V., McCalley, J.D., Anderson, P.M., and Fouad, A.A. *Power System Control and Stability*, IEEE Press Series on Power and Energy Systems, 3e. New York, NY: Wiley. ISBN: 13-978-1119433712.

Other Controls Used in the Electrical Grid

Altuve Ferrer, H.J. and Schwweitzer, E.O. III, (2010). *Modern Solutions for Protection, Control and Monitoring of Electric Power Systems*. Pullman, WA: Schweitzer Engineering Laboratories. ISBN: 13-978-0-9725026-3-4.

Fink, D.G. and Beaty, H.W. (1987). *Standard Handbook for Electrical Engineers*. New York, NY: McGraw-Hill, Inc.

Ibrahim, M.A. (2016). *Protection and Control for Power System*. Createspace Independent Publishing Platform, Amazon. ISBN: 13-9781537023809.

Kundur, P. (1993). *Power System Stability and Control*. New York, NY: McGraw-Hill, Inc. ISBN: 13-978-0070359581.

Kundur, P. (1994). *Power System Stability and Control*. New York, NY: McGraw-Hill, Inc.

Lewis Blackburn, J. and Domin, T.J. (2014). *Protective Relaying: Principles and Applications*, 4e. Boca Raton, FL: CRC Press, Taylor & Francis Group. ISBN: 13-978-1439888117.

Saadat, H. (2011). *Power System Analysis*, 3e. PSA Publishing LLC. ISBN: 978-0-9845438-6-1.

Wood, A.J., Wollenberg, B.F., and Sheblé, G.B. (2014). *Power Generation, Operation, and Control*, 3e. Hoboken, NJ: Wiley. ISBN: 978-0-471-79055-6.

5

Communication Architecture

Chris Dyer

Senior Project Engineer, POWER Engineers, Inc., Meridian, ID, USA

Objectives

This chapter summarizes the history of power system communication, reviews the different components typically used, identifies the major topologies, and introduces emerging concepts that will continue to stress the reliance of system operation on high-speed communication systems.

5.1 Introduction

Early power system architectures were designed around a central generation facility and local loads. Typically, if there were a system malfunction the power system would "black-out" and the outage would last until the source of the failure was determined and fixed. Customers were relatively accustomed to these outages and, even during switching operations or other system reconfigurations, loads were often offline as field operators communicated via technology of the time and manually switched in and out sections of the system.

As technology allowed advances in system protection and communications, these practices evolved along with customer expectations and lack of tolerance for long-term outages. System protection systems were soon combined with telephony systems for rudimentary remote control schemes to allow for operation from a centralized location providing quicker response and action. Further advancements, such as the ability to introduce digital data on traditional voice systems (e.g. MOdulater/DEModulator [MODEM]), allowed for further implementation of control systems that provided two-way communication and the modern supervisory control and data acquisition (SCADA) system was born. In the 1960s and 1970s, manufacturers began developing remote terminal units (RTUs), which brought limited discrete data from remote outstations to a centralized control facility. Data included both digital state data (digital inputs) and digitized analog data representing power system measurements analog inputs. The systems also provided control outputs at the outstation that could be operated remotely (digital outputs [1]). Operators could now, in near real time, monitor the condition of the power system throughout the grid, send a control to switch in/out sections of the grid, and receive near instantaneous feedback on the success of these operations. Soon there were several manufacturers and vendors competing heavily for the new market; custom communication protocols were usually developed along with the proprietary RTUs and energy management systems (EMS) such that other vendors RTUs

Resilient Control Architectures and Power Systems, First Edition.
Edited by Craig Rieger, Ronald Boring, Brian Johnson, and Timothy McJunkin.
© 2022 The Institute of Electrical and Electronics Engineers, Inc. Published 2022 by John Wiley & Sons, Inc.

would not communicate on a different vendor's system. This was not sustainable and soon efforts to develop a standardized, open protocol began.

The next advance in SCADA and communication technology began in earnest in the 1980s with the availability of powerful, reasonably priced microprocessors. Manufacturers were taking advantage of these products and increasing the functionality of system components like protective relays, meters, voltage regulators, and other control devices. The added functionality soon included communication capabilities and the industry began to coalesce around the earlier effort to create common communication protocols for utility use. The primary example of this in the North American market is the distributed network protocol (DNP), which has been adopted as IEEE Standard 1815. The industry began to incorporate those adopted standard protocols in various devices, now known as intelligent electronic devices (IEDs), such that information gathered during the normal course of these devices functions can easily be shared within the SCADA system. RTUs began to evolve into communication gateways or "data concentrators," which, rather than having discrete input/output (I/O) connections, now had a variety of communications ports to gather and transmit data between local IEDs and the centralized control center. Additional advances into the 1990s would soon see computer platforms that allowed for the replacement of traditional "mimic-boards" in operations center, which provided a graphical interpretation of the power grid, to be replaced by customizable computer screens that could be quickly modified and adapted as necessary with minimal effort compared to past systems. These graphical user interfaces (GUIs) were expensive, but as technology progressed, soon were cost-effective for deployment as local interfaces in substations and smaller portions of the grid.

As the desire and need for more information from remote areas of the grid grew, the communications system that served the data evolved. Communications systems advanced from the 1200 bits per second (bps) (common even in the 1990s), to transmission rates in the giga-bits per second (Gbps) (common in modern "trunk" or "backhaul" communication links). Even localized substation communication in modern substations are routinely designed around high-speed Ethernet communication systems that connect IEDs to each other and data concentrators.

As systems have progressed over the decades to provide more and more data, system operators, engineers, and other users have become highly dependent upon the data to provide reliable system performance, customer service, and market functions. Interruptions in the SCADA systems or communication have become highly intolerable. These requirements have led to continued advancement in engineering and designs of the systems to provide redundancy and reliability to the communication and data gathering systems, as well as the electric power system itself. Entire management systems are now dedicated solely to ensuring the continued operation of the communication system that is the backbone of this data collection empire.

This chapter will further discuss the communication systems and its makeup.

5.2 Communication Media

Electric power utility systems use a variety of methods of telecommunications to transmit data for purposes as wide ranging as business enterprise services, voice and email data, high-speed data for system protection relays, operational data for grid management purposes, and a vast range of engineering data to maintain infrastructure and apparatus throughout the electric power grid. This telecommunication system is transmitted typically in three different ways on three basic communication mediums. A communication medium can be defined as any substance through which data are transmitted. Typically, this does not define the means of transmission; however, in our discussion, the means and physical substance will be discussed together. We will focus on three dominant mediums used in electric power system communications: copper wire, fiber-optic cable, and radio-frequency (RF) communications. The emphasis in this chapter will be on digital data transmission rather than analog data transmission. Analog data transmission, such as amplitude modulation (AM) or frequency modulation (FM), are used on occasion in power systems communication, but the focus here will be on digital transmission. In digital transmission systems, data is binary, existing in two discrete states that are transmitted across the mediums discussed below.

5.2.1 Copper Wire

While copper wire may not be the first communication medium used by people, it is the most widely used throughout history, even into the modern era. Copper wire is used to transmit electricity wherein the data are inserted in various manners. We will focus on the predominant signal types associated with copper wire used in utility communications. The two predominant signal types utilize voltage references to differentiate the binary data being sent. In other words, one voltage represents a "1" and the second voltage represents a "0." Often signals of this nature on a copper wire are colloquially referred to in the utility industry as "serial" because the data are sent one bit at a time down the wire with different methods (protocols/handshaking) to indicate the beginning and end of a message. Various standards have been developed to define how these messages are to be constructed, including the electric signals involved and how to differentiate the binary bits on the wire.

5.2.1.1 Telecommunication Industry Association (TIA)/Electronic Industries Association (EIA) RS-232

The Electronic Industries Association (EIA) developed the Recommended Standard (RS)-232 and the standard has since been updated and revised by the Telecommunications Industry Association. The three prefixes for the standard are often still used interchangeably and referred to in literature [2]. The Telecommunication Industry Association (TIA) RS-232 standard describes a signaling technique with two data signals, a transmit and a receive signal, referenced to a ground signal. The standard describes the electrical characteristics to adhere to the standard so that transmitting and receiving equipment can differentiate between a "0" and a "1" in binary. Typically, the voltage on the wire is between 12 and 15 Vdc with high impedance resulting in a low current and a negligible voltage drop across the signal wire. The standard does not dictate a standard cable length but does provide for a maximum capacitance that the connecting communication circuit can tolerate. Typical "rule of thumb" practice is for a single run to be no longer than 50 feet under ideal conditions without interference. The standard was developed to be a point-to-point communication connection so communications between multiple devices on one "bussed" cable is not intended. Work arounds have been developed but are not suggested due to the lack of reliability in that configuration.

5.2.1.2 Twisted Pair (TIA RS-485)

The RS-485 standard describes a signaling technique that is also voltage referenced, like the RS-232 standard. This signal requires at a minimum two wires. The difference in voltage between the two wires is the measured voltage to identify the signal. A common/ground wire is generally used, but technically not necessary. With a 2(3)-wire configuration such as this, only one-way data transmission is possible, referred to as "half-duplex." If two devices are on an RS-485 system with 2(3) wires, one device can transmit and the other device receives. Only when that transmission is complete can the other device transmit a response back to the first device because the same signal wires are used.

To alleviate this a 4(5)-wire option is available with two sets of differential signals, one for transmitting and one for receiving. This allows for a "full-duplex" scenario where two devices can transmit simultaneously. The typical use case for RS-485 wiring is to use a twisted pair with conductors that are twisted within the cable sleeve to reduce electromagnetic interference. This, combined with the differential signal, helps RS-485 achieve higher bandwidth communication over longer cable runs as compared with RS-232.

Cable markings in RS-485 can be confusing since the signal is a differential voltage between two references. Commonly one signal cable is referred to as "A" and the other "B." The binary states are often then referred as "1" when B is a higher voltage in reference to A and "0" when A is a higher voltage in reference to B. Other references seen in industry are "+" or "TX+/RX+" for "B" and "−" or "TX−/RX−" for "A" [3]. As with RS-232, these are often not used or agree with different vendors/manufacturers and documentation should be consulted for specific devices.

5.2.1.3 Twisted Pair (Ethernet [10Base-T])

The Ethernet standard is defined in IEEE 802 to describe the common local area network (LAN) technology used for computing systems today. Ethernet systems have been in use in utility business networks for several years and in the past decade have become common in Operational Technology networks, including substation networks.

The original Ethernet standard utilized coaxial cable with limited bandwidth. This system has become largely obsolete. The most common Ethernet cabling in use today is twisted-pair copper cable and fiber-optic cables. The most common wiring pattern for twisted-pair Ethernet cables uses eight wires twisted in four pairs of differential signaling wires terminated using RJ-45 termination plugs (see Figure 5.1).

The terminations are defined by the TIA 568 standard. The difference between T568A and T568B termination assignments is the transmit and receive cables are swapped. Most twisted-pair Ethernet cables use T568A on both ends with "cross-over" functions in the connected devices. In the case a true cross-over cable is required, the T568A and T568B can be used together to create this function. In more recent years, increased bandwidth has introduced the 100Base-T and 1000Base-T variants using physical cabling with higher tolerance for noise immunity. More recently power has been introduced in unused pairs in Power over Ethernet (PoE) standards.

5.2.2 Fiber-Optic Cable

Fiber-optic cable is similar in application as copper electrical cabling and is often installed alongside other copper cables, though it follows a stricter procedural installation practice due to the nature of the cable. Rather than carrying data signals using electricity as copper cables do, fiber-optic cables

Figure 5.1 RJ-45 termination plugs.

are constructed of multiple plastic or glass optic cables that carry light as the signaling method. The cables are bundled and protected in a variety of ways depending upon applications. Fiber-optic cable has an advantage over copper cable in certain applications. Since the data are being transmitted using light and not an electrical signal, the typical cause of signal degradation (electromagnetic noise) is eliminated. Fiber cables can safely transmit data through high noise environments without introducing complicated designs or shielding to protect the signal. In addition, the optical cable is immune to ground fault rise situations that are common in a substation yard during a grid fault incident. Fiber cables can be installed between equipment in the substation yard and the control building without designing intricate protection systems to prevent voltage variation issues during a fault event. The two main types of optic cables used are classified as single-mode and multi-mode. The cables differ in size and light propagation patterns.

Single-mode fiber typically has a fiber core with a diameter of 9 μm and a cladding of diameter 125 μm. The signal is transmitted in wavelengths of typically 1310 and 1550 nm. Multi-mode fiber usually has a core diameter of either 50 or 62.5 μm and a cladding of 125 μm. Multi-mode fiber, due to its larger core, supports more than one propagation mode, as opposed to a single mode [4]. The core differences also create different dispersion behaviors. These factors lead to two major differences between the two types of fiber.

Multi-mode fiber ends up having a higher degradation of signal and cannot run without repeaters/boosters for more than a mile or so. However, due to the light characteristics involved in the larger core size, the transmitter electronics are of a lower cost. Due to this, typical practice seen in industry is to use multi-mode fiber optic in short-run installations, such as inside a building, plant, or substation yard. Single-mode fiber is more often used in longer, backhaul applications, such as between substations and between substation and control centers.

Fiber-optic cables must be installed and pulled through ducts and cable trays very carefully, always keeping the "maximum allowed bending radius" of the cable (see Figure 5.2). This information is provided by the cable manufacturer and must be adhered to very closely.

5.2.2.1 Optical Ground Wire (OPGW)

Ground or shield wires are common in many transmission lines of varying voltage levels. The wire is intended to assist in "shielding" the power conductors from overhead lightning strikes by providing a quick path to ground. Optical ground wires (OPGWs) took the standard shield wire conductor and embedded a fiber-optic cable made up of multiple individual optical cables in the core (see Figure 5.3). These cables run the length of the transmission line along with the power conductors. Utilities will use these OPGW cables to facilitate communications between substations for SCADA,

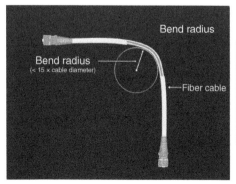

Figure 5.2 Maximum bending radius.

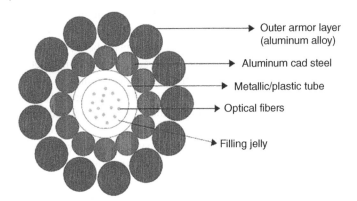

Outer armor layer (aluminum alloy)

Aluminum cad steel

Metallic/plastic tube

Optical fibers

Filling jelly

Figure 5.3 Individual optical cables in the core. Source: Ref. [5].

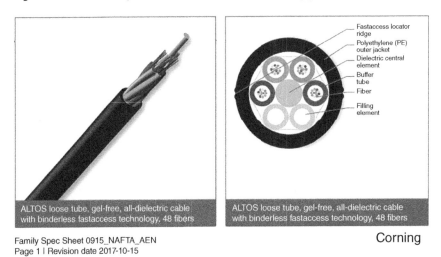

Fastaccess locator ridge

Polyethylene (PE) outer jacket

Dielectric central element

Buffer tube

Fiber

Filling element

ALTOS loose tube, gel-free, all-dielectric cable with binderless fastaccess technology, 48 fibers

ALTOS loose tube, gel-free, all-dielectric cable with binderless fastaccess technology, 48 fibers

Family Spec Sheet 0915_NAFTA_AEN
Page 1 I Revision date 2017-10-15

Corning

Figure 5.4 All-dielectric self-supporting cable. Source: Ref. [6].

protection, and a variety of other services. These cables must be sized accordingly for their applications to protect the fiber optics and the cable itself from thermal damage.

5.2.2.2 All-Dielectric Self-Supporting (ADSS) Cables

Another backhaul solution is the All-Dielectric Self-Supporting (ADSS) Cable. These are often used in distribution systems and are "underhung" below the power conducting cable (see Figure 5.4). These cables are designed with strength members along with the fiber-optic cables, cladding, and jackets to allow them to be "self-supporting" between tower spans. There are other types of all-dielectric fiber-optic cables that can be put into ducts or even direct bury into the ground.

5.2.2.3 Underground Cables

Burying the fiber cables often provides more security, though potentially for a higher price. The two major options for underground cables are a non-ruggedized cable that is pulled through a protective conduit and "direct-buried" cable, which is jacketed in armored shielding to protect from the rigors of being directly buried in the ground without a conduit. In addition, there are special cables designed to be installed underwater for submarine applications that are sealed to prevent water penetration.

5.2.2.4 Splice Box

Fiber-optic cable, regardless of whether single-mode or multi-mode is used and regardless of what type of installation application is used, will only be produced in reasonable length bundles. To span distances, the cable must be spliced at certain points. Unlike copper cable, these splices are not trivial, and the glass fibers must be exposed during the splice. Splice boxes are space where the spliced ends, themselves protected after the splicing, are stored to protect them further. In addition, excess cable is often stored along with splice boxes. For aerial installations, the cable is generally run down the tower structure and the splice boxes are located at the base of the towers. For underground and submarine installations, the splice boxes are often located in vaults that are themselves buried with access hatches as necessary. The other reason for splicing the fiber-optic cable is if there are sharp turns of the cable route through which the fiber-optic cable cannot be safely pulled. In that case, it needs to be pulled in two sections and spliced at the turning point.

There are generally two methods of splicing fiber that are utilized: fusion and mechanical splicing. As tools and techniques have improved over the past decade, fusion splices have become cost-effective and are preferable due to the reduced attenuation of the splice. In a simple mechanical splice, the ends of the glass strands in a fiber-optic cable are aligned and mechanically held in place by an assembly. The fibers themselves are not joined; they are nearly adjacent such that light passes through one end into the other. Mechanical splicing is usually used only as a temporary solution (see Figure 5.5).

However, fusion splices effectively "weld" the two ends of the glass fibers together; therefore, it requires more technique to align, heat, and fuse. Typical values vary depending upon source, but the attenuation of a fusion splice can be conservatively approximated at 0.3 dB. However, well-spliced joints should provide lower attenuation.

5.2.2.5 Fiber-optic Terminations

A fiber-optic cable must be terminated to be connected to an end device, such as an IED. Unlike copper cable, bare fiber cannot connect to a terminal lug. Fiber-optic terminations come in a variety of styles and sizes that have been standardized over the years. The most common ones are noted below [7]:

- Standard connector (SC): Single fiber termination, rugged, inexpensive, common in legacy equipment (see Figure 5.6).
- Straight tip (ST): Single fiber termination, secure, rugged (see Figure 5.7).
- Lucent connector (LC): Also colloquially referred to as "Little Connector," it is a smaller form factor for high-density I/O modules. Single fiber termination, but often paired with another connector and fashioned as a "duplex" connector (see Figure 5.8).
- Mechanical transfer registered jack (MT-RJ): A smaller form factor for high-density I/O modules (see Figure 5.9). It is often paired for duplex connections.

Figure 5.5 Mechanical splicing.

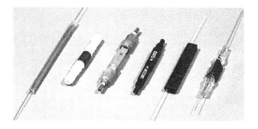

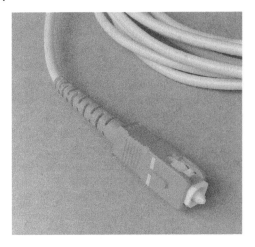

Figure 5.6 Standard connector.

Figure 5.7 Straight tip.

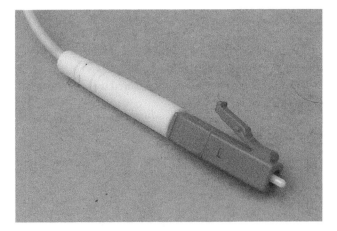

Figure 5.8 Lucent connector.

Figure 5.9 Mechanical transfer registered jack.

Figure 5.10 Patch panel.

5.2.3 Patch Panel

Most backhaul or trunk fiber contains numerous fiber cables. Typical end devices will require only two fiber-optic cables (transmit and receive). The trunk needs a termination spot where the fiber pairs can be broken out to individual devices. This often occurs at "patch panels." A patch panel provides one side that the multi-fiber cable can be terminated to, typically with fusion splices like other splices intermediate to the cable (see Figure 5.10). The other side of the patch panel will break out each individual fiber optic to a termination socket. Numerous terminations for fiber are not discussed here, but the terminations protect the end of the fiber and are designed so that patch cables can be installed and removed repeatedly without damaging the fiber cable itself.

5.2.3.1 Patch Cables

The final run from the patch panel to the end device is accommodated by what is commonly referred to as a "patch cable" or "patch cord." These can be either simplex cable with one strand of fiber in a protective jacket or duplex fiber, which essentially consists of a pair of single strand fibers, individually jacketed, but constructed such that the jackets are connected for the run of the patch cable with terminations that act as one unit with two terminations. These cables are used for a single communication between devices with both a transmit fiber and a receive fiber. They will be terminated at both ends with the appropriate terminations for the patch panel and the end device. It is not necessary that these are the same and often are not. The important item is that the fiber type be the same from end to end through to the light-transmitting device to the receiving device. A multi-mode cable cannot be mixed with a single-mode cable without first passing through an intermediate device, which would retransmit the signal.

5.2.3.2 Fiber-optic Loss Calculations

Optical power loss in a fiber-optic cable can be determined by summing the attenuation of the cable, splices, and any terminations. Measurement of optical power is expressed in units of dBm. As a decibel (dB) is a unit of power in reference to another level on a logarithmic scale, dBm is used to describe absolute power with a reference of 1 mW. Loss measurements are described in dB and are shown as negative as they are less than the reference power [8]. When designing fiber circuits, the fiber-optic power budget must be calculated to ensure there is enough light power to successfully traverse the fiber circuit. The basic information required is:

- Minimum transmit power (in dBm) – The minimum amount of light power generated by the transmitter equipment
- Minimum receiver sensitivity (power in dBm) – The minimum amount of light power at the receiving equipment necessary to register the signal and operate without error.

To calculate the power budget simply subtract these two values:

- Power budget = Minimum transmit power – Minimum receiver sensitivity.

To determine if a circuit design is adequate without repeaters, the total attenuation of a dB of the circuit must be calculated. This is a simple exercise of adding up the losses along the circuit. Typical "rule of thumb" losses can be used, but detailed attenuation specifications should be sought for the equipment and material used in each design. Fiber-optic cable attenuation, splice loss, and connectors losses must all be considered. To determine if the design is adequate, subtract the losses from the power budget. If the value is greater than zero, the design should be valid.

- Power budget – Total loss ≥0.

It is even better if the power budget is at least 3 dB higher than the total loss; this would ensure enough optical power over the years and aging of the fiber and connectors, less-than-perfectly clean connectors, etc. [4].

5.2.4 Radio-Frequency (RF) Communications

RF communications generally encapsulate various communication methods that utilize radio transmission of data via electromagnetic waves of various frequencies (see Figure 5.11). Different transmission and connection patterns are also considered.

5.2.4.1 Microwave

Backhaul or trunk communications are often accomplished with fiber-optic OPGW in modern systems, but there is still a variety of utilities that utilize microwave transmission due to existing infrastructure or to provide route diversity. Microwave transmission frequency is measured in gigahertz (GHz) and can support high-bandwidth data transmission. Antennas are directional, focused, require line-of-sight (LoS), and care must be taken in the design to ensure path continuity.

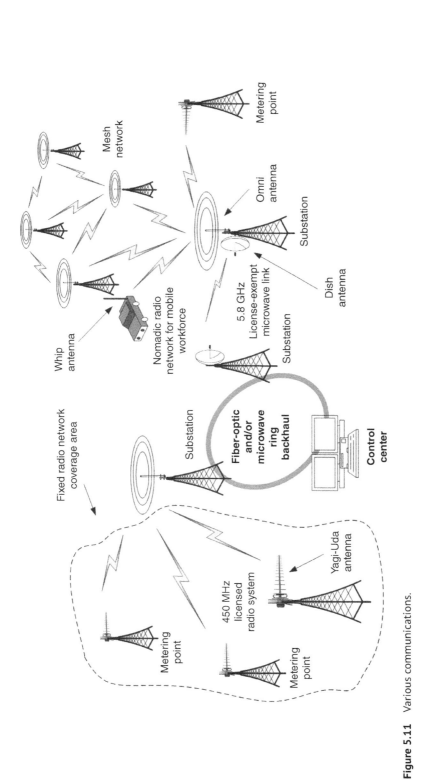

Figure 5.11 Various communications.

5.2.4.2 VHF/UHF Radio

Very-high frequency (VHF) and ultra-high frequency (UHF) are LoS and non-LoS radio transmissions with frequencies measured in megahertz (MHz). VHF is typically 30–300 MHz and UHF 300–3000 MHz [9]. VHF/UHF can be directional (point-to-point) or omni-directional (point to multi-point or broadcast) depending upon the antenna configurations. Bandwidths are not as high as microwave data transmission but are often used for SCADA and distribution automation (DA) functions where devices are spread across an area making wired communications impractical. Many frequencies require fixed licenses to be obtained by the owners to prevent numerous transmissions at the same frequency leading to interference.

5.2.4.3 Spread-Spectrum Communication

Spread-spectrum communication is a special form of unlicensed radio transmission that utilizes "frequency hopping." The transmitter and receiver are both coded with randomly generated frequencies that are switched at the same time. The frequency hopping helps to prevent interference with other systems or eavesdropping. Spread-spectrum radios are often in the VHF/UHF frequency range but also in the microwave bands and commonly used for SCADA and automation purposes.

5.2.4.4 Mesh Communication Networks

Radios can also be deployed in mesh network topologies, usually at short distances. Typically, low-powered radios are used in this type of architecture. They are deployed such that each individual radio can be received by two or more other radios, and messages are relayed through the network as necessary to arrive at the appropriate destination radio.

5.2.4.5 Radio Propagation and Path Studies

When looking to design and deploy a radio system, it is necessary to understand the requirements of the system and the path between transmitter and receiver. Radio systems, especially those of higher frequencies (VHF, UHF, and microwave) prefer if not absolutely require LoS between the two nodes. Some systems can operate under partially obstructed conditions, but it is not the ideal case when designing a system. One of the initial tasks is to perform a Path Study between the expected locations of the two nodes to determine potential obstacles and provide initial design considerations, such as tower heights and antenna placement.

Path studies also includes radio power budget calculation and determining if the obtained fade margin is sufficient for the radio link. At least a 20-dB fade margin is required for lower frequency SCADA radio systems and at least 30 dB for microwave systems in the GHz frequency bands [4].

The same basic steps as those taken in fiber-optic design are considered, which identify the power specifications of the transmitter/receiver and the attenuation (losses) along the radio link. Losses would occur at terminations, connections, cables, and "free-space" propagation loss.

5.2.5 Local Area Networks

A LAN has colloquially come to refer to a collection of devices contained in a physical space, such as a small building, or department that are connected via Ethernet. This differentiates LANs from wide area networks (WANs) that carry across multiple locations, such as a campus or an entire corporation. Though Ethernet, both wired and Wi-Fi, is the most common technology used for LANs, the term could be used for systems that are interconnected via other communication systems, especially in power utility substation networks, which could conceivably consist of modern and legacy technologies in a hybrid design. We will explore LANs in three different environments and compare the technologies and uses in electric power systems, the business enterprise networks, the control center operation enterprise network, and the remote outstation networks.

5.2.5.1 Business Enterprise Networks

Most of our discussion about control systems communication networks will be focused on the operational network for an electric power system; however, there will generally be a business enterprise functioning alongside the operational enterprise system. The business needs in terms of communication are similar to those of the business networks in other organizations. Modern business networks have Information Technology (IT) departments solely dedicated to ensuring their reliability, security, and function. Reliability is typically designed with various levels of redundancy in equipment, infrastructure, and data storage. Recent trends have data storage being outsourced to "the cloud" with strict security protocols placed on access. Interactions with the operations enterprise is very limited with the primary exception being Customer Information Systems (CIS). Customer data are required on the business networking for obvious billing and marketing functions. In most modern utilities, customer energy usage information is obtained through a communication system to a "smart meter" at the customer site, which measures both the quantity of energy used and the time of use. This information is typically collected and transmitted over an operational network. In Advanced Distribution Management Systems (ADMS), smart-meter information can be used by the Outage Management Systems (OMSs) to determine if there is a system failure. In addition, customer contact information in the business network is often shared to the OMS. The data are shared between the systems and strict firewalls must be maintained to ensure customer information security as well as reliability of data.

5.2.5.2 Operational Enterprise Networks

Electric power systems and the North American grid have grown to be one of the most complex "machines" in the world, and due to the nature of electricity, the power consumed must be balanced continually by the power that is being generated. This operational nightmare is the job of electric power utility system operators in the generation, transmission, and distribution spaces. We will focus on the transmission and distribution spaces in this discussion. EMS and Distribution management systems (DMS) have been developed to help facilitate the operation of the system.

Energy Management Systems (EMS) EMS must actively monitor thousands upon thousands of discrete data points coming in from outstations and data created locally. An EMS is made up of a variety of disparate components working in concert together. These components can include:

- The SCADA component of the EMS is the data acquisition element that collects data from the outstations of concern for that operational area. The data are generally collected using standard SCADA protocols, such as DNP, via the front-end processor (FEP), and initially processed in a real-time database (RTDB). The data are forwarded to other modules as necessary.
- Generation management applications help to coordinate the necessary generation based upon real-time and scheduled demand. The application can consist of subsystems such as:
  - Load forecasting: Application that predicts expected load in various time frames.
  - Generator commitment: Application that schedules generation with commitments based upon type of generation (baseline, peaking, balancing intermittent renewables, etc.).
- Transmission management applications that assist in operating the ever-changing transmission system configuration using real-time data and manually entered data. These applications can consist of:
  - Network topology
  - State estimation
  - Power flow optimization.

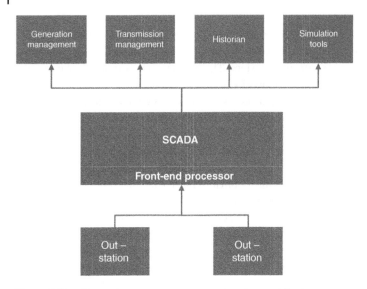

Figure 5.12 Illustrative energy management system architecture.

- Offline applications engineering applications that do not require real-time analysis including:
  - System simulation
  - Historian
  - System analysis
  - Training [9].

These systems can be obtained from a single vendor or be multi-vendor with a common data translation interface that allows for information to be passed directly. The high-level architecture is shown in Figure 5.12.

The EMS is commonly on a LAN isolated from other networks including the business network. Operators often need to work on separate isolated networks: one to operate the system and one for corporate systems (e.g. email). Due to a focus on reliability and availability, especially in larger utilities, the EMS systems are redundant with an architecture that could look like in Figure 5.13.

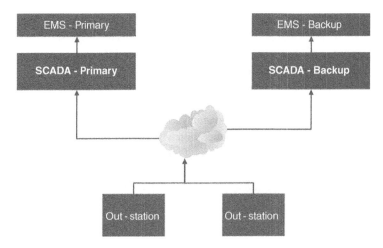

Figure 5.13 Redundant energy management system architecture.

The architecture shown in Figure 5.13 could be exaggerated even further to provide diversification in communication redundancy and physical separation. Architectures like this would come at a higher cost than others and could be required for larger systems that are vital to the security and reliability of the bulk electric system (BES).

Distribution Management Systems (DMS) The DMS is the energy management system for an electric power system's lower voltage distribution system. While there may be similarities in architecture for basic system operation functions, the distribution system introduces challenges of its own and has additional functions not typically seen in the transmission system. Real-time information for the distribution system is not as complete in some networks and systems must have the capability to react and operate with less data in some cases. In addition, the sheer number of connections in the distribution space is greater than the transmission including looped "doubly-fed" systems with two sources, and radial systems, both with taps for customer loads both large commercial and smaller residential. In addition to a central operation center like an EMS, the distribution system will occasionally contain automation systems that operate semi-independently from the central operation for certain activities. These actions are monitored and can be reversed as necessary, but automation can often provide system recovery in a shorter timeframe than can be obtained with manual intervention alone.

DMSs often have the following subsystems:

- SCADA systems similar to that in the EMS to collect data from the field for use in the DMS.
- Advanced metering infrastructure (AMI) including "smart meters" that have the capability of not only communicating energy usage information for billing back to the utility, but also providing real-time visibility of each load and system status at that load.
- OMSs that assist in recovery from system disturbances. Modern OMS will have ties to the AMI system to assist in determining the location or extent of an outage and can assist in direct customer outreach.
- Customer Information Systems (CISs) contain detailed customer information for billing and contact purposes and are considered one of the more confidential data in the utility Information and Operation Technology systems.
- Asset management systems provide a utility with visibility of high-value assets in their system including power transformers, circuit breakers, and other substation infrastructure. In modern architectures, these can be enhanced with condition-based monitoring, which provides real-time data on assets (e.g. transformer oil temperature and dissolved gas composition) that can aid in ongoing maintenance of assets as well as determining critical emergency loading conditions of the system.
- Geographic information systems (GISs) provide geospatial data for the distribution systems aiding in asset management and outage management.

The subsystems are connected via the distribution operational enterprise LAN, which is secured and separate from the business networks like the EMS LAN. An example architecture is shown in Figure 5.14 [9].

The DMS typically has some level of redundancy but may not be as diverse and complex as the EMS redundancy requirements. Security is still a priority for control systems of any size due to the potential influence of external actors as evidenced by the 2015 cyberattack on the Ukrainian power grid's distribution system. During that attack, hackers were able to gain access to the operational technology network via an unsophisticated spear phishing attack. Once inside the network the hackers were able to remotely operate the system and interrupt the electrical power supplied to

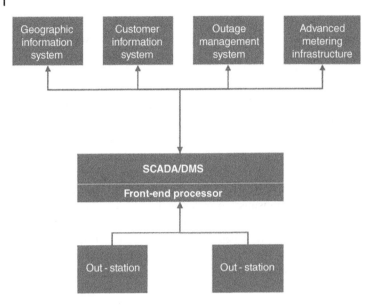

Figure 5.14 Illustrative distribution management system architecture.

customers and modify firmware in communication devices, slowing the eventual recovery of the power system.

Distribution Automation (DA) Beyond the distribution control center, communications between devices in the field contributes to what is generically called distributed automation and entails any automatic functions that occur based upon system inputs and dynamic changes in the system that is monitored, detected, and acted upon without manual intervention.

Numerous schemes could be implemented in the name of DA. A routine Fault Location, Isolation, and System Restoration (FLISR) with corresponding communication and data requirements is described below to illustrate.

Consider a distribution system resembling the architecture shown in Figure 5.15.

It is assumed the two feeders on buses A and B are operated radially with a Normally Open (NO) sectionalizing switch at S3. It is further assumed that under normal conditions, the two feeders can

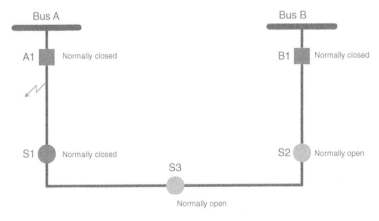

Figure 5.15 Distribution system example.

support the loads along the distribution line from B1 to A1 radially (i.e. substation B can support the entire line if breaker A1 is open and vice versa). The DA operation begins with fault detectors at substation A and sectionalizing switch S1 sensing the fault located between A1 and S1. The protective relays at substations A and B operate to isolate the fault by opening breakers A1 and B1. Communication systems between the three sectionalizing switches and the two substations use the information drawn from IEDs at all five locations to determine the location of the fault, verify the configuration of substation B is "normal" and able to carry the loads between S1 and S3, and automatically reconfigure the feeder by opening sectionalizing switch S1, closing sectionalizing switch S3, then finally reclosing breaker B1. The DMS system is notified, the changes in the system configuration and all alarms associated with the incident are displayed for the operator while crews are dispatched to the faulted line and customers affected by the disturbance are sent automatic messages with contact information drawn from the CIS.

Communication subsystems between the sectionalizing switches and the two substations could conceivably be designed using ADSS fiber-optic cables run with the distribution lines or possibly low-power LoS radios using spread spectrum. Fiber-optic cables would be inherently more secure, but solutions with encrypted messages in either medium would be considered secure by best practices. Communications between Substations A and B and the DMS would use traditional SCADA system communication paths either using a traditional serial SCADA signal or one of the networked systems discussed earlier.

5.2.5.3 Remote Outstation Networks

A site where any data is collected remotely from the central management systems is considered an "outstation" and can consist of any number of facilities. This discussion will focus on the evolution and topology of networks in electric power utility substations or switching stations, but networks like those described below are scalable and could appear in smaller forms in distribution reclosers, regulators, and other applications as discussed earlier.

Outstation networks of the past, as discussed earlier, were primarily an RTU with hard-wired I/O using a single communication connection, typically a modulated signal transmitted on owned or leased telephone circuits to a control center as shown in Figure 5.16.

As the technology advanced and IEDs were available with communication options to extract even more data, gateway devices were created to concentrate the data from the IEDs and send a summarized database to the control center. Early data concentrators developed unique ways to also collect hardware I/O-containing data that was not available via IEDs. Communications to IEDs varied, but typically involved RS-232 or RS-485 serial communications links using a standard utility protocol or, occasionally, a legacy protocol as necessary. An example of the basic architecture is shown in Figure 5.17.

In rare cases where an owner felt they needed additional reliability in the data from IEDs *AND* if supported by the IEDs, redundant communication links could be made to a single IED. The data would be collected by the gateway as two disparate databases and would either have to be served to the master station this way or logic developed in the gateway could serve one set of data to the master and transfer to the second set of data based upon a loss of communication to the IED on the primary channel. Figure 5.18 shows an example.

The next evolution of outstation networks began when Ethernet was introduced to the station. Communication management, reliability, and bandwidth increased as the technology progressed for substation Ethernet equipment. The network topology most often used is a star topology with an Ethernet switch (shown in Figure 5.19). If equipment supported dual Ethernet interfaces a redundant scheme could also be developed with two independent star networks (see Figure 5.20).

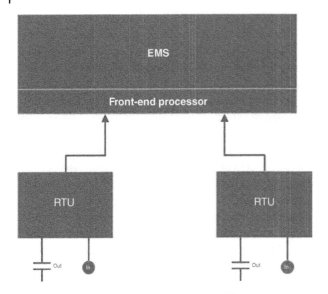

Figure 5.16 Legacy RTU communication architecture.

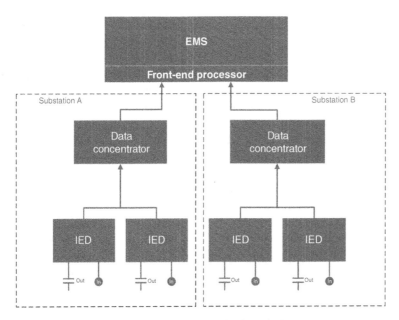

Figure 5.17 Data concentrator/IED communication architecture.

Many substations evolve from legacy devices to modern Ethernet devices without a wholesale replacement of equipment. This necessitates a hybrid approach that can support traditional RS-232/RS-485 serial communication and Ethernet equipment in the same network. An example of this is shown in Figure 5.21.

The next evolution of outstation communication architecture is being driven by the IEC-61850 standard. The standard provides data constructs that are object oriented and discoverable with the intent that the engineering design process for substations gains efficiency and substations of the future can be "digitized" further. The goal most often discussed when considering an IEC-61850

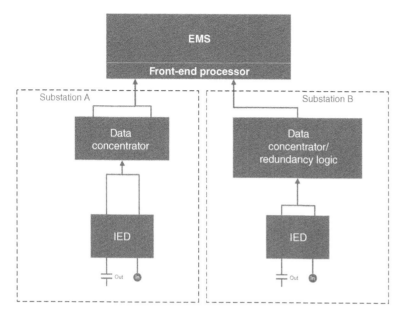

Figure 5.18 Data concentrator/IED redundant communication architecture.

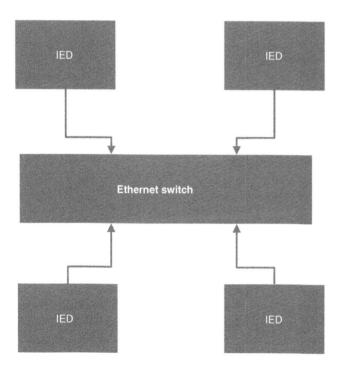

Figure 5.19 Ethernet star topology architecture.

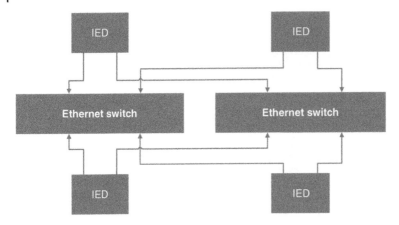

Figure 5.20 Redundant Ethernet star topology architecture.

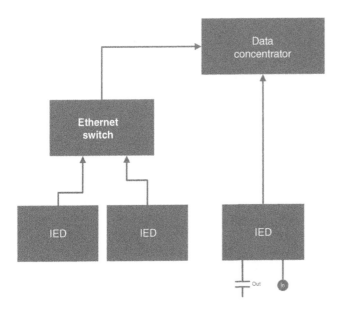

Figure 5.21 Hybrid communication architecture.

design is the replacement of signaling between devices in a substation with traditional control logic using voltage sensing and hardware I/O to one with digitized signals that are shared between devices on an Ethernet network. One goal is to reduce the sheer number of control wire cabling required in a substation, but other advantages are increased visibility of signals and the ability to maintain and test the system without lengthy power outages that affect customers. An example of this is a simple auxiliary relay that is used to multiply contact outputs. A system protection relay would be expected to operate closing a contact on the relay, which completes the circuit to assert voltage on the input of the auxiliary relay. The auxiliary relay would, in turn, close numerous contacts, thereby sending the signal from the original relay to several destinations. This is shown logically in Figure 5.22.

In a digital substation, the Generic Object-Oriented Substation Event (GOOSE) protocol could be used with intelligent devices. The source of the signal (the protective relay from the traditional

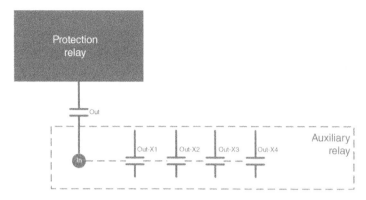

Figure 5.22 Auxiliary relay contact multiplication.

example) would "publish" a GOOSE signal containing the data that initiates the same action represented by the contact closure in the previous example. Receiving devices would be on the same Ethernet network and would "subscribe" to the publishing relay message. The message would be presented and repeated as per the IEC-61850 standard and received by the various destinations that would then act based on the scheme under consideration.

A different protocol described by the IEC-61850 standard is Sampled Values. Sampled Values would replace the copper cabling that supplies the power system measurement signals from the system to IEDs that are monitoring it. For a traditional system, these would be supplied by current and voltage transformers (CTs/VTs) that transform the high voltage/current signals on the power system to levels that can be input to protective relays, meters, etc. In a digital substation, a "Merging Unit" would be directly connected to the CTs/VTs and digitize the analog signal. Fiber-optic cable would connect the Merging Unit in the substation yard to the Ethernet network in the substation control enclosure. IEDs that would need the data would subscribe to the data stream via the connected network. Figure 5.23 shows an example of a system using GOOSE and Sampled Values.

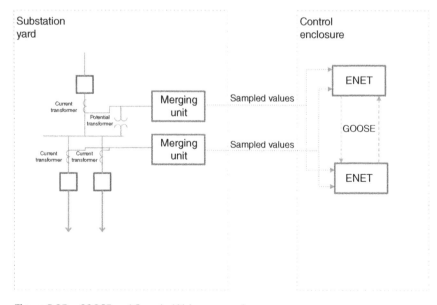

Figure 5.23 GOOSE and Sampled Values example.

The digital substation briefly described above would require a high-availability communication network. Redundancy in Ethernet networks has evolved over the years, along with other aspects of the communication architecture. Rapid Spanning Tree Protocol (RSTP) is a redundancy scheme that has been deployed in SCADA systems in recent years. The Ethernet protocol does not support looped networks, so interconnecting network switches to increase the number of paths possible for a signal for redundancy would cause a broadcast storm. RSTP prevents that by scanning the network, identifying the loops, and automatically shutting off ports, as necessary. Should the RSTP logic detect a change in the system (broken switch, cable disconnected, etc.), it would immediately scan the system for the changes and reroute the system, turning ports on and off as necessary to facilitate communications between all devices around the communication outage. RSTP is effective in SCADA systems where the scan intervals are often measured in seconds. In a system employing IEC-61850, however, the message requirements are measured in power cycles (i.e. milliseconds) and RSTP would not resolve quickly enough to maintain the necessary messaging streams for protective relay and other high-speed functions. Updated redundancy protocols have been developed to improve on RSTP include the Parallel Redundant Protocol (PRP) and the High-availability Seamless Redundancy (HSR). Both provide near-continuous redundancy with differing topologies.

HSR is designed as a ring with each device on the network being a node on the ring. Each device would have two connections. Any data packet sent from a source device would be simultaneously sent out each port (in terms of a ring one message could be considered traveling "clockwise" while the other is traveling "counter-clockwise"). The packets would travel along the ring until reaching their destination device. The destination device would receive one package and process it. If the second packet then arrived at the destination device, it would determine it was a duplicate and discard it. If there was a break in the ring such that one packet never arrived, there would be no delay aside from latency for travel along the ring.

PRP is structured in a more traditional manner in a star topology with an Ethernet switch. Each device on the network would still have two independent communication ports. The two ports would be connected to identical, isolated networks. Like the HSR network, a device would send two identical packets: one from each port and onto the two networks. The receiving device would receive one of the two packets first and process it, discarding the second packet if it is received on the second network. In the event of a network failure or connection loss on one of the networks, the receiving device would only receive one packet. Since both packets are sent simultaneously in both cases, there is no latency if there is an N-1 outage condition on the networks.

One consideration for these advanced networks is the maintenance required. Networks of this complexity in the Operational Technology environment are likely to be maintained like those operated in the IT environment seen in the business and enterprise networks. As substations are no longer isolated by legacy serial communications links and are now connected into a larger network, consisting of the control center, outstation, and maintenance networks, security also becomes a higher consideration.

5.2.6 Backhaul Communications

A "trunk" or "backhaul" communication system is one that connects separate localized networks. As discussed elsewhere in this chapter, legacy systems were often connected to the control center via modulated analog signals using technology from legacy telephone systems. As the communications demands increased and the communications technology turned toward digital communications, these systems evolved. Larger electric power utilities, especially transmission system operators, have utilized the availability of OPGW on utility-owned infrastructure to provide portions of the backhaul system. Other mediums include microwave and leased space from telecommunications corporations providing digital channels to outstations. In larger stations with reliability being a large concern, diversification of communication path and technology is often desired and redundant links are provided via different media and technologies.

Two technologies that have been used in backhaul communications are time division multiplexing (TDM) and multiprotocol label switching (MPLS). TDM is the technology developed for digitized voice communications, which divides a communication channel into channels divided by 64 kbps labeled DS0. T1 (or DS1) consists of 24 DS0s multiplexed into a single stream. When used for data, the DS0 channels can be combined to increase the bandwidth of a data channel. For example, two DS0s combined provide a bandwidth of 128 kbps. Synchronous optical networks (SONET) are high-rate TDM systems originally deployed for voice over fiber-optic systems. Similar multiplexing concepts apply [10].

The second technology that is becoming more prevalent in backhaul systems is MPLS. MPLS is a routing technique at the "switching level" of the open systems interconnection (OSI) model. MPLS utilizes path "labels," which are a 32-bit identifier that neither fits in Layer 2 or Layer 3 of the OSI model and is often referred to as Layer 2.5, though there is no official Layer 2.5 of the model. Therefore, the labels direct the packet that does not require laborious routing tables in a full Layer 3 implementation [11].

5.2.7 Emerging Technologies and Other Considerations

A final consideration in terms of electric power utility communication systems are the demands of technologies that are emerging in the industry. Much of this expansion is a result of new technologies that allow generation adjacent to load centers rather than the traditional model of centralized generation with radial feeds to loads. As photovoltaic (PV) becomes more commonplace on residential and commercial properties "behind the meter" and electric vehicles (EVs) become more prevalent, communications between these systems and the larger distribution system will be required for stability, resilience, and safety.

Microgrid controllers and Distributed Energy Resource Management (DERM) controllers will facilitate the coordination with various resources, including loads, energy storage, and renewable generation on a smaller scale. These devices will require high-availability communication channels to facilitate data transfer, sharing current state and ready to react to a system change. For example, a microgrid controller in a system, similar to that shown in Figure 5.24, could be operating in a campus environment such as a university. Inside the microgrid boundaries, there are numerous loads that have been prioritized, a Battery Energy Storage System (BESS), PV generation, and a

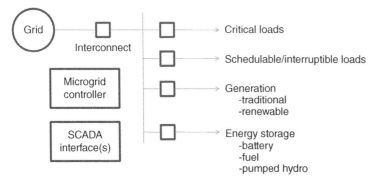

Figure 5.24 Microgrid.

synchronous generator that also is used to supply steam to the campus. The connection to the larger grid is normally closed and the system operates as a component of the outside grid. Should a disturbance be detected by the microgrid controller, several actions could take place. The controller could determine that a separation from the grid at large was necessary, what the load demand is inside the microgrid, what generation is available, including storage, and execute a load-shedding scheme based on lowest prioritization to reduce the demand to be within the available generation inside the microgrid and then separate from the external grid. While disconnected from the grid, the controller maintains communication with all the available generation sources, storage, and loads. The synchronous generation is supplemented by the PV, which is aided by the BESS to maintain a steady generation source regardless of changing weather patterns (i.e. cloud-cover-reducing PV generation) and balances generation with load. These units can be separated by several feet up to miles. The communication between them is essential to maintain operation and the reliability and security of this system and others will be challenged by outside sources and will require design solutions to maintain.

5.3 Summary

Since the advent of electric power generation and delivery data has been key to successful operation. Data systems including the monitoring and analysis have evolved immensely over the years. Monitoring, transmission, and analysis of data has evolved over the years. The communication systems supporting the transmission of the data have also advanced to provide the vast quantities that are required to operate efficiently. From the now primitive telemetry systems in the early days of electronic communication to the systems today that can provide operations and post-event analysis. This chapter was intended to provide a high-level review of these systems.

To design a truly resilient architecture, one must understand all points of weakness that may be targeted. In our current era of cybersecurity awareness, the communication system is key in an interconnected system. Not only must one know the modern communication schemes used but also understand how these modern systems are often interconnected with legacy systems and what risks is posed by both.

For further reading see the texts referenced in this chapter, noted below.

Thoughtful Questions to Ensure Comprehension

1 Explain some of the benefits of introducing communication systems into the electric power system?

2 How did SCADA systems improve operations for electric power utilities?

3 What advantages were gained with the introduction of common communication protocols within the utility industry?

4 Compare the two standards RS-485 and RS-232. What is common between the two standards? Identify a situation when RS-485 would be selected over RS-232.

5 What advantages does fiber-optic cable have over copper cabling for communication systems? Are there any disadvantages?

6 What is the purpose of a fiber-optic patch panel?

7 In what applications using radio-frequency communications would microwave communications be a better choice than VHF radio frequencies?

8 What is the purpose of using spread-spectrum communications?

9 What are some things that must be considered when performing a radio propagation path study?

10 Describe the differences and similarities between an Energy Management System (EMS) and a Distribution Management System (DMS).

11 Describe the benefits gained by utilizing Distribution Automation schemes such as FLISR.

12 What is the major disadvantage of an Ethernet star topology network? What is one method of alleviating that disadvantage?

13 What is one benefit of using IEC-61850 communication protocols such as GOOSE and Sampled Values?

14 Why is considerable effort put into redundancy mechanisms such as Parallel Redundant Protocol and High-availability Seamless Redundancy for Sampled Value and GOOSE communication systems?

References

1 Daniel, E. and Nordell, P. (2008). *Substation Communication History and Practice*. IEEE.
2 Wikipedia (n.d.). Wikipedia: RS-232. Retrieved from Wikipedia. https://en.wikipedia.org/wiki/RS-232 (accessed 31 August 31 2021).
3 Wikipedia (n.d.). Wikipedia: RS-485. Retrieved from Wikipedia. https://en.wikipedia.org/wiki/RS-485 (accessed 31 August 31 2021).
4 Lehpamer, H. (2016). *Introduction to Power Utility Communications*. Artech House.
5 What is OPGW? – Optical Ground Wire, November 8, 2017, Electrical Concepts. https://electricalbaba.com/what-is-opgw-optical-ground-wire/.
6 ALTOS® Loose Tube, Gel-Free, All-Dielectric Cables with Binderless* FastAccess® Technology, 12-72 Fibers, Revision date 2017-10-15.
7 Wikipedia (n.d.). Wikipedia: optical fiber connector. Retrieved from Wikipedia. https://en.wikipedia.org/wiki/Optical_fiber_connector (accessed 31 August 31 2021).
8 The Fiber Optic Association, I. (n.d.). Guide to fiber optics & premises cabling – topic: testing fiber optic neworks: understanding "dB". Retrieved from The Fiber Optic Association, Inc. https://www.thefoa.org/tech/ref/testing/test/dB.html (accessed 31 August 31 2021).
9 Thomas, M. (2015). *Power System SCADA and Smart Grids*. CRC Press.
10 Budka, K.C., Deshpande, J.G., and Thottan, M. (2014). *Communication Networks for Smart Girds*. Springer.
11 Rouse, M. (n.d.). DEFINITION – multiprotocol label switching (MPLS). Retrieved from SearchNetworking. https://searchnetworking.techtarget.com/definition/Multiprotocol-Label-Switching-MPLS (accessed 31 August 31 2021).

Part III

Disciplinary Fundamentals

Resilient Control Architectures and Power Systems, First Edition.
Edited by Craig Rieger, Ronald Boring, Brian Johnson, and Timothy McJunkin.
© 2022 The Institute of Electrical and Electronics Engineers, Inc. Published 2022 by John Wiley & Sons, Inc.

6

Introducing Interdisciplinary Studies

Craig Rieger

National and Homeland Security, Idaho National Laboratory, Idaho Falls, ID, USA

Objectives

The objectives of this chapter are to establish the need for interdisciplinary education and to present initiatives bringing interdisciplinary studies to resilient control systems. In particular, this chapter establishes the role of the resilient control systems for the power grid course and Grid Game in promoting interdisciplinary perspectives.

6.1 Introduction

Over the years of having advanced my own education and moving from a practitioner to a researcher, I have seen the nuances of opinion on what education should provide. That is, what should a student be exposed to before graduation? If you go through an engineering track, you are exposed to curricula that have evolved slowly over recent decades. These curricula include aspects that are unrelated to the discipline, historically integrated to provide a broader "education" as compared to vocational training. This Renaissance man's perspective, in contrast, goes back centuries.

The Renaissance man's perspective suggests a liberal education is beneficial in developing critical thinking beyond a single subject area. Whether this is being achieved in the university environment today is worthy of discussion on its own, but I will suggest that to solve the complex problems of today, education that exposes students to the complexity of the issues is critically important. Issues of cybersecurity, control-system autonomy, and human interaction with autonomy crosscut a range of disciplines from the liberal arts, sciences, and engineering. Knowing the issues indicates another important perspective, which is establishing the challenges to inspire the next generation to pursue a science–technology–engineering–math (STEM) education.

Some educators and agencies have long recognized the potential value of such education, and yet the nuances involved are dependent on the problem set. From a practitioner standpoint, engineers will often work with technicians, but did their education provide them a perspective on each other's value and contribution to a team? This is unlikely. Yet the dynamic established by multidisciplinary engineering teams could have benefited from the perspective provided ahead of entering the job market. By using a technician to take design input from the engineer, which helps to success-fully build a bench scale pilot, a positive understanding of a professional dynamic is established.

Resilient Control Architectures and Power Systems, First Edition.
Edited by Craig Rieger, Ronald Boring, Brian Johnson, and Timothy McJunkin.

Many engineers and scientists have worked on projects with different disciplines, but did they effectively communicate? Was there a clear understanding of contributions to recognize equally the importance of each other's contribution (an interdisciplinary team)? Establishing this dynamic is even more important with teams that cross not only engineering boundaries but also the boundaries between science and engineering and the underlying disciplines.

Today's highly technological environment has created some natural challenges to the next generation of bright young minds. The desire to advance to a *Star Wars* or *I Robot* level of autonomy is upon us, fulfilling an ongoing desire to reduce the cost of production and efficiency. However, the pathway to achieve resilience while developing these systems is ever present. It is not only important to ensure these new control systems will be threat-resilient (from cyberattack, hurricanes, human error, control system failure, etc.) but also to achieve resilience within our current complex infrastructures and industry. Resilient solutions are not possessed by any one person or discipline, but in a combination. The effectiveness in providing teams that will successfully address this resilience challenge will also be an educational challenge and a mindset change for how we educate and how we build teams.

6.2 The Pathway to an Interdisciplinary Team

Organizational champions have developed initiatives to interdisciplinary teaming through funding motivators, such as the critical resilient interdependent infrastructure systems and processes (CRISP) program area of the National Science Foundation (NSF). The resilient control systems for the power grid course was established for just this purpose and builds upon an initiative within Idaho National Laboratory (INL) and any partner university institutions. Through research at INL and its collaborators, a group of senior researchers was established to represent engineering and science disciplines, and a particular field of study within each discipline. This team was developed from in-house resources because the initiative was tasked with building upon the existing capability. This capability was to enhance and evolve the current to a forward leaning strategy that advanced the reputation of INL. This team achieved this goal in establishing an international research area that recognized the limitations of current designs to human and complex failures. Through several years of research, the wealth of interactions and subsequent efficiencies of interactions were developed, which resulted in mutual respect. However, a mature career in several of the individuals and the challenge to relearn an already comfortable pallet of research focus also limited the ability to adapt.

As one outcome of the initiative was to develop the interdisciplinary perspective for early careers, the resilient control systems course was developed within the research community. This community included those that had already participated in the prior and ongoing research, as well as others who have worked in this and related areas nationally. To achieve a dynamic that addresses the various contributions, one might consider the challenges of achieving resilience in a control system design. This includes not only the designers but also those that would create the diverse challenges to these designs to ensure the effectiveness to threats. To recognize the resilience challenges of the role, whether cognitive, control, cyber, or in the power system chosen as the domain for this text, the interdisciplinary team needs to communicate on the individual roles. If this dynamic were created within a game, the individual and the team would score. The motivator would ensure that the desire to improve one's self, as well as understand what the team can do to support improvement to one contributor that may be falling short, results in a learning experience. As learning to design is benefited by considering failures, the game would include the opportunity to perform both the designer and test/scoring role. In much the same way the cybersecurity community performs red/blue exercises, so would the dynamic of this game be established.

The Grid Game was developed to provide the individual perspective of this experience. Discussed in a later part of this book, the student can develop a perspective on the unique challenges that cross the disciplinary seams. Partnering this game with the individual challenges in each discipline, the student can consider more directly how they would approach methods and techniques to develop the appropriate ability to adapt to threats. Additional design perspectives are provided on interdependency studies, architectures, and the like to further enhance the perspectives on the various challenges that are broader than the primary focus of the book. Within the resilience community, the focus is not just on technological considerations but also the policies and processes that governments must consider when adapting effectively to the many resilience challenges.

Further Reading

Hall, P. and Weaver, L. (2001). Interdisciplinary education and teamwork: a long and winding road. *Medical Education* 35 (9): 867–875.

Lattuca, L.R., Trautvetter, L.C., Codd, S.L. et al. (2011). Working as a team: enhancing interdisciplinarity for the engineer of 2020. 2011 ASEE Annual Conference and Exposition (June 2011).

McJunkin, T.R., Rieger, C.G., Johnson, B.K. et al. (2015). Interdisciplinary education through "Edutainment": electric grid resilient control systems course. 122nd ASEE Annual Conference and Exposition, ASEE.

Taajamaa, V., Westerlund, T., Guo, X. et al. (2014). Interdisciplinary engineering education – practice based case. Fourth Interdisciplinary Engineering Design Education Conference, 31–37.

7

Cybersecurity

Daniel Conte de Leon[1], Georgios M. Makrakis[2], and Constantinos Kolias[2]

[1] Computer Science, University of Idaho, Moscow, Idaho USA
[2] Computer Science, University of Idaho, Idaho Falls, Idaho USA

Objectives

The objectives of this chapter are to describe fundamental concepts of security and cybersecurity within the context of Industrial Control Systems (ICSs) and Cyber–Physical Systems (CPSs) and their control subsystems, to identify the fundamental properties of a cybersecure system and the design principles and techniques used to help ensure such properties, to describe current threats that target ICSs and CPSs and how to apply methods and techniques used to counteract or mitigate these threats, and to overview processes and techniques used for security management, risk assessment, threat mitigation, and incident detection and response.

7.1 Introduction

Security and *safety* are two essential properties of all control systems. The goals of both security and safety are to maximize the protection of a Cyber–Physical System (CPS) or Industrial Control System (ICS). In the case of security, the primary goal is to protect subjects and objects that are deemed within the system boundary. In the case of safety, the primary goal is to protect subjects and objects outside of the system boundary. Hence, the definitions of security and safety rely and depend on the determination of system boundaries. System boundaries may change depending on the scope and object of the evaluation.

In an ICS, security can be classified between *physical security* and *cybersecurity*, though the latter is dependent on the former. Security and cybersecurity are not achievable without adequate physical security. However, adequate physical security does not necessarily result in a secure system.

Cybersecurity is a property of a system intended to ensure the following three fundamental properties: *confidentiality*, *integrity*, and *availability*, while enabling and facilitating the system's objectives. Due to the nature of ICSs or CPSs, and the physical systems or processes they control, lack of optimal cybersecurity in an ICS or CPS could result in safety violations, hazards, accidents, or catastrophes. Essentially, in an ICS or CPS, *loss of cybersecurity* could result in the *loss of security and safety*. Note that this means that in an ICS or CPS, loss of confidentiality or integrity would likely lead to loss of service (loss of availability) and loss of safety. Many standards, processes, and techniques exist for minimizing the negative effects of cybersecurity events and breaches in order

Resilient Control Architectures and Power Systems, First Edition.
Edited by Craig Rieger, Ronald Boring, Brian Johnson, and Timothy McJunkin.
© 2022 The Institute of Electrical and Electronics Engineers, Inc. Published 2022 by John Wiley & Sons, Inc.

to reduce the consequences on the system or process being controlled. We introduce some of these processes and techniques in this chapter.

7.2   Systems and Control Systems

7.2.1   Systems, Subsystems, and Analysis Boundaries

Systems can be natural, man-made, or hybrid. *Control systems* are artificial or man-made subsystems created with the purpose of controlling some other system or process. A CPS or ICS is a collection of interconnected physical, control, computing, and networking devices with a common objective. Such objective is usually the control and management of a physical or chemical process. Examples of these systems are the power grid, chemical plants, water treatment facilities, and manufacturing plants. Examples of common control system devices are REMOTE TERMINAL UNITS (*RTUs*), INTELLIGENT ELECTRONIC DEVICES (*IEDs*), and REAL-TIME AUTOMATION CONTROLLERS (RTAC), among many others.

7.2.2   System Subjects and Objects

Within a system, *subjects* are system entities that can process information and events and initiate or take actions. *Objects* are system resources that subjects act upon. For example, in an ICS, an *operator* and an *IED* are subjects. In fact, a device may actually implement multiple digital control processes or system subjects, for example, an IED with relay and metering capabilities or a firewall and a router process operating within the same digital device. Notice that certain devices or processes may act as subjects or objects depending on the context. Also, which objects and subjects are considered to be internal versus external to the system depends on a sometimes relative or not clearly specified boundary of analysis.

Subjects and objects may be *tangible* or *intangible*. Examples of *tangible subjects* are people, living entities, and assets capable of acting upon other objects or the environment. Examples of *intangible subjects* are digital or analog control processes implemented by or within devices, such as processes within IEDs or Programmable Logic Controllers (PLCs). *Tangible objects* are all physical assets, such as switchgear, electrical conductors, pipes, and valves. *Intangible objects* may be data or information in any form or other physical world intangible objects such as light or current. Data or information is an intangible object of the control system. Light or current is an intangible object of the controlled physical system.

Subjects and objects may be *permanent* or *transient*. The determination of which category these entities (subjects and objects) belong to may depend on the subsystem and the objectives of the analysis. Examples of permanent subjects are human users or operators. Examples of transient subjects are transient or temporary system processes, some of which may have a longer lifespan than others (from microseconds or less to years). Examples of permanent objects are physical devices such as digital relay devices, switchgear, and valves. Examples of transient objects are data and representations of system states, measurements, events, and actions.

7.2.3   Subject Communication and Cyber Systems

System subjects, such as human operators or digital processes, must communicate in order to implement the system's goals. Human subjects communicate using spoken or written language or diagrams written in a multiplicity of languages. Digital devices or processes communicate using digital Machine-to-Machine (M2M) communication protocols. Analog processes communicate using analog signals, some as simple as voltage difference, and some more complex, such as frequency-based information carrier protocols. Humans and machines or devices communicate through specialized visualization and control input and output (I/O) devices, such as Control Panels or Human Machine Interfaces (HMI). New speech synthesis and recognition devices are also increasing as a Human-Machine communication medium.

Transient objects may be further classified as *consumable* or *ephemeral*. Examples of consumable objects are *events* and *actions*. Examples of ephemeral, transient objects are non-stored events. Events and actions are very important types of objects in the realm of digital control systems. Transient data and information may also be classified as consumable or ephemeral. In an ICS, physical system events and actions have a corresponding digital representation.

The addition of digital communications to a control system is what makes the control system a *cyber system* and hence the controlled physical system a modern ICS or CPS.

7.3 Fundamental Cybersecurity Objectives: The CIA Triad

Confidentiality, *integrity*, and *availability* are three essential properties of a cybersecure system and are commonly referred to as the *CIA triad*. A fourth essential property is *privacy*.

The objectives for the confidentiality, integrity, and availability properties are outlined in the NIST standard FIPS 199 [1]. FIPS 199 defines *loss of security*, referring to the loss of cybersecurity, precisely through the partial or complete loss of one of these three properties. The process of securing a system involves corrective actions to ensure all three properties. Nevertheless, when taking into account the specific requirements of the application at hand, these properties, and their corresponding implementation techniques, may act antagonistically. For example, while the use of new or additional authentication requirements increases confidentiality, it might indirectly reduce the availability of information for a certain group of users. In other situations, some security objectives must be prioritized over others due to inherent system limitations. For example, achieving confidentiality, at the device level, may prove difficult when dealing with devices with very limited computational resources. System design and implementation choices must be made carefully and consciously and must always be compatible with system objectives, functionality, limitations, and cybersecurity requirements. It is important to consider that in a modern ICS or CPS, loss of cybersecurity properties could propagate, intentionally or accidentally, into *loss of visibility* or *loss of control* of an industrial process, that may subsequently result in injury, loss of critical services, or even loss of life [2].

7.3.1 Confidentiality

Within the context of cybersecurity, the property of confidentiality states that knowledge of system subjects or objects must be restricted to those subjects with permission resulting from a need-to-know or need-to-access. Implementing confidentiality relies on specifying and enforcing this access to enable the required functions but no more. Only subjects or users that should have access to certain objects should be given such access while the rest must be denied access. This follows what is called the *Principle of Least Privilege (PoLP)*, which is described in further detail in subsequent sections in this chapter.

The three primary techniques used to implement the property of confidentiality are as follows: (i) *cryptography*, (ii) *authentication*, and (iii) *access control*. In the context of confidentiality, cryptographic techniques are usually used for encrypting data during processing, while in transit, or at rest. This is done to ensure only authorized (key-bearing) subjects can read such data. Passwords or two-factor authentication are used to ensure that only authorized users have access to systems or data (system subjects and objects). An access-control list (ACL) is one method that can be used to define which users (subjects) may access a resource (object).

7.3.2 Integrity

Within the context of cybersecurity, *integrity* is a property of a system or subsystem that ensures the absence of unauthorized behaviors, additions, removals or deletions, modifications, or fabrications to, or of, any subject or object. Integrity applies to all subjects and objects within the system's boundary. Safety is concerned with the integrity of subjects and objects outside the system's boundary.

When referring to *tangible subjects*, integrity aims to ensure that humans and devices are who they claim to be and that are acting within their intended and expected behavior. This applies to human subjects and also analog and digital devices. Integrity also aims to ensure that nonhuman tangible subjects, such as control devices, have not been tampered with, modified, added, or removed. For example, a PLC has not been replaced without this action being approved beforehand by authorized personnel.

When referring to *intangible subjects*, integrity aims to ensure that processes are valid and are executing within their intended and expected behavior. When integrity refers to processes, it aims to assure that the intended operations are performed without any divergence from specification due to intentional or unintentional modification events. For example, the firmware of a PLC, or the firmware, software, or configuration data of an IED, has not been modified or replaced, neither statically nor dynamically.

When referring to *tangible objects*, integrity aims to ensure that devices, including their components and connectors, have not been tampered with, modified, added, or removed without prior approval. For example, hardware devices or serial cables have not been moved or disconnected without approval.

When referring to *intangible objects* such as data or information, integrity aims to ensure that data or devices have not been modified, fabricated, deleted, held, delayed, or replayed without valid authorization. For example, configuration parameters for a digital protective relay or the value of a coil in an ICS protocol message have not been altered unintentionally or maliciously. Supply chain security also applies to the integrity of firmware and software updates. In the case of data or information, integrity must be ensured across three axes: in use, in transit, or at rest (in storage).

The two primary techniques for achieving system subject and object integrity are *cryptography*, and *replication and redundancy*. An example of the use of cryptography to ensure object integrity is digital signatures (DS), which may be used to ensure the message sender's validity. An example of the use of replication and redundancy is the automatic backup of configuration files.

There are three dimensions to integrity with respect to system implementation stages: before, during, and after device or subsystem installation and commissioning. Integrity before installation and commissioning lies within the scope of the early phases of *supply chain security*. Integrity, during installation, lies within the scope of *supplier and contractor security and cybersecurity*. Integrity, after commissioning, is within the realm of *operational security and cybersecurity*. These realms fall within what is known as *security management*.

7.3.3   Availability

Within the context of cybersecurity, availability is a property of a system or subsystem that ensures that system functions are provided without unexpected interruptions. In a modern ICS or CPS, availability of control functions and data are necessary to ensure the uninterrupted operation of the system. Partial or complete loss of availability on the control subsystem may lead to system-wide loss of service or safety hazards or accidents.

The three primary techniques for achieving subject and object availability are *redundancy and replication*, *access control*, and *cryptography*. An example of a technique for ensuring availability using redundancy and replication is redundant design. An example of ensuring availability using access control is restricting process execution to authorized known processes (subjects).

7.4   Fundamental Cybersecurity Techniques

In this section, we describe the most common techniques and mechanisms used to achieve different aspects of the three fundamental security properties of confidentiality, integrity, and availability described in Section 7.3. Table 7.1 summarizes this discussion.

7.4.1   Cryptography

Cryptography refers to a wide range of mathematical techniques that are applied to increase the level of security of a system. Different types of cryptography are applied to provide confidentiality, integrity, and availability. Existing cryptographic mechanisms can be divided into three major families: (i) symmetric cryptography, (ii) asymmetric cryptography, and (iii) secure hash functions.

Some examples of applications that fall into each one of these cryptographic families are (i) symmetric encryption, (ii) asymmetric encryption, (iii) MESSAGE AUTHENTICATION CODES, and (iv) Digital signatures.

Symmetric cryptography refers to a set of mathematical functions that may provide the expected output only if a certain key is used. In this case, the same key is used for *encryption* and *decryption*, hence the name symmetric. This key is assumed to be shared among all parties a priori (in advance) and must be kept secret. If the symmetric key is stolen, then confidentiality may be lost. Securely generating, sharing, and maintaining symmetric encryption keys in a digital system is challenging. In modern systems, these tasks or functions are implemented using other cryptographic techniques, primarily public key (asymmetric) cryptography and digital certificates.

Asymmetric cryptography, also called public-key cryptography, is based on the existence of a pair of keys per communicating entity, one of which must remain private at all times, while the other can be freely shared with other parties, potentially even adversaries. In general, asymmetric cryptography operations are considered more resource-intensive. Hence, it is usually used for authentication and the exchange of symmetric keys. Another challenge associated with asymmetric cryptography is the need to verify that a certain public key corresponds to a particular subject and not an adversary who simply claims to be that subject. Moreover, since private keys may be compromised, a system for their management and revocation must be available. This is a hard problem that, in practice, is tackled by relying on what is called a PUBLIC KEY INFRASTRUCTURE (PKI).

Secure hash functions receive a message as input and produce a fixed-length fingerprint of that message, namely, the digest, as an output. While the possibilities for input are infinite, the number of unique digests that can be produced is finite and depends on the size of the digest. For example, when the digest length is 128-bits, there may be 2^{128} unique fingerprints. It is, therefore, possible that two different messages may lead to the same digest or fingerprint, though statistically very unlikely; When this happens, it is called a *collision*.

Secure hash functions should have the following properties: (i) while it is easy to calculate the digest of a message, it is computationally infeasible to calculate the original message given its digest; (ii) given a message or its digest it is computationally infeasible to find a different message that will have the same digest; and (iii) it is computationally infeasible to find a pair of messages with the same digest. When secure hash functions have these properties, they may be safely used as building blocks for integrity and authentication techniques.

7.4.1.1 Symmetric Encryption

Symmetric encryption aims to render messages or data incomprehensible to parties that do not possess the secret key. The reader should recall that symmetric encryption, as any other symmetric cryptographic operation, relies on the existence of known secrets (keys), which are shared among all parties. Due to their efficiency and robustness, symmetric encryption algorithms, also called *symmetric ciphers*, are considered best suited for ensuring the confidentiality of large messages, files, or communication sessions, for example, files stored on a hard-drive or most messages exchanged during network communications.

The most common modern cipher used for symmetric encryption is the Advanced Encryption Standard (AES) [3] that was created to fill the inefficiencies of the older Data Encryption Standard (DES). Other symmetric encryption chiphers are described in [4]. The older DES and 3DES symmetric encryption standards [5] may also be found in use but should be phased out and replaced by AES or another modern cipher.

A challenge with symmetric encryption is key management. In an ICS, many parties share the same key, thus increasing its exposure and the risk of system compromise. For example, often based on the location of devices, the keys do not rotate regularly, or are built into the software, in order to increase operational efficiency, without proper protection. For the latter, such protection can be provided with the use of secure Integrated Circuits (ICs) such as the Trusted Platform Modules (TPMs) [6].

7.4.1.2 Asymmetric Encryption

In asymmetric approaches, encryption is done by the sender using the receiver's public key, while for the operation of decryption, the receiver is using their own private key. The reader should notice that even if adversaries are in possession of the public key of a user, they will only be able to encrypt and send messages to that user. The receiver may require proof regarding the authorship of the message, and if not satisfied, they will reject the message.

In practice, asymmetric encryption is used to encrypt small messages and, more specifically, keys for symmetric encryption so they can securely be transferred across an untrusted channel. This happens because asymmetric encryption is considerably less computationally efficient than symmetric encryption. Well-known asymmetric encryption algorithms are RSA [7] and Elliptic Curve algorithms [8, 9].

A PKI is a set of entities, documents, and protocols for supporting operations that involve asymmetric cryptography. Today, PKI is the dominant solution regarding the dissemination and management of public keys for applications such as the Internet. PKI relies on external entities, namely certificate authorities (CAs), whose primary role is to guarantee the validity of public keys and their unique correspondence with users or subjects. This is accomplished by means of digital documents called *authentication certificates*. Authentication certificates contain the user or subject identifier, the corresponding public key, among other fields, and are cryptographically signed by the certificate authority. One fundamental requirement of this model is that all communicating subjects must trust the CA, who is a trusted third party.

An example of implementing integrity is the use of secure hash functions to calculate a digest over a firmware image for an industrial control device such as a PLC or IED. Before using such a firmware image, the digest of the image file can be checked against the digest provided by the vendor to verify that the firmware is not modified by an adversary or corrupted in transit.

An example of implementing both integrity and confidentiality using secure hash functions and encryption is as follows: the sender creates a message, calculates the message digest, appends it to the original message, encrypts the new message, and transmits the hashed and encrypted message. The receiver may then decrypt the message, separate the message from the digest, independently calculate the digest of the message, and compare the calculated digest with the received digest. Even if just one bit of the original message has been altered, the sent and calculated digests should not match. In this case, it should be computationally infeasible for an attacker to modify the message without the receiver noticing the lack of integrity due to the differing message digests.

7.4.1.3 Digital Signatures

Digital signatures are cryptographic methods that, combined with key management, provide authentication regarding the origin of a message as well as its integrity. The most common types of digital signatures rely on asymmetric cryptography. In this case, they can also ensure non-repudiation as the subject that signed the message may not be able to deny the action.

To generate a digital signature, the sender first calculates the digest of a message (which is typically much smaller in length). Then, the sender encrypts the digest with its own private key. Signing a message is the process of attaching its resulting digital signature. The sender then may send the signed message to the receiver.

To verify the authenticity and validity of the message, the receiver first calculates the digest of the received message. Then, the receiver uses the sender's public key to decrypt the signature that was appended by the sender. If the calculated and retrieved digests match, then it is unlikely that the message was forged by another entity, or it has been altered in any way. Moreover, since the message was signed with the sender's private key (which must remain secret at all times), the sender may not deny the authorship of the message unless they can prove that the public key certificate was lost or stolen.

As specified in the Digital Signature Standard by NIST [10], the main signature algorithms are the Digital Signature Algorithm (DSA), the RSA DSA, and the Elliptic Curve Digital Signature Algorithm (ECDSA).

Digital signatures can be applied in ICSs as a means to check that changes to application logic or software updates come from a trusted source. As an example, all application logic and software updates should be verified by a PLC before starting the execution of the updated logic or perform the software update. Furthermore, a device that sends information to a control center can also include a singed part of the data to avoid malicious forgeries by third-party entities.

7.4.2 Authentication and Identity

Authentication refers to mechanisms used to (i) verify a subject's (human or process) identity to a system or (ii) to prove the valid origin of a transmitted message.

There are three general categories for verifying a subject's identity:

- Something a user knows – It is usually implemented in the form of a password, passphrase, or personal identification number (PIN). This approach is based on keeping the token secret and hard to guess.

- Something a user has – It is based on the physical ownership of tokens such as a card or mobile device. This approach is based on items that tend to be unique, hard to forge, and highly personal to the users.
- Something a user is – Verifies the users' identity based on some biometric characteristics (e.g. face morphology, fingerprint, iris). This approach assumes that such characteristics are overly complex and hard to counterfeit.

To prove the origin of a message, a MAC is the most common method used in real-life scenarios. A MAC is a digest that gets appended to the original message similar to a digital signature. Unlike digital signatures, a MAC relies on symmetric cryptography; hence, MACs are produced and verified with the same symmetric keys. For this reason, the receiver of a message with a valid MAC is assured the sender must be in possession of the shared key.

7.4.3 Authorization and Access Control

Authorization determines the type and level of privileges that users have upon the system's resources. In this context, resources may be files or software, and the privileges refer to potential actions that subjects may perform, such as read, write, or execute. Secure systems typically adopt the *PoLP*, which dictates that subjects must only be granted the minimum amount of privileges that are required to perform their task. Naturally, authorization assumes that a stage of authentication and user identification has taken place.

Authorization at the network level is achieved by integrated systems such as MS Active Directory (AD); while at the OS level, it is typically achieved via ACLs or capabilities. Due to space reasons, these approaches are not discussed in this chapter. Indirectly, authorization may be applied to protect data integrity.

The permissions given to a subject are defined by the access control policies. The main categories of access control policies are as follows:

- Discretionary access control (DAC) – Access to objects is based on the identity of subjects and the access rules that define what those subjects are allowed to do. Moreover, a subject with a certain access permission has the right to hand over or delegate that permission to other subjects.
- Mandatory access control (MAC) – Access to objects is based on the sensitivity of the information that those objects contain and the authorization of the subjects that want to access such information. An administrator centrally defines and controls all permissions, and users (subjects) do not have the ability to override or delegate them.
- Role-based access control (RBAC) – Access to objects is based on the specific roles that subjects possess in the system and on rules that declare what type of access is allowed for each role. Roles are structured into a hierarchy and may apply to a single or many subjects.
- Attribute-based access control (ABAC) – Access is based on the attributes of the subject, the resource to be accessed (object), and the actions the subject will perform to the accessed resource, plus sometimes other system or environmental attributes such as time-of-day.

One example of the use of access control in ICS environments is the allowance to use an HMI only if the user has used its badge to enter the room (use of credentials and physical security). A second example is a permission for a user to handle or manage a process only from a specific controller device.

Table 7.1 Security mechanisms and their properties.

	Confidentiality	Integrity	Availability
Cryptography	•	•	•
Authentication	•	•	•
Access control	•	•*	•
Accountability		•	

*Access control and authorization.

7.4.4 Accountability

Accountability includes a number of mechanisms to ensure that the actions of a subject of a system can be traced back solely to that subject and no other. The term is associated with non-repudiation, which gives emphasis to the fact that subjects must be unable to deny their actions. Accountability is essential to enable security audit procedures [11] and forensic analysis after a system has been compromised.

Accountability requires that at least the following activities are performed: (i) identification, i.e. bonding a user with a unique identifier when interacting with the system, (ii) authentication, i.e. verifying that the identifier uniquely corresponds to the actual subject, and (iii) logging of all subject activity, i.e. monitoring and maintaining detailed information about the way subjects interact with the system.

7.4.5 Redundancy and Replication

Redundancy and replication techniques should be used to ensure both availability and integrity. This applies to both system subjects and objects and to hardware, software, and data.

In ICS and CPS systems, it is common to use hardware redundancy to ensure availability in the occurrence of failures, whether accidental, environmental, or malicious. In some systems, replication is mandated by regulations.

In safety-critical systems, sensor replication must be used to make systems resilient to sensor failures or misreadings. Voting or agreement techniques are then used to decide on the most probable value among the readings from multiple redundant sensors.

In addition to hardware redundancy, process and data replication should also be used to ensure system integrity. For example, frequent, scheduled, off-device, and off-site backups of configuration files should be maintained. If an integrity violation is discovered, then configuration files may be reinstated from known good backups.

7.5 Threats, Vulnerabilities, and Attacks

7.5.1 Definitions

A *threat* is any subject, event, or condition that could cause or lead to damage to a system object or subject or harm the system as a whole. A threat can be intentional (e.g. malicious activity) or a random occurrence. Threats and their potential impact on a system should be identified through risk assessment. Then, a set of policies and procedures should be developed and applied for their elimination or mitigation. This is called risk management.

A *vulnerability* is defined by NIST [12] as "A flaw or weakness in system security procedures, design, implementation, or internal controls that could be exercised (accidentally triggered or intentionally exploited) and result in a security breach or a violation of the system's security policy." The Common Vulnerability Scoring System (CVSS) [13] is an open standard for classifying security vulnerabilities according to their severity level. The Mitre Corporation maintains a list of known vulnerabilities in its Common Vulnerabilities and Exposures [14] database, which is organized following the CVSS. Examples of sources of vulnerabilities are hardware misconfigurations, software-related bugs, and lack of proper training of human subjects.

An *attack* is an action that exploits a vulnerability in the system. An attack aims to inflict loss of confidentiality, integrity, or availability to the system or reduce the privacy of its users. Attacks could cause leakage of sensitive information, unauthorized access to system resources, altering of processes or data, or service unavailability.

The terms threat and attack are often wrongfully used interchangeably. An important difference between a threat and an attack is that the former constitutes a potential danger to the system, while the latter is an orchestrated sequence of actions with the intention to cause damage.

7.5.2 Common Types of ICS Vulnerabilities

Vulnerabilities exist in modern ICS environments primarily due to the high level of architectural complexity, which involves the integration of both analog and digital components. Like any system, ICSs evolve over time with new components being added and old components becoming obsolete. Therefore, it is natural that new vulnerabilities are created over time. Indeed, vulnerabilities may be embedded at the component level (e.g. software and hardware), or may be the outcome of improper or rushed design decisions (e.g. the way system components interact with each other). This section describes the most common types of vulnerabilities observed in ICS environments, along with examples of attacks that exploited such vulnerabilities.

7.5.2.1 Human Related

The human element of systems may be a point of weakness. Untrained employees may knowingly or unknowingly bypass existing security measures or violate established security policies (e.g. the use of weak credentials). Moreover, untrained employees may fall victim to social engineering attacks such as email phishing. Through such practices, an attacker may be granted permission to critical resources or may achieve the installation of unauthorized and malicious software. For example, during the Stuxnet incident [15, 16], employees uploaded pictures from the operating environment during the President's Ahmadinejad visit at the Natanz facilities. Close examination of the monitors in pictures revealed critical information about the SCADA systems. Another example of human factor exploitation can be seen with the BlackEnergy malware [17], used in the 2015 Ukrainian power grid cyberattacks, and the Havex malware [18], which was used for industrial espionage. In both these cases, the attackers relied on phishing techniques to deceive employees into opening a malicious document that contained the first stage of the malware as an email attachment.

Moreover, the possibility exists that some employees may become malicious insiders. For example, in the case of Stuxnet, the exact knowledge of the targeted PLC models is attributed to a malicious insider who was leaking information. Similarly, in the Maroochy incident [19], it was a former employee who was responsible for the attack. The employee had privileged information about the setup of the targeted water facilities, which allowed him to remotely issue malicious commands.

Automated tools exist to assist the prevention of some human-related attacks, such as anti-phishing tools. However, since attacking practices constantly evolve, existing automated solutions cannot be fully trusted to perform with high accuracy. An effective and continuous security training program in combination with detailed auditing of activities is necessary.

7.5.2.2 Software or Firmware Based

Software and firmware lie at the heart of most modern systems. However, software and firmware can be infested with bugs, some of which may lead to security vulnerabilities. A bug is a flaw in the logic of the software or firmware that under certain conditions causes the corresponding application (or, as a result, the entire system) to show unexpected behavior and provide unwanted outcomes. Software or firmware bugs may be found in adopted third-party tools and libraries and devices, in-house developed applications, or communication protocols. Typically, when bugs are discovered, they are addressed by eliminating or correcting the erroneous module at the code level. While in some cases correcting is as easy as performing a software or firmware updates, in other cases, correcting is a time-consuming and error-prone process that involves reconfiguration of the system. The biggest challenge regarding bugs lies in their timely discovery. Particularly, a zero-day (also known as 0-day) is a vulnerability that remains unknown or for which the vendor has yet to develop a solution (patch). These are the most catastrophic types of bugs as there is no readily available remedy, which in turn provides aggressors with a window of opportunity for exploitation (zero-day attack).

The Stuxnet malware took advantage of alternative zero-day vulnerabilities to increase its chances of penetrating and proliferating inside the target network, including USB flash drives (CVE-2010-2568), a Windows Print Spooler service (CVE-2010-2729), a Windows Server service (CVE-2008-4250), a Windows keyboard layout loading (CVE-2010-2743), and WinCC and PCS 7 SCADA system (CVE-2010-2772) vulnerabilities. Another example is the Duqu malware, whose main purpose is the leakage of valuable information from within highly protected critical infrastructures. In the incidents that were analyzed, it is presumed that the malware-infected the target host through malicious code embedded in Microsoft Word documents that exploited a zero-day vulnerability in the Windows TrueType font parsing library (CVE-2011-3402), which in turn enabled remote code execution (RCE).

7.5.2.3 Policies and Procedures

Policies and procedures are an essential part of a security program, and the lack of adequate policies and procedures may lead to organizational or system vulnerabilities. For example, a common set of required procedures is patch management and system upgrades.

Software and firmware bugs often get fully eliminated or at least mitigated with appropriate patches or system upgrades. Patching usually refers to small-scale changes to the problematic code. System upgrades usually entails more extensive changes to multiple components of the system along with the addition of modules that provide new functionality.

Despite being highly desirable, the process of applying bug fixes in a complex system is not a simple task. In further detail, the patch itself may negatively impact the performance of a system and generate incompatibility. Moreover, applying a system update may require significant downtime (e.g. a system reboot) for the corresponding changes to take effect. For this reason, a comprehensive patch management process must be in effect. Such frameworks involve the phases of (i) monitoring for available updates for all components of the system, (ii) evaluating the importance and prioritizing the most critical among them, (iii) testing the effects of the updates before deployment, (iv) deploying the updates in a fashion that ensures the minimum down-time and loss of service to its legitimate users, (v) measuring their impact to the performance of the system, and finally (vi) following a sequence of actions as a fallback plan in case the application of a patch generates unwanted behavior.

The example of the WannaCry [20] malware is indicative of the importance of an effective patch management policy to exist in critical infrastructures. WannaCry is a worm that affected Windows computers by encrypting local files and then requiring ransom for their decryption. The malware exploited an underlying vulnerability in Microsoft's SMB v1 protocol implementation. Interestingly, Microsoft had already addressed the vulnerability, but the affected systems were still unpatched 1 month after. Other critical infrastructures that were impacted by WannaCry are the National Health Service (NHS) of the United Kingdom, the Spanish telecommunications company Telefonica, and the U.S. delivery service, FedEx.

7.5.3 Attack Stages and the Cyber Kill Chain

There are some common misconceptions about how an attack in an ICS actually happens. A successful cyberattack is usually the outcome of a complex and time-consuming multistage process. In general, attackers first begin with a reconnaissance stage when they learn as much as possible about their target system. Then they attempt to exploit the system. The last step is the execution of the intended malicious actions. By then, usually, attackers have been active within the target system for long periods of time and have gained access to many devices within the system.

The Cyber Kill Chain is a cyberattack model that dissects an attack into seven discrete stages. Each stage involves a series of actions with a well-defined outcome as part of an attack. The Cyber Kill Chain model was introduced by Lockheed Martin [21] with the purpose of understanding and communicating critical attack aspects toward enabling the determination of adequate defenses for each stage.

In this section, we describe the stages in the Cyber Kill Chain model using the Stuxnet incident as an exemplar. The phases of the Cyber Kill Chain are detailed in the following subsections.

7.5.3.1 Reconnaissance

During the reconnaissance stage, an attacker or penetration tester, known colloquially as a pentester, attempts to assess a target environment in a stealthy manner in order to enumerate existing devices, networks, users, and vulnerabilities. With this information at hand, an attacker or pentester may then orchestrate a course of action for reaching its objectives. This information is gathered using any available method of discovery but with the objective of avoiding detection. The most common method used at this stage is called Open Source Intelligence (OSINT). OSINT is the use of publicly available resources to learn details about a system—for example, the number of an organization's employees through social media. In the case of the Stuxnet attack, with the help of a malicious insider, attackers identified the specific PLC brands and models in the target infrastructure, namely Siemens S7 315 and 417 PLCs.

7.5.3.2 Weaponization

Using information derived during reconnaissance, an attacker or pentester may use and tailor exploits or malware with the objective of circumventing system defenses. In the case of Stuxnet, it is believed that the malware modules were based on the Tilde-D or "Tilded" framework [22].

Actually, three iterations of the malware were found in public [15, 23], each having different goals and exploiting different vulnerabilities. More specifically, in its last version, the Stuxnet malware made use of four zero-day vulnerabilities.

7.5.3.3 Delivery
This phase involves all actions performed by an attacker or pentester to deliver the developed malware to the victim. Stuxnet, for example, is believed to have been delivered initially via USB drives. After initial infection, Stuxnet spread to other devices inside the target network via several other zero-day vulnerabilities. Interestingly, to maintain a low profile, after infecting three devices, it was programmed to erase itself. Moreover, after a specific date (i.e. 24 June 2012), the malware ceased any attempt to infect new targets.

7.5.3.4 Exploitation
This is usually a short-lived phase involving actions of exploitation of specific vulnerabilities. In the Stuxnet example, after its initial delivery to the target, the malware took advantage of hard-coded credentials embedded in specific versions of the WinCC software commonly used for configuring Siemens PLCs.

7.5.3.5 Installation
In this step, an attacker or pentester attempts to achieve persistence by installing copies of the malware in several places or devices within the target system. It also installs components needed for taking targeted malicious actions, sometimes also referred to as the payload. Stuxnet ensured persistence by installing itself on multiple workstations by capitalizing on many alternative zero-days. The actual installation aiming to deliver the payload was done against the specific PLCs identified during the reconnaissance stage.

7.5.3.6 Command and Control
The next stage involves establishing and maintaining a line of communication with the malware for control and data funneling. Malware may remain dormant and stealthy until instructed to perform a malicious task. These instructions originate from dedicated malicious entities known as Command and Control (C&C) servers. While such (C&C) servers are reachable through the Internet in the case of Stuxnet, the target of the malware resided in a network isolated environment deprived of direct communication with external networks. Such networks are also known as "air-gapped." To circumvent this challenge, the malware authors chose to compromise external contractor companies and relied on the conventional physical interaction of the employees of the two networks to carry their commands, thus breaking the "air-gap."

7.5.3.7 Actions on Objectives
The final phase involves actions performed by the attacker or pentester to accomplish their intended objective. In an ICS system, this may include taking control of the system at the PLC and IED levels allowing attackers or pentester complete control of the system operations with the highest potential consequences. In the case of Stuxnet, the malware was used to alter the rotation frequency of centrifuges, thus damaging the equipment.

7.5.3.8 ICS Cyber Kill Chain
The Cyber Kill Chain has been adapted to the ICS domain by Lee and Assante [24]. The concept of the *ICS Cyber Kill Chain* is introduced to enable the defenders to understand the ICS-custom cyberattacks in greater detail, as the more generic model of the Cyber Kill Chain does not focus on

the peculiarities of these environments. The ICS Cyber Kill Chain consists of two stages. In Stage 1, the attacker performs espionage operations with the aim of acquiring knowledge and ensuring future access to an ICS or CPS. When this is complete, Stage 2 takes place, where the attacker develops an ICS specific attack based on the information gathered from Stage 1 and executes this attack. Stage 2 may not immediately follow Stage 1, as a delay for processing the discoveries of Stage 1 might occur.

The main differences between the traditional Cyber Kill Chain and the ICS Cyber Kill Chain are as follows: (i) the attackers that target ICS or CPS usually do not have the primary goal of information exfiltration. Instead, they focus on the disruption or destruction of the control process; (ii) the attackers can acquire the information from the IT environment by utilizing common and well-tested techniques. However, as each ICS or CPS has specific characteristics and configurations, the attack that is created in Stage 2 must be tailored to these environments to maximize its results.

7.6 Secure System Design Principles

In this section, we introduce secure design principles that must be carefully taken into consideration when designing, modifying, maintaining, and configuring a cyber-enabled ICS or CPS. Several of the principles described below are adapted from a seminal article by Saltzer and Schroeder entitled, *The Protection of Information in Computer Systems* [25]. Others are new or an evolution or enhancement of the original principles.

As the reader will observe, and as mentioned previously, the maximization of these principles is antagonistic. For a given system, maximizing a given principle may diminish another. Hence, the challenge for system engineers and users is to design, maintain, configure, and operate a system in such a manner that it maximizes, to the highest possible extent, all the secure design principles. For this, many engineering trade-offs will need to be carefully studied, analyzed, and implemented.

7.6.1 Continuous Improvement

ICS systems and their environments are dynamic entities. Threats to a system's security and safety change every day due to internal and external actions and factors. It is imperative to design and perform a continuous risk assessment process and take corrective actions within the system and the organization to improve cybersecurity and safety on a daily basis. Corrective actions of today may not be optimal or sufficient tomorrow. For example, a newly disclosed vulnerability in an ICS device could result in system risks changing drastically from one day to the next. Several authors have introduced variations of this principle [26, 27]. These continuous assessment and improvement processes are called *Security Management* and are further described in Section 7.7.

7.6.2 Defense in Depth

Systems must be designed, implemented, maintained, and configured to provide multiple and independent layers of security for each provided security function in a system. If security layers are designed and implemented appropriately, then an attacker or pentester will need to break through, or bypass, multiple security layers to be successful. For example, using strong authentication where the user must present a secret plus a hardware token in order to authenticate is a form of defense in depth strategy. Another form of defense in depth strategy is segregating an ICS network using an Onion-style design, which is called *network micro-segmentation*. Careful consideration must be placed on the property of layer independence. For example, using multiple firewalls of the same type or brand will likely not provide any additional security layers. The principle of defense in depth may be considered an expansion [26, 27] of the principle of *Complete Mediation* described by Saltzer and Schroeder [25].

7.6.3 Least Privilege

The PoLP determines that subjects in the system must be granted and configured with the minimum (least) number and level of privilege (permissions) needed to carry their assigned tasks and not more. If a subject holds multiple roles for different sets of tasks, it is best to separate their authentication on a subject-role basis rather than allowing the subject to operate always with the most privileged role. This principle helps reduce the potential damage from an attacker or pentester that has compromised a device or is impersonating a user or process (subject). For example, if a human user is logged on to a remote engineering workstation with rights to remotely configure an ICS when such user, most of the time, only needs to view the status of the ICS, this would be a violation of this principle. This condition may allow an attacker that has compromised the engineering workstation to configure the ICS even though such functions are rarely used by the legitimate user. An approach that would mitigate such an attack would be for the user (subject) to have, at least, two separate authentication credential sets, one for ICS-view and one for ICS-manage. In other words, permissions assigned to a subject are the least needed set to accomplish that role or set of tasks.

7.6.4 Validated Design and Implementation

System designers, owners, and operators must ensure that multiple stakeholders review system designs, implementations, and configurations. Ensuring the review of a system's design and implementation by different people with diverse backgrounds and from diverse disciplines and roles is needed. This review provides validation and may discover additional threats or vulnerabilities that need to be mitigated. If the system (software or hardware) is manufactured by a vendor, designers, owners, and operators must ensure that subsystem functions, security functions, and interfaces are well-documented and well-understood. This principle does not imply that the designs, implementations, and configurations of a system must be made public, but rather that numerous, rich, and varied expert reviews benefit a system's design and rich and expert validation is essential. Validated Design and Implementation also implies that the security of a system must not depend on secret devices, algorithms, or functions. For example, using an unpublished encryption algorithm would violate this principle. For example, the AES symmetric cipher has been validated by many experts. This principle was introduced as *Open Design* by Saltzer and Schroeder [25].

7.6.5 Fail-safe Defaults

The Fail-safe Defaults principle refers to designing and implementing a system in such a way that default actions of a system or a subsystem, in response to unexpected or unforeseen events, are to revert to or maintain a secure state. This is usually well-understood within ICS when the safety of a subsystem or device or the environment is at risk. For example, in the power grid, automated relays cut power when the lines are overloaded to prevent line damage. However, when the trade-off is between confidentiality and availability, the optimal default action may be harder to discern and describe because many more variables may be involved. Furthermore, in some ICSs, lack of availability may lead to critical or catastrophic accidents. This principle was introduced by Saltzer and Schroeder with the same name [25]. Smith called a variation of this principle *Deny by Default* [27].

7.6.6 Separation of Duties

For tasks that are critical to the operation of the system or that may have major consequences if performed incorrectly (intentionally or accidentally), the Separation of Duties principle ensures that at least two or more stakeholders are needed. This could be implemented asynchronously, for example, by the creation, review, and approval workflows, or synchronously, for example, by ensuring that two stakeholder subjects perform the task together and that both must present their authentication credentials. System designers, owners, and operators must also ensure that such stakeholders are enabled and empowered to act independently. This principle helps ensure that critical tasks are carefully performed and also mitigates insider threats of high consequence. This principle also applies to nonhuman subjects in the system. If the critical task or action is performed or initiated by an automated system, then both redundancy and subsystem independence are needed. This principle was introduced as *Separation of Privilege* by Saltzer and Schroeder [25].

7.6.7 Psychological Acceptability

The principle of psychological acceptability states that system user interfaces, including HMI, and interfaces used for security and protection, must be designed in a manner that facilitates the usage and utility of the system or subsystem (maximize ease of use) while maintaining, or helping maintain, security (confidentiality, integrity, and availability) (maximize security). Human errors are minimized when the language, processes, and model of the system's user interfaces match the user's language and mental models of the system and its processes [25]. This principle also applies to application programming interfaces (APIs) or any other form of M2M communication protocol, which is ultimately programmed or configured by humans. A later version of this principle by Saltzer and Kaashoek was described as *least astonishment* [28].

7.6.8 Modularization

The principle of *modularization* states that the architecture, design, implementation, and configuration of a system must be as modular as possible and must ensure that the interfaces between subsystems or modules are simple, well-defined, well-documented, and well-understood by all stakeholders. Subsystem interfaces for HMI, M2M, hardware, software, analog/digital, and communication protocols should be simple and stable enough that replacing different subsystems or modules is always successful and does not lead to unintended consequences. This principle is a variation, combination, and enhancement of the principles of *Simplicity*, *Economy of Mechanism*, and *Least Common Mechanism* presented by Saltzer and Schroeder [25].

7.6.9 Accountability

The principle of *accountability* states that logs of creation, modifications, and accesses to system subjects and objects must be recorded in a separate, add-only subsystem. For example, it is usual for ICS control actions and feedbacks to be logged on a subsystem called the *Historian*. Cyber-initiated events must also be logged. Such recording enables analysis, correlation, and discovery of attempted or successful intrusions or mishaps. In this chapter, we describe Accountability as a fundamental technique that is essential in implementing secure CPSs and provide additional details in Section 7.4.4. In the Saltzer and Schroeder article of 1975 [28], an earlier version of this principle was called *Compromise Recording*.

7.7 Approaches for Threat and Risk Assessment and Mitigation

A system will always be exposed to threats and risks, which could be external or internal, malicious, or benign. Some threats and their corresponding risks may be catastrophic to the system's functions; others may coexist with the system and be managed. Some threats and risks exist for the lifetime of the system, and others are temporary or recurring. Analyzing the security threats and risks to a system and mitigating them falls within the field of *Risk Management*, which is a subarea of *Security Management*.

The field of risk management includes principles, knowledge, techniques, and organizational processes for identifying, reducing and managing threats and risks and their potential consequences. There are many standards for risk management that describe the processes and tasks needed for effective threat and risk treatment and mitigation. In general, most standards describe a continuous risk management cycle comprised of the primary stages of (i) *assess*, (ii) *select*, (iii) *implement*, and (iv) *monitor*.

The NIST Risk Management Framework defines four major continuous phases for risk management: *framing*, *assessment*, *response*, and *monitoring* [29]. Each of these phases defines a set of inputs, outputs, and tasks. In this section, we describe risk management based primarily on the phases as detailed by the NIST Risk Management Framework (NIST SP 800-39 [29]).

7.7.1 Risk Framing, Legal, and Compliance

During the *Framing* phase, an organization must develop and document an understanding of all regulations, drivers, and conditions that may influence risk-based decisions. *Inputs* to this phase include, but are not limited to, laws and regulations, an organization's goals and objectives, and strategic plans. The *output* of this phase should be a comprehensive risk management strategy. *Tasks* in this phase include: (i) *identify assumptions that affect how risk is assessed, responded to, and monitored within the organization*, (ii) *identify constraints on the conduct of risk assessment, risk response, and risk monitoring activities within the organization*, (iii) *identify the level of risk tolerance for the organization*, and (iv) *identify priorities and trade-offs considered by the organization in managing risk* [29].

All organizations work within a legal and regulatory environment. In particular, the operations and functions of organizations managing critical infrastructures are regulated by laws as decreed by legislative bodies and also executive power decrees, rules, and regulations. Different government agencies may have authority over the operations of different infrastructure sectors. For example, in the United States, the Electric Power Industry is regulated primarily by the Federal Energy Regulatory Commission (U.S. FERC), and medical devices are regulated by the Food and Drug Administration (U.S. FDA). State or province-level and local agencies may also regulate the operations of organizations in a way that affects the design, implementation, and configuration of their CPSs.

Laws, rules, and regulations often mandate security or cybersecurity controls, techniques, or procedures. Other cybersecurity techniques and processes are applied on a voluntary basis. It is essential for system managers, designers, owners, operators, and manufacturers to know the applicable laws and regulations and how they may affect ICS or CPS systems and processes. It is also important to know that these laws and regulations are usually designed as the minimum requirement. Security controls, techniques, and procedures, beyond those mandated by regulations, should be implemented in order to achieve the desired level of security. Relying on compliance for security is very likely not an optimal approach to keep an organization's systems secure. Threat and risk assessment, and the corresponding mitigation implementations, procedures, and processes must not only be based on applicable laws and regulations but also on the particularities of the system being protected and the risk of loss or damage to the system or the environment.

7.7.2 Risk Assessment

During the *assessment* phase, an understanding of all threats and risks to all subjects, objects, and subsystems is developed and documented. Based on the latter plus the functions and criticality of the objects, subjects, and subsystems, a determination of organizational (mission, functions, image, and reputation) risks is also made. *Inputs* to this phase include, but are not limited to, risk framing documents, organizational policies and procedures, risk assessment methods, and detailed subsystem descriptions. The *Output* of this phase, is the documented determination of risk to the organization and also each and all subsystems, subjects, and objects. *Tasks* in this phase include: (i) *identify threats to and vulnerabilities in [...] systems and the environments in which the systems operate*, and (ii) *determine the risk to organizational operations and assets, individuals, [and] other organizations [...]* [29].

7.7.3 Risk Response or Treatment

During the *response* phase, a detailed plan for how to handle risks is developed. In addition, corrective actions are taken. During this stage, an organization identifies, evaluates, determines, and implements risk response actions for all subsystems, subjects, and objects. These actions may be to accept, avoid, mitigate, share, or transfer each risk. *Inputs* to this phase include, but are not limited to, risk framing documents, organizational policies and procedures, and detailed risk assessment, including threats, vulnerabilities, and their potential consequences. The *outputs* of this phase are (i) a detailed risk response plan and (ii) the implementation of said plan. *Tasks* in this phase include: (i) *identify alternative courses of action to respond to risk[s] determined during the risk assessment*, (ii) *evaluate alternative courses of action for responding to risk[s]*, (iii) *decide on the appropriate course of action for responding to risk[s]*, and (iv) *implement the course of action selected to respond to risk[s]* [29].

7.7.4 Risk Monitoring

During the *monitoring* phase, detailed measurement, and understanding of the performance of the organization for mitigating risk and risk-affecting changes are developed. During this stage, an organization measures and documents: (i) how well the risk response plan is being implemented, and (ii) how internal or external changes to subsystems, subjects, and objects affect risk. *Inputs* to this phase include, but are not limited to, detailed risk response plan, framing documents, organizational policies and procedures, subsystem risk and change measurement methods and tools. The *output* of this phase is a detailed and documented evaluation of (i) risk treatment performance and (ii) risk-affecting organizational and environmental changes. *Tasks* in this phase include: (i) *develop a risk monitoring strategy for the organization that includes the purpose, type, and frequency of monitoring activities*, and (ii) *monitor organizational [...] systems and environments of operation on an ongoing basis to verify compliance, determine the effectiveness of risk response measures, and identify changes* [29].

7.7.5 Security Management and Continuous Improvement

The plans, procedures, policies, processes, tools, and data resulting from the risk management processes described in the previous subsections must change and adapt to the organization and according to external events. This is essential and consistent with the secure design principle of Continuous Improvement (Section 7.6.1).

Changes that may significantly affect the security posture of an organization may come from many different types of events. These include the following. (i) Addition, change, or removal of subjects, including human users, operators, or administrators and device processes, for example, the hiring or addition of a new operator, whether internal or from a partner organization. (ii) Addition, change, and removal of system objects, for example, changing a relay configuration value or connecting to a new partner network. (iii) The occurrence of external or environmental events or action, for example, the discovery of a critical vulnerability for a device commonly used within the organization or the accidental splicing of a fiber path from a substation to the control center. (iv) Changes in laws, regulations, or internal strategic plans.

In order to successfully manage and adapt to change, an organization must implement procedures for designing, implementing, recording, observing, and measuring changes and their effects. This helps to ensure that when the status of the system changes, security processes change accordingly and also that these changes do not diminish the security posture of the organization. For example, if an employee leaves the organization, their access to control systems should be removed, and their permissions must be immediately reassigned to one or more other employees. Similarly, if a new network-capable device is connected to the control network, its functions and risks must be well-analyzed and documented, and its configuration must optimize service levels while maximizing security. Planning and measurement implement the principle of *Continuous Improvement*. Approval implements the principles of *Separation of Privilege* and *Validated Design and Implementation*. Recording implements *Accountability*.

7.8 Approaches for Incident Detection and Response

7.8.1 Incident and Intrusion Detection

Intrusion Detection Systems (IDS) are hardware devices or software that monitor a network or a system and provide indicators of malicious activity or policy violations. Such tools analyze the header and/or the payload of network packets that arrive at a specific location in the network. Typical IDS implementations usually provide alerts along with additional information about incidents but do not attempt to eliminate the threat. More advanced systems involve the incorporation of INTRUSION PREVENTION SYSTEMS (IPS) that act proactively to identify threats and block them.

Intrusion detection is based on the assumption that the observable events resulting from malicious actions will differ from those resulting from legitimate subjects. However, a precise distinction between malicious and normal behavior is not always feasible, especially in the presence of novel advanced attacks. Also, observable behaviors of benign users or subjects may change over time or due to exceptional conditions. Because of this, an IDS may result in what is called *false negatives*, which is intrusive behavior being misclassified as benign or normal. An IDS may also generate *false positives*, that is, intended behavior misclassified as anomalous. Too many false positives may result in rendering the IDS non-trustworthy and leading the system or network administrator to overlook or reject many of these alerts. A well-configured IDS should be able to detect a substantial percentage of intrusions or attempts while keeping the number of false positives at relatively low levels. The challenge for IDS maintainers and operators is to maximize the detection rate while minimizing the false alarm rate.

IDS may be classified into Host-based Intrusion Detection Systems (HIDS) or Network-based Intrusion Detection Systems (NIDS) according to the location of their sensors on a host or the network.

7.8.1.1 Host-Based IDS

A *Host-based IDS* (HIDS) monitors the activity of a system at various levels within a host or device. For example, the status of specific files, the type of services running, the system call traces, or the network traffic entering or leaving the device. Because of its location within the system, an HIDS is able to detect all intrusion attempts that target a specific host or device, including insider threats. Furthermore, an HIDS is capable of analyzing encrypted network traffic since data is first decrypted upon arriving at the target host or device. A disadvantage of an HIDS is the additional management and maintainability costs.

7.8.1.2 Network-Based IDS

A *Network-based IDS* (NIDS) [30] is a type of IDS that is deployed close to critical network components such as firewalls, switches, or routers. An NIDS may monitor and analyze network traffic belonging to a multiplicity of networking and application communication protocols, for example, ICS protocols such as DNP3 and Modbus. An NIDS, depending on its placement on the network, may provide wider visibility of a subsystem's network. Moreover, today the extensive adoption of encryption by various applications and protocols hinders the ability of an NIDS to analyze all data traveling through the network.

7.8.1.3 Distributed or Hybrid IDS

A *Distributed Intrusion Detection Systems* (DIDS) is a distributed and collaborative system that utilizes detectors at both device and network levels. By combining the advantages of HIDS and NIDS [31], DIDS may effectively discover threats in a system not detectable by either an HIDS or NIDS.

7.8.1.4 Signature Detection

An alternative categorization is based on the intrusion analysis method and it distinguishes IDS as *signature detection IDS* or *anomaly detection IDS* [32].

To perform detection, a *signature detection IDS* attempts to match device and network activity with known signatures of attacks. Such IDS assume that signatures have been produced a priori, usually manually, by analyzing malicious network traffic or spurious behavior inside an OS.

A signature-based IDS can detect malicious behavior with high accuracy and low false positives. Their weakness comes when a novel or zero-day attack is introduced, which would not correspond to known signatures or heuristic rules. One of the most widely used IDS of this category is Snort [33].

7.8.1.5 Anomaly Detection

Anomaly detection IDS create models or profiles of normal activity. Model construction can be performed by statistical analysis, knowledge-based, or machine learning-based approaches [34, 35]. If a deviation from expected (normal) behavior is found, then an alert is generated.

An anomaly-based IDS does not need the maintenance and update of attack signatures. The advantage of an anomaly-based IDS is their increased potential for discovering novel/zero-day attacks. The disadvantage is their high false-positive rate, which may require further attention.

7.8.2 Incident Response

Incident Response (IR) is a set of methodologies and procedures that are used for handling cyber-security incidents. It aims to limit the damage from potential incidents and creates an environment that will assist the transition back to normal operations. Moreover, the information gathered from past incidents can be used to create stronger defense measures and better preparation for future incidents.

A comprehensive IR plan is the most effective way to respond to an incident. According to the NIST Computer Security Incident, Handling Guide (SP 800-61) [36], there are four fundamental phases that should be followed: (i) *preparation*, where the IR team and IR plan are formed along with the appropriate training; (ii) *detection and analysis* where the monitoring of attack vectors and signs of an incident take place together with the prioritization of which incidents should be addressed first; (iii) *containment, eradication, and recovery*, where all the necessary steps for miti-gation and recovery of the affected parts of the systems are performed; and (iv) *post-incident activity* in which the organization reviews lessons form an incident and retains evidence for legal and orga-nizational policy reasons.

The creation of an IR plan for ICS or CPS should consider any potential life, safety, and property impacts related to both the incident outcomes and the actions taken by the IR team. This should also comply with current procedures and safety mechanisms that are already in place. Another important aspect is information sharing among organizations such as the Cybersecurity and Infras-tructure Security Agency (CISA), or the Information Sharing and Analysis Centers (ISACs), which could assist other organizations (in the same or different sectors) in learning from past incidents and creating more effective defenses and IR plans [37].

7.9 Summary

In this chapter, we described the fundamental ICS and CPS cybersecurity concepts, and how a cybersecure system is designed and maintained to provide such properties. We outlined the most prominent threats (intentional or random occurrences) that ICS and CPS face, and the methods and techniques that are necessary to mitigate these threats. The analysis of security management, risk assessment, threat mitigation, and incident detection and response aims to provide the neces-sary apparatus for the individuals that want to understand, create or improve the security posture of organizations that are involved with ICS and CPS. The common cybersecurity techniques and mechanisms should take into consideration not only the data exchanged between various systems but also the systems' objectives, functionality and limitations with regards to the controlled pro-cesses, to ultimately avoid safety violations, hazards, accidents, or catastrophes.

7.10 Thoughtful Questions to Ensure Comprehension

1. What is the difference between Authentication and Authorization?

2. How can Accountability be achieved in an ICS environment (e.g. manufacturing plant) for both cyber and physical assets?

3. Should the organizations follow the CVSS strictly in order to evaluate the criticalness of a vulnerable part of their system?

4. Can tools provide all the required security for an organization or should work in combination with policies and procedures?

5. What are some common ICS consequences that origin from an unauthorized access to an ICS?

6. Which stages of the Cyber Kill Chain model can an IDS possibly detect?

7. What is the most appropriate type of IDS for detecting novel attacks?

Further Reading

General information about computer security can be found in [38]. Industrial protocols and their related network topologies are discussed in [39]. The subject of Cybersecurity particularly in Industrial Control settings is discussed in [40].

References

1 PUB FIPS (2004). *199, Standards for Security Categorization of Federal Information and Information Systems.* Gaithersburg, MD: Computer Security Division, Information Technology Laboratory, National Institute of Standards and Technology.

2 Stouffer, K. (2011). Guide to industrial control systems (ICS) security. *NIST Special Publication* 800 (82): 16.

3 Nechvatal, J., Barker, E., Bassham, L. et al. (2001). Report on the development of the advanced encryption standard (AES). *Journal of Research of the National Institute of Standards and Technology* 106 (3): 511.

4 Ferguson, N., Schneier, B., and Kohno, T. (2010). Cryptography Engineering. *Design Princi.*

5 U.S. National Bureau of Standards (1977). FIPS-46: Data Encryption Standard.

6 Trusted Computing Group (2011). TPM main part 1 design principles. *TCG White Paper.*

7 Rivest, R.L., Shamir, A., and Adleman, L. (1978). A method for obtaining digital signatures and public-key cryptosystems. *Communications of the ACM* 21 (2): 120–126.

8 Miller, V.S. (1986). Use of elliptic curves in cryptography. In: Williams H.C. (eds) *Advances in Cryptology – CRYPTO '85 Proceedings. CRYPTO 1985. Lecture Notes in Computer Science*, 218 417–426. Berlin, Heidelberg: Springer, 10.1007/3-540-39799-X_31.

9 Koblitz, N. (1987). Elliptic curve cryptosystems. *Mathematics of Computation* 48 (177): 203–209.

10 Kerry, C.F. and Gallagher, P.D. (2013). Digital Signature Standard (DSS). *FIPS PUB 186-14.*

11 Shirey, R. (2007). Internet Security Glossary, Version 2. *Technical report, RFC 4949*, The IETF Trust, Fremont, CA, August 2007.

12 Grance, T., Hash, J., Peck, S. et al. (2002). Security Guide for Interconnecting Information Technology Systems: Recommendations of the National Institute of Standards and Technology. *Technical report, NIST Special Publication 800-47.* Gaithersbur, MD: National Institute of Standards and Technologies.

13 FIRST (2020). Common Vulnerability Scoring System.

14 MITRE (2020). CVE-Common Vulnerabilities and Exposures.

15 Falliere, N., Murchu, L.O., and Chien, E. (2011). W32. Stuxnet dossier. *White Paper, Symantec Corp., Security Response* 5 (6): 29.

16 Langner, R. (2013). A technical analysis of what Stuxnet's creators tried to achieve, p. 37.

17 DHS (2016). Cyber-attack against Ukrainian critical infrastructure | CISA.

18 DHS (2014). ICS focused malware | CISA.

19 Abrams, M.D. and Weiss, J. (2016). Malicious Control System Cyber Security Attack Case Study: Maroochy Water Services, Australia, May 2016.

20 Microsoft Security (2017). WannaCrypt Ransomware worm targets out-of-date systems, May 2017.

21 Hutchins, E.M., Cloppert, M.J., Amin, R.M. et al. (2011). Intelligence-driven computer network defense informed by analysis of adversary campaigns and intrusion kill chains. *Leading Issues in Information Warfare & Security Research* 1 (1): 80.

22 Kuznetsov, I. and Gostev, A. (2011). Stuxnet/Duqu: The evolution of drivers.

23 McDonald, G., Murchu, L.O., Doherty, S., and Chien, E. (2013). Stuxnet 0.5: The missing link, p. 18.

24 Assante, M.J. and Lee, R.M. (2015). *The Industrial Control System Cyber Kill Chain*, 1. SANS Institute InfoSec Reading Room.

25 Saltzer, J.H. and Schroeder, M.D. (1975). The protection of information in computer systems. *Proceedings of the IEEE* 63 (9): 1558–2256.

26 Lamb, C.C. (2015). A Survey of Secure Architectural Principles. *Tech. Report SAND2015-5657*, Sandia National Laboratories, Albuquerque, NM.

27 Smith, R.E. (2012). A contemporary look at Saltzer and Schroeder's 1975 design principles. *IEEE Security and Privacy* 10 (06): 20–26.

28 Saltzer, J.H. and Kaashoek, M.F. (2009). *Principles of Computer System Design*. Morgan Kaufmann - Elsevier.

29 NIST Joint Task Force Transformation Initiative (2011). Managing information security risk: organization, mission, and information system view, March 2011.

30 Mukherjee, B., Heberlein, L.T., and Levitt, K.N. (1994). Network intrusion detection. *IEEE Network* 8 (3): 26–41.

31 Snapp, S.R., Brentano, J., Dias, G. et al. (1991). DIDS (distributed intrusion detection system)-motivation, architecture, and an early prototype.

32 Porras, P. (1993). Stat–A state transition analysis tool for intrusion detection.

33 Roesch, M. (1999). Snort: Lightweight intrusion detection for networks. *Lisa*, Volume 99, pp. 229–238.

34 Garcia-Teodoro, P., Diaz-Verdejo, J., Maciá-Fernández, G., and Vázquez, E. (2009). Anomaly-based network intrusion detection: techniques, systems and challenges. *Computers & Security* 28 (1–2): 18–28.

35 Creech, G. (2014). Developing a high-accuracy cross platform Host-Based Intrusion Detection System capable of reliably detecting zero-day attacks. PhD thesis. Canberra, Australia: University of New South Wales.

36 Cichonski, P., Millar, T., Grance, T., and Scarfone, K. (2012). Computer security incident handling guide. *NIST Special Publication* 800 (61): 1–147.

37 DHS (2009). Recommended Practice: developing an Industrial Control Systems cybersecurity incident response capability.

38 Stallings, W. and Brown, L. (2018). *Computer Security: Principles and Practice*. Upper Saddle River, NJ: Pearson Education.

39 Knapp, E.D. and Langill, J.T. (2014). *Industrial Network Security: Securing Critical Infrastructure Networks for Smart Grid, SCADA, and other Industrial Control Systems*. Syngress.

40 Flaus, J.-M. (2019). *Cybersecurity of Industrial Systems*. Wiley.

8

Control Theory

Desineni S. Naidu

Electrical Engineering, University of Minnesota-Duluth, Duluth, Minnesota, USA

Objectives

The objectives of this chapter are to provide an overview of linear open- and closed-loop, deterministic, and stochastic optimal control systems for both regulation and tracking, including calculus of variations, Lagrange and Hamiltonian methods, and the Pontryagin principle, and of recent results on nonlinear closed-loop optimal control systems for both regulation and tracking including state-dependent Riccati equations (SDRE).

8.1 Introduction

From the classical/basic control point of view, modeling is generally done using differential and/or difference equations and the corresponding Laplace and Z-transform domains, respectively. The analysis is done by various tools such as Routh, Nyquist, Root Locus, Bode plot, and Nichols chart. The third ingredient of synthesis or design is important for control system engineering because design deals with meeting the performance specifications or simply "satisfying the customer." Similarly from the modern/advanced control point of view [1], modeling is done in terms of state variables using Lagrange method, analysis is mostly done for linear and nonlinear systems using Lyapunov and Popov methods.

The classical (conventional) control theory concerned with single input and single output (SISO) is mainly based on Laplace transform theory and its use in system representation in Laplace transform. The modern or advanced control theory concerned with multiple inputs and multiple outputs (MIMO) is based on state variable representation in terms of a set of first-order differential (or difference) equations. The modern/advanced control theory dictates that all the state variables should be fed back after suitable weighting and the input $\mathbf{u}(t)$ to the plant is determined by the controller (consisting of error detector and compensator) driven by system states $\mathbf{x}(t)$ and reference signal $\mathbf{r}(t)$, all or most of the state variables are available for feedback control, and it depends on well-established matrix theory, which is amenable for large-scale computer simulation.

Resilient Control Architectures and Power Systems, First Edition.
Edited by Craig Rieger, Ronald Boring, Brian Johnson, and Timothy McJunkin.
© 2022 The Institute of Electrical and Electronics Engineers, Inc. Published 2022 by John Wiley & Sons, Inc.

8.1.1 Formal Statement of Optimal Control Problem

The main objective of optimal control is to determine control signals that will cause a process (plant) to satisfy some physical constraints and at the same time extremize (maximize or minimize) a chosen performance criterion (performance index or cost function). We are interested in finding the optimal control $\mathbf{u}^*(t)$ (* indicates optimal condition) that will drive the plant P from initial state to final state with some constraints on controls and states and at the same time extremizing the given performance index J.

The formulation of optimal control problem requires a mathematical description (or model) of the process to be controlled (generally in state variable form), a specification of the performance index, and a statement of boundary conditions and the physical constraints on the states and/or controls. The optimal control problem is to find the optimal control $\mathbf{u}^*(t)$ ($*$ indicates extremal or optimal value) that causes the linear time-invariant plant (system):

$$\dot{\mathbf{x}}(t) = \mathbf{A}\mathbf{x}(t) + \mathbf{B}\mathbf{u}(t) \tag{8.1}$$

to give the trajectory $\mathbf{x}^*(t)$ that optimizes or extremizes (minimizes or maximizes) a performance index:

$$J = \mathbf{x}'(t_f)\mathbf{F}\mathbf{x}(t_f) + \int_{t_0}^{t_f} [\mathbf{x}'(t)\mathbf{Q}\mathbf{x}(t) + \mathbf{u}'(t)\mathbf{R}\mathbf{u}(t)]dt, \tag{8.2}$$

where $\mathbf{R}$ is a positive definite matrix, $\mathbf{Q}$ is a weighting matrix, which can be positive semi-definite, and $\mathbf{F}$ is a positive semi-definite matrix and state and control vectors $\mathbf{x}(t)$ and $\mathbf{u}(t)$ and other matrices $\mathbf{A}, \mathbf{B}$ are of appropriate dimensionality.

or which causes the nonlinear system:

$$\dot{\mathbf{x}}(t) = \mathbf{f}(\mathbf{x}(t), \mathbf{u}(t), t) \tag{8.3}$$

to give the state $\mathbf{x}^*(t)$ that optimizes the general performance index:

$$J = S(\mathbf{x}(t_f), t_f) + \int_{t_0}^{t_f} V(\mathbf{x}(t), \mathbf{u}(t), t)dt \tag{8.4}$$

with some constraints on the control variables $\mathbf{u}(t)$ and/or the state variables $\mathbf{x}(t)$. The final time t_f may be fixed, or free, and the final (target) state may be fully or partially fixed or free. Thus, we are basically interested in finding the control $\mathbf{u}^*(t)$, which when applied to the plant described by (8.1) or (8.3), gives an optimal performance index J^* described by (8.2) or (8.4). In other words, optimal control problem is the optimization of an integral function subject to the constraints in the form of differential equations and the associated boundary conditions.

8.2 Deterministic Linear Systems

8.2.1 Open-Loop Optimal Control of Linear Systems

In this section, we approach the optimal control system by variational techniques, and in the process introduce the Hamiltonian function, which was used by Pontryagin and his associates to develop the famous Minimum Principle [2].

Here we consider the optimal control system where the performance index is of general form containing a final (terminal) cost function in addition to the integral cost function. Such an optimal control problem is called the Bolza problem. Consider the nonlinear, time-varying plant shown in

Eq. (8.3) and the performance index consisting of both terminal cost and integral cost functions shown in Eq. (8.4) the given boundary conditions as:

$$\mathbf{x}(t_0) = \mathbf{x}_0; \qquad \mathbf{x}(t_f) \text{ is free and } t_f \text{ is free,} \tag{8.5}$$

where $\mathbf{x}(t)$ and $\mathbf{u}(t)$ are n- and r-dimensional state and control vectors, respectively, and $S(\mathbf{x}(t_f), t_f)$ is the terminal cost function. Using calculus of variations [3], the optimal control $\mathbf{u} * (t)$ is given by:

$$\left(\frac{\partial \mathcal{H}}{\partial \mathbf{u}}\right)_* = 0, \tag{8.6}$$

where the Pontryagin or Hamiltonian $\mathcal{H}$ function is given by:

$$\mathcal{H}(\mathbf{x}(t), \mathbf{u}(t), \lambda(t), t) = V(\mathbf{x}(t), \mathbf{u}(t), t) + \lambda'(t)\mathbf{f}(\mathbf{x}(t), \mathbf{u}(t), t), \tag{8.7}$$

and the optimal state $\mathbf{x}^*(t)$ and optimal costate $\lambda^*(t)$ are solved as a two-point boundary value problem (TPBVP) consisting of the $2n$ differential equations:

$$\dot{\mathbf{x}}^*(t) = +\left(\frac{\partial \mathcal{H}}{\partial \lambda}\right)_*, \qquad \dot{\lambda}^*(t) = -\left(\frac{\partial \mathcal{H}}{\partial \mathbf{x}}\right)_*. \tag{8.8}$$

In the previous control relation shown in Eq. (8.6), the state and the costate relations (8.8) are solved for optimal control $\mathbf{u}^*(t)$, optimal state $\mathbf{x}^*(t)$, and costate $\lambda^*(t)$ using the given initial condition $\mathbf{x}_0$ and the final condition:

$$\left[\mathcal{H}^* + \frac{\partial S}{\partial t}\right]_{t_f} \delta t_f + \left[\left(\frac{\partial S}{\partial \mathbf{x}}\right)_* - \lambda^*(t)\right]'_{t_f} \delta \mathbf{x}_f = 0 \tag{8.9}$$

to be evaluated depending upon the nature of the final conditions. In solving the TPBVP arising due to the state and costate relations shown in Eq. (8.8), and then substituting in the control relation shown in (8.6), one gets only the open-loop optimal control. Here, one has to construct or realize an open-loop optimal controller (OLOC) and in many cases it is very tedious. Also, changes in plant parameters are not taken into account by the OLOC. This prompts us to think in terms of a closed-loop optimal control (CLOC) (i.e. to obtain optimal control $\mathbf{u}^*(t)$ in terms of the state $\mathbf{x}^*(t)$). This CLOC will have many advantages such as sensitive to plant parameter variations and simplified construction of the controller. The CLOC systems are discussed in the next section.

8.2.2 Closed-Loop Optimal Control of Linear Systems

In this section, we present the CLOC of linear plants or systems with quadratic performance index or measure. This leads to the linear quadratic regulator (LQR) system dealing with state regulation, output regulation, and tracking. Broadly speaking, we are interested in the design of optimal linear systems with quadratic performance indices. We discuss the plant and the quadratic performance index with particular reference to physical significance. This helps us to obtain some elegant mathematical conditions on the choice of various matrices in the quadratic cost functional. Thus, we will be dealing with an optimization problem from the engineering perspective.

Consider a linear, time-varying (LTV) system:

$$\dot{\mathbf{x}}(t) = \mathbf{A}(t)\mathbf{x}(t) + \mathbf{B}(t)\mathbf{u}(t), \qquad \mathbf{y}(t) = \mathbf{C}(t)\mathbf{x}(t), \tag{8.10}$$

with a cost functional (CF) or performance index (PI):

$$J(\mathbf{u}(t)) = \frac{1}{2}\left[\mathbf{z}(t_f) - \mathbf{y}(t_f)\right]'\mathbf{F}(t_f)\left[\mathbf{z}(t_f) - \mathbf{y}(t_f)\right]$$

$$+ \frac{1}{2}\int_{t_0}^{t_f}\left[\left[\mathbf{z}(t) - \mathbf{y}(t)\right]'\mathbf{Q}(t)\left[\mathbf{z}(t) - \mathbf{y}(t)\right] + \mathbf{u}'(t)\mathbf{R}(t)\mathbf{u}(t)\right]dt, \tag{8.11}$$

where, $\mathbf{e}(t) = \mathbf{z}(t) - \mathbf{y}(t)$. If our objective is to keep the state $\mathbf{x}(t)$ near zero (i.e. $\mathbf{z}(t) = 0$ and $\mathbf{C} = I$), then we call it state regulator system. In other words, the objective is to obtain a control $\mathbf{u}(t)$, which takes the plant from a nonzero state to zero state. This situation may arise when a plant is subjected to unwanted disturbances that perturb the state (for example, sudden load changes in an electrical voltage regulator system, or a sudden wind gust in a radar antenna positional control system). If we try to keep the output or state near a desired state or output, then we are dealing with a tracking system. We see that in both state and output regulator systems, the desired or reference state is zero and in tracking system the error is to be made zero.

8.2.3 Finite-Time Linear Quadratic Regulator: Time-Varying Case

For a finite-time LTV quadratic regulator, the result is summarized below [3]. Given a LTV plant:

$$\dot{\mathbf{x}}(t) = \mathbf{A}(t)\mathbf{x}(t) + \mathbf{B}(t)\mathbf{u}(t) \tag{8.12}$$

and a quadratic performance index:

$$J = \frac{1}{2}\mathbf{x}'(t_f)\mathbf{F}(t_f)\mathbf{x}(t_f) + \frac{1}{2}\int_{t_0}^{t_f} \left[\mathbf{x}'(t)\mathbf{Q}(t)\mathbf{x}(t) + \mathbf{u}'(t)\mathbf{R}(t)\mathbf{u}(t)\right] dt, \tag{8.13}$$

where $\mathbf{u}(t)$ is not constrained, t_f is specified, and $\mathbf{x}(t_f)$ is not specified, $\mathbf{F}(t_f)$ and $\mathbf{Q}(t)$ are $n \times n$ symmetric, positive semidefinite matrices, and $\mathbf{R}(t)$ is $r \times r$ symmetric, positive definite matrix, the optimal control is given by:

$$\mathbf{u}^*(t) = -\mathbf{R}^{-1}(t)\mathbf{B}'(t)\mathbf{P}(t)\mathbf{x}^*(t) = -\mathbf{K}(t)\mathbf{x}^*(t), \tag{8.14}$$

where $\mathbf{K}(t) = \mathbf{R}^{-1}(t)\mathbf{B}'(t)\mathbf{P}(t)$ is called Kalman gain and $\mathbf{P}(t)$, the $n \times n$ symmetric, *positive definite* matrix (for all $t \in [t_0,\ t_f]$), is the solution of the matrix differential Riccati equation (DRE):

$$\dot{\mathbf{P}}(t) = -\mathbf{P}(t)\mathbf{A}(t) - \mathbf{A}'(t)\mathbf{P}(t) - \mathbf{Q}(t) + \mathbf{P}(t)\mathbf{B}(t)\mathbf{R}^{-1}(t)\mathbf{B}'(t)\mathbf{P}(t) \tag{8.15}$$

satisfying the final condition $\mathbf{P}(t = t_f) = \mathbf{F}(t_f)$. The optimal state is the solution of:

$$\dot{\mathbf{x}}^*(t) = \left[\mathbf{A}(t) - \mathbf{B}(t)\mathbf{R}^{-1}(t)\mathbf{B}'(t)\mathbf{P}(t)\right]\mathbf{x}^*(t), \tag{8.16}$$

and the optimal cost is $J^* = \frac{1}{2}\mathbf{x}^{*\prime}(t)\mathbf{P}(t)\mathbf{x}^*(t)$. The optimal control $\mathbf{u}^*(t)$, shown in (8.14), is linear in the optimal state $\mathbf{x}^*(t)$. The solution of the matrix DRE (8.15) is not readily available with MATLAB* and hence a MATLAB-based program is developed [4] for solving the matrix DRE based on the analytical solution of matrix DRE [5].

8.2.4 Infinite-Interval Regulator System: Time-Invariant Case

For the time-invariant case of the infinite-interval regulator system, the result is summarized below. For a controllable, linear, time-invariant plant (8.1) and the infinite interval cost functional:

$$J = \frac{1}{2}\int_0^\infty \left[\mathbf{x}'(t)\mathbf{Q}\mathbf{x}(t) + \mathbf{u}'(t)\mathbf{R}\mathbf{u}(t)\right] dt, \tag{8.17}$$

the optimal control is given by:

$$\mathbf{u}^*(t) = -\mathbf{R}^{-1}\mathbf{B}'\overline{\mathbf{P}}\mathbf{x}^*(t) = -\overline{\mathbf{K}}\mathbf{x}^*(t), \tag{8.18}$$

where $\overline{\mathbf{K}} = \mathbf{R}^{-1}\mathbf{B}'\overline{\mathbf{P}}$, and $\overline{\mathbf{P}}$, the $n \times n$ constant, positive definite, symmetric matrix, is the solution of the nonlinear, matrix algebraic Riccati equation (ARE):

$$-\overline{\mathbf{P}}\mathbf{A} - \mathbf{A}'\overline{\mathbf{P}} + \overline{\mathbf{P}}\mathbf{B}\mathbf{R}^{-1}\mathbf{B}'\overline{\mathbf{P}} - \mathbf{Q} = 0, \tag{8.19}$$

the optimal trajectory is the solution of $\dot{\mathbf{x}}^*(t) = \left[\mathbf{A} - \mathbf{B}\mathbf{R}^{-1}\mathbf{B}'\overline{\mathbf{P}}\right]\mathbf{x}^*(t)$.

Figure 8.1 Implementation of the closed-loop optimal control: infinite final time.

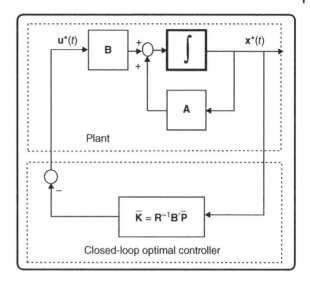

The implementation is shown in Figure 8.1. An interesting property that an optimal control system can be solved and implemented as an OLOC configuration or a CLOC configuration. From the implementation point of view, the closed-loop optimal controller is much simpler in terms of hardware or software than the open-loop optimal controller. With a closed-loop configuration, all the advantages of conventional feedback are incorporated.

8.2.5 Linear Quadratic Tracking System: Finite-Time Case

In tracking (trajectory following) systems, we require that the output of a system tracks or follows a desired trajectory in some optimal sense. Thus, we see that this is a generalization of *regulator* systems in the sense that the desired trajectory for the regulator is simply the zero state. For the LTV system shown in Eq. (8.12) with the desired output $\mathbf{z}(t)$, the error $\mathbf{e}(t) = \mathbf{z}(t) - \mathbf{y}(t)$, and the performance index:

$$J = \frac{1}{2}\mathbf{e}'(t_f)\mathbf{F}(t_f)\mathbf{e}(t_f) + \frac{1}{2}\int_{t_0}^{t_f}\left[\mathbf{e}'(t)\mathbf{Q}(t)\mathbf{e}(t) + \mathbf{u}'(t)\mathbf{R}(t)\mathbf{u}(t)\right]dt, \tag{8.20}$$

then the optimal control $\mathbf{u}^*(t)$ is given by:

$$\mathbf{u}^*(t) = -\mathbf{K}(t)\mathbf{x}^*(t) + \mathbf{R}^{-1}(t)\mathbf{B}'(t)\mathbf{g}(t), \tag{8.21}$$

where the $n \times n$ symmetric, positive definite matrix $\mathbf{P}(t)$ is the solution of the nonlinear, matrix DRE:

$$\dot{\mathbf{P}}(t) = -\mathbf{P}(t)\mathbf{A}(t) - \mathbf{A}'(t)\mathbf{P}(t) + \mathbf{P}(t)\mathbf{E}(t)\mathbf{P}(t) - \mathbf{V}(t) \tag{8.22}$$

with final condition $\mathbf{P}(t_f) = \mathbf{C}'(t_f)\mathbf{F}(t_f)\mathbf{C}(t_f)$ and the nth order $\mathbf{g}(t)$ is the solution of the linear, non-homogeneous vector differential equation:

$$\dot{\mathbf{g}}(t) = -[\mathbf{A}(t) - \mathbf{E}(t)\mathbf{P}(t)]'\,\mathbf{g}(t) - \mathbf{W}(t)\mathbf{z}(t) \tag{8.23}$$

with final condition $\mathbf{g}(t_f) = \mathbf{C}'(t_f)\mathbf{F}(t_f)\mathbf{z}(t_f)$, where $\mathbf{E}(t)$, $\mathbf{V}(t)$, and $\mathbf{W}(t)$ are defined appropriately, the optimal state (trajectory) is the solution of the linear state equation:

$$\dot{\mathbf{x}}^*(t) = [\mathbf{A}(t) - \mathbf{E}(t)\mathbf{P}(t)]\,\mathbf{x}^*(t) + \mathbf{E}(t)\mathbf{g}(t). \tag{8.24}$$

8.2.6 Gain Margin and Phase Margin

We know that in classical control theory, the features of gain and phase margins are important when evaluating the system performance with respect to robustness to plant parameter variations and uncertainties. The engineering specifications often place lower bounds on the phase and gain margins. Here, we interpret some of the classical control features such as gain margin and phase margin for the CLOC system [6]. In the single-input case, it is shown [3]:

$$r + |\mathbf{c}'[j\omega\mathbf{I} - \mathbf{A}']^{-1}\mathbf{b}|^2 = r|(1 + \overline{\mathbf{k}}[j\omega\mathbf{I} - \mathbf{A}]^{-1}\mathbf{b})|^2. \tag{8.25}$$

The previous result may be called another version of the Kalman equation in frequency domain. The previous relation implies that $|(1 + \overline{\mathbf{k}}[j\omega\mathbf{I} - \mathbf{A}]^{-1}\mathbf{b})|^2 \geq 1$. Here, we can easily recognize that for a single-input, single-output case, the optimal feedback control system is exactly like a classical feedback control system with unity negative feedback and transfer function as $G_o(s) = \overline{\mathbf{k}}[s\mathbf{I} - \mathbf{A}]^{-1}\mathbf{b}$. Thus, the frequency domain interpretation in terms of gain margin and phase margin can be easily done using Nyquist, Bode, or some other plot of the transfer function $G_o(s)$.

8.2.7 Gain Margin

We recall that the gain margin of a feedback control system is the amount of loop gain (usually in decibels) that can be changed before the closed-loop system becomes unstable. Let us now apply the well-known Nyquist criterion to the unity feedback. Here, we assume that the Nyquist path is clock-wise (CW) and the corresponding Nyquist plot makes counter-clock-wise (CCW) encirclements around the critical point $-1 + j0$. According to Nyquist stability criterion, for closed-loop stability, the Nyquist plot (or diagram) makes CCW encirclements as many times as there are poles of the open-loop transfer function $G_o(s)$ lying in the right half of the s-plane.

Further, we note that $\left|(1 + \overline{\mathbf{k}}[j\omega\mathbf{I} - \mathbf{A}]^{-1}\mathbf{b})\right| \geq 1$ or $|1 + G_o(j\omega)| \geq 1$ implies that the distance between the critical point $-1 + j0$ and any point on the Nyquist plot is at least 1. This condition implies that the Nyquist plot of $G_o(j\omega)$ is constrained to avoid all the points inside the unit circle (centered at $-1 + j0$). Note that the number of encirclements of the critical point $-1 + j0$ is the same and the number of encirclements of any other point inside the unit circle. Thus, it is clear that the closed-loop optimal system has infinite gain margin. Let us proceed further to see if there is a lower limit on the gain factor.

Now, if we multiply the open-loop gain with some constant factor β, the closed-loop system will be asymptotically stable if the Nyquist plot of $\beta G_o(j\omega)$ encircles $-1 + j0$ in a CCW direction as many times as there are poles of $\beta G_o(s)$ in the right-half plane. This means that the closed-loop system will be stable if the Nyquist diagram of $G_o(j\omega)$ encircles the critical point $-(1/\beta) + j0$ the same number of times as there are open-loop poles in the right-half plane. Consequently, there is a lower limit of $\beta > \frac{1}{2}$. Thus, we have an infinite gain margin on the upper side and a lower gain margin of $\beta = 1/2$.

8.2.8 Phase Margin

Let us first recall that the phase margin is the amount of phase shift in CW direction (without affecting the gain) through which the Nyquist plot can be rotated about the origin so that the gain crossover (unit distance from the origin) passes through the $-1 + j0$ point. Simply, it is the amount by which Nyquist plot can be rotated CW to make the system unstable. Consider a point P at unit distance from the origin on the Nyquist plot. Since we know that the Nyquist plot of an optimal regulator must avoid the unit circle centered at $-1 + j0$, the set of points, which are at unit distance from the origin and lying on Nyquist diagram of an optimal regulator, are constrained to lie on the portion on the circumference of the circle with unit radius and centered at the origin.

Here, we notice that the smallest angle through which one of the admissible points A_n on (the circumference of the circle centered at origin) the Nyquist plot could be shifted in a CW direction to reach $-1 + j0$ point, is $60°$. Thus, the closed-loop optimal system or LQR system has a phase margin of at least or a minimum of $60°$. It is easily seen that the maximum phase margin corresponding to the point A_x is $(30 + 180 + 30) = 240°$.

8.3 Pontryagin Principle and HJB Equation

The Pontryagin principle is now summarized below. Given the plant as:

$$\dot{\mathbf{x}}(t) = \mathbf{f}(\mathbf{x}(t), \mathbf{u}(t), t), \tag{8.26}$$

the performance index as:

$$J = S(\mathbf{x}(t_f), t_f) + \int_{t_0}^{t_f} V(\mathbf{x}(t), \mathbf{u}(t), t) dt, \tag{8.27}$$

and the boundary conditions as $\mathbf{x}(t_0) = \mathbf{x}_0$ and t_f, $\mathbf{x}(t_f) = \mathbf{x}_f$ are free, to find the optimal control, form the Pontryagin $\mathcal{H}$ function:

$$\mathcal{H}(\mathbf{x}(t), \mathbf{u}(t), \lambda(t), t) = V(\mathbf{x}(t), \mathbf{u}(t), t) + \lambda'(t) \mathbf{f}(\mathbf{x}(t), \mathbf{u}(t), t), \tag{8.28}$$

minimize $\mathcal{H}$ w.r.t. $\mathbf{u}(t)(\leq \mathbf{U})$ as:

$$\mathcal{H}(\mathbf{x}^*(t), \mathbf{u}^*(t), \lambda^*(t), t) \leq \mathcal{H}(\mathbf{x}^*(t), \mathbf{u}(t), \lambda^*(t), t), \tag{8.29}$$

and solve the set of $2n$ state and costate equations:

$$\dot{\mathbf{x}}^*(t) = \left(\frac{\partial \mathcal{H}}{\partial \lambda}\right)_* \quad \text{and} \quad \dot{\lambda}^*(t) = -\left(\frac{\partial \mathcal{H}}{\partial \mathbf{x}}\right)_*, \tag{8.30}$$

with appropriate the boundary conditions.

8.3.1 The Hamilton–Jacobi–Bellman (HJB) Equation

In this section, we present an alternate method of obtaining the CLOC, using the *principle of optimality* and the Hamilton–Jacobi–Bellman (HJB) equation. First we need to state Bellman's principle of optimality [7] – *any portion of the optimal trajectory is optimal.* Alternatively, the optimal policy (control) has the property that no matter what the previous decisions have been (i.e. controls), the remaining decision must constitute an optimal policy. Consider the nonlinear plant (8.3) and the performance index as:

$$J(\mathbf{x}(t_0), t_0) = \int_{t_0}^{t_f} V(\mathbf{x}(t), \mathbf{u}(t), t) dt. \tag{8.31}$$

Now, we provide the alternative approach, called HJB approach, and obtain a control law as a function of the state variables, leading to CLOC. This is important from the practical point of view in implementation of the optimal control. Let us define a scalar function $J^*(\mathbf{x}^*(t), t)$ as the minimum value of the performance index J for an initial state $\mathbf{x}^*(t)$ at time t such that:

$$J^*(\mathbf{x}^*(t), t) = \int_{t}^{t_f} V(\mathbf{x}^*(\tau), \mathbf{u}^*(\tau), \tau) d\tau. \tag{8.32}$$

In other words, $J^*(\mathbf{x}^*(t), t)$ is the value of the performance index when evaluated along the optimal trajectory starting at $\mathbf{x}(t)$. Here, we used the principle of optimality in saying that the trajectory from t to t_f is optimal. However, we are not interested in finding the optimal control for specific initial state $\mathbf{x}(t)$, but for any unspecified initial conditions. Thus, our interest is in $J(\mathbf{x}(t_0), t_0)$ as a function of $\mathbf{x}(t_0)$ and t_0. Now consider:

$$
\begin{aligned}
\frac{dJ^*(\mathbf{x}^*(t), t)}{dt} &= \left(\frac{\partial J^*(\mathbf{x}^*(t), t)}{\partial \mathbf{x}^*}\right)' \dot{\mathbf{x}}^*(t) + \frac{\partial J^*(\mathbf{x}^*(t), t)}{\partial t}, \\
&= \left(\frac{\partial J^*(\mathbf{x}^*(t), t)}{\partial \mathbf{x}^*}\right)' \mathbf{f}(\mathbf{x}^*(t), \mathbf{u}^*(t), t) + \frac{\partial J^*(\mathbf{x}^*(t), t)}{\partial t}.
\end{aligned}
\tag{8.33}
$$

Let us introduce the Hamiltonian as:

$$
\mathcal{H} = V(\mathbf{x}^*(t), \mathbf{u}^*(t), t) + \left(\frac{\partial J^*(\mathbf{x}^*(t), t)}{\partial \mathbf{x}^*}\right)' \mathbf{f}(\mathbf{x}^*(t), \mathbf{u}^*(t), t).
\tag{8.34}
$$

Using (8.34), we have:

$$
\frac{\partial J^*(\mathbf{x}^*(t), t)}{\partial t} + \mathcal{H}\left(\mathbf{x}^*(t), \frac{\partial J^*(\mathbf{x}^*(t), t)}{\partial \mathbf{x}^*}, \mathbf{u}^*(t), t\right) = 0; \ \forall \ t \in [t_0, t_f)
\tag{8.35}
$$

with boundary condition from (8.32) as:

$$
J^*(\mathbf{x}^*(t_f), t_f) = 0, \quad \text{or} \quad J^*(\mathbf{x}^*(t_f), t_f) = S(\mathbf{x}^*(t_f), t_f)
\tag{8.36}
$$

if the original performance index (8.31) contains a terminal cost function. Equation (8.35) is called the Hamilton–Jacobi equation. Since this equation is the continuous-time analog of Bellman's recurrence equations in dynamic programming [8], it is also called the HJB equation. The HJB equation, in general, is a nonlinear partial differential equation in J^*, which can be solved for J^*. Once J^* is known, its gradient $J_{\mathbf{x}}^*$ can be calculated and the optimal control $\mathbf{u}^*(t)$ is obtained. Often, the solution of HJB equation is very difficult.

8.4 Stochastic Linear Systems

We describe optimal state estimation problem leading to the celebrated Kalman filter for continuous-time systems in stochastic environment. The Kalman filter is basically an optimal state estimator minimizing the mean-squared error between the true value and estimated value of the state of a stochastic system composed of a process model and measurement model subjected to process and measurement disturbances. There is the extended Kalman filter (EKF) for linearized models of the nonlinear systems. Consider optimal estimation and control with finite horizon and incomplete state information.

8.4.1 Optimal Estimation

Let us suppose that the entire state $\mathbf{x}(t)$ is not available but only the output $\mathbf{y}(t)$ is available or measurable [9]:

$$
\dot{\mathbf{x}}(t) = \mathbf{A}(t)\mathbf{x}(t) + \mathbf{B}(t)\mathbf{u}(t) + \mathbf{B}_w(t)\mathbf{w}(t); \quad \mathbf{y}(t) = \mathbf{C}(t)\mathbf{x}(t) + \mathbf{v}(t),
\tag{8.37}
$$

where $\mathbf{w}(t)$ and $\mathbf{v}(t)$ are process and measurement (white, Gaussian) random noises with zero mean (i.e. $\overline{\mathbf{w}}(t) = \overline{\mathbf{v}}(t) = 0$) and covariances $\mathbf{Q}_w(t)$ and $\mathbf{R}_v(t)$, respectively, and assumed to be uncorrelated,

Figure 8.2 Continuous-time LQG regulator – finite-horizon: summary.

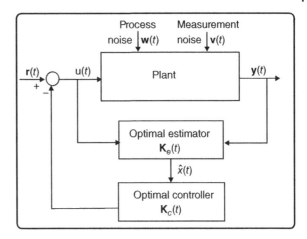

In order to find the best estimate $\hat{\mathbf{x}}(t)$ and the corresponding covariance matrix $\mathbf{P}_e(t)$, first, the filter (estimate) equations are as follows:

$$\dot{\hat{\mathbf{x}}}(t) = \mathbf{A}(t)\hat{\mathbf{x}}(t) + \mathbf{B}(t)\mathbf{u}(t) + \mathbf{K}_e(t)[\mathbf{y}(t) - \mathbf{C}(t)\hat{\mathbf{x}}(t)]; \quad \hat{\mathbf{x}}(t_0) = \overline{\mathbf{x}}(t_0), \tag{8.38}$$

where $\mathbf{K}_e(t)$, the optimal Kalman estimator (filter) gain, is obtained as $\mathbf{K}_e(t) = \mathbf{P}_e(t)\mathbf{C}'(t)\mathbf{R}_v^{-1}(t)$ and $\mathbf{P}_e(t)$ is the solution of the matrix DRE:

$$\dot{\mathbf{P}}_e(t) = \mathbf{A}(t)\mathbf{P}_e(t) + \mathbf{P}_e(t)\mathbf{A}'(t) + \mathbf{B}_w(t)\mathbf{Q}_w(t)\mathbf{B}'_w(t)$$
$$-\mathbf{P}_e(t)\mathbf{C}'(t)\mathbf{R}_v^{-1}(t)\mathbf{C}(t)\mathbf{P}_e(t); \quad \mathbf{P}_e(t_0) = \mathbf{P}_{e0}. \tag{8.39}$$

Let us also note that the estimation error covariance $\mathbf{P}_e(t)$ is independent of the measurements and the control and is deterministic.

8.4.2 Optimal Control

Now, the problem described by (8.38) constitutes a standard LQR problem. The problem of optimal control is reformulated to minimize the performance measure w.r.t. the admissible control, subject to the condition imposed by the filter. Now, using the results of linear quadratic gaussian (LQG) regulator except that the state is now the optimal estimate $\hat{\mathbf{x}}(t)$:

$$\mathbf{u}^*(t) = -\mathbf{R}^{-1}(t)\mathbf{B}'(t)\mathbf{P}_c(t)\hat{\mathbf{x}}(t) = -\mathbf{K}_c(t)\hat{\mathbf{x}}(t), \tag{8.40}$$

where, $\mathbf{K}_c(t) = \mathbf{R}^{-1}(t)\mathbf{B}'(t)\mathbf{P}_c(t)$, is the Kalman (controller) gain and $\mathbf{P}_c(t)$ is the solution of the matrix DRE:

$$-\dot{\mathbf{P}}_c(t) = \mathbf{P}_c(t)\mathbf{A}(t) + \mathbf{A}'(t)\mathbf{P}_c(t) + \mathbf{Q}(t)$$
$$-\mathbf{P}_c(t)\mathbf{B}'(t)\mathbf{R}^{-1}(t)\mathbf{B}(t)\mathbf{P}_c(t); \quad \mathbf{P}_c(t_f) = \mathbf{F}(t_f). \tag{8.41}$$

The summary of the LQG regulator problem is shown in Figure 8.2. Here, we see that the original plant is subjected to input noise $\mathbf{w}(t)$ and measurement noise $\mathbf{v}(t)$.

8.5 Deterministic Nonlinear Systems

Many nonlinear control design techniques exist, but each has benefits and weaknesses. Most of them are limited in range of applicability and use of certain nonlinear control techniques for a specific system that usually demands choosing between different factors (e.g. performance, robustness, optimality, and cost). Some of the well-known nonlinear control techniques are feedback linearization, adaptive control, nonlinear predictive control, sliding mode control, and approximating sequence of Riccati Equations. One of the highly promising and rapidly emerging techniques for nonlinear optimal controllers designing is the state dependent Riccati equation (SDRE) technique. Although a number of other methods exist for stabilization of nonlinear systems, the SDRE-based approach involves imitating standard LQR design for linear systems. The SDRE-based techniques are among the few successful approaches that have important properties, such as applicability to a large class of nonlinear systems, allowing the controller designer to make a trade-off between control effort and state errors, and its systematic formulation [10].

8.5.1 Finite-Horizon Regulation and Tracking for Nonlinear Systems

The nonlinear system considered in this chapter is in the form:

$$\dot{\mathbf{x}}(t) = \mathbf{f}(x) + \mathbf{g}(x)\mathbf{u}(t), \qquad \mathbf{y}(t) = \mathbf{h}(\mathbf{x}). \tag{8.42}$$

This nonlinear system can be expressed in a state-dependent linear-like form:

$$\dot{\mathbf{x}}(t) = \mathbf{A}(x)\mathbf{x}(t) + \mathbf{B}(\mathbf{x})\mathbf{u}(t), \qquad \mathbf{y}(t) = \mathbf{C}(\mathbf{x})\mathbf{x}(t), \tag{8.43}$$

where $\mathbf{f}(x) = \mathbf{A}(x)\mathbf{x}(t)$, $\mathbf{B}(x) = \mathbf{g}(x)$, and $\mathbf{h}(x) = \mathbf{C}(x)\mathbf{x}(t)$.

The goal is to find a state feedback optimal control law of the form $\mathbf{u}(\mathbf{x}) = -\mathbf{K}\mathbf{x}(t)$, that minimizes a cost function given by Naidu [3]:

$$\mathbf{J}(\mathbf{x}, \mathbf{u}) = \frac{1}{2}\mathbf{x}'(t_{\mathrm{f}})\mathbf{F}\mathbf{x}(t_{\mathrm{f}}) + \frac{1}{2}\int_{t_0}^{t_{\mathrm{f}}} \left[\mathbf{x}'(t)\mathbf{Q}(\mathbf{x})\mathbf{x}(t) + \mathbf{u}'(\mathbf{x})\mathbf{R}(\mathbf{x})\mathbf{u}(\mathbf{x}) \right] dt, \tag{8.44}$$

where $\mathbf{Q}(\mathbf{x})$ and $\mathbf{F}$ are symmetric positive semi-definite matrices, and $\mathbf{R}(\mathbf{x})$ is a symmetric positive definite matrix. Moreover, $\mathbf{x}'\mathbf{Q}(\mathbf{x})\mathbf{x}$ is a measure of state accuracy and $\mathbf{u}'(\mathbf{x})\mathbf{R}(\mathbf{x})\mathbf{u}(\mathbf{x})$ is a measure of control effort.

8.5.2 Finite-Horizon Regulator

To minimize the above cost function (8.44), a state feedback control law is given as:

$$\mathbf{u}(\mathbf{x}) = -\mathbf{K}\mathbf{x}(t) = -\mathbf{R}^{-1}(\mathbf{x})\mathbf{B}'(\mathbf{x})\mathbf{P}(\mathbf{x})\mathbf{x}(t), \tag{8.45}$$

where $\mathbf{P}(\mathbf{x}, t)$ is a symmetric, positive-definite solution of the differential SDRE. Strictly speaking it could be called the state-dependent differential Riccati equation (SD-DRE), of the form:

$$-\dot{\mathbf{P}}(\mathbf{x}) = \mathbf{P}(\mathbf{x})\mathbf{A}(\mathbf{x}) + \mathbf{A}'(\mathbf{x})\mathbf{P}(\mathbf{x}) - \mathbf{P}(\mathbf{x})\mathbf{B}(\mathbf{x})\mathbf{R}^{-1}(\mathbf{x})\mathbf{B}'(\mathbf{x})\mathbf{P}(\mathbf{x}) + \mathbf{Q}(\mathbf{x}), \tag{8.46}$$

with the final condition $\mathbf{P}(\mathbf{x}, t_{\mathrm{f}}) = \mathbf{F}$. The resulting differential SD-DRE-controlled trajectory becomes the solution of the state-dependent closed-loop dynamics:

$$\dot{\mathbf{x}}(t) = [\mathbf{A}(\mathbf{x}) - \mathbf{B}(\mathbf{x})\mathbf{R}^{-1}(\mathbf{x})\mathbf{B}'(\mathbf{x})\mathbf{P}(\mathbf{x})]\mathbf{x}(t). \tag{8.47}$$

As the differential SDRE is a function of $(\mathbf{x}, t)$, we do not know the value of the states ahead of present time step. Consequently, the state-dependent coefficients cannot be calculated to solve (8.46) with the final condition by backward integration from t_{f} to t_0. To overcome this problem, an approximate analytical approach is used [11] that converts the original nonlinear differential Ricatti equation to a linear differential Lyapunov equation (DLE) that can be solved in closed form at each time step. In order to solve the DRE (8.46), one can follow the following steps at each time step [12]:

1. Solve the ARE to calculate the steady-state value $\mathbf{P}_{ss}(\mathbf{x})$:

$$\mathbf{P}_{ss}(\mathbf{x})\mathbf{A}(\mathbf{x}) + \mathbf{A}'(\mathbf{x})\mathbf{P}_{ss}(\mathbf{x}) - \mathbf{P}_{ss}(\mathbf{x})\mathbf{B}(\mathbf{x})\mathbf{R}^{-1}(\mathbf{x})\mathbf{B}'(\mathbf{x})\mathbf{P}_{ss}(\mathbf{x}) + \mathbf{Q}(\mathbf{x}) = 0. \tag{8.48}$$

2. Use changing-of-variables procedure and assume that:

$$\mathbf{K}(\mathbf{x}, t) = [\mathbf{P}(\mathbf{x}, t) - \mathbf{P}_{ss}(\mathbf{x})]^{-1}. \tag{8.49}$$

3. Calculate the value of $\mathbf{A}_{cl}(\mathbf{x})$ as:

$$\mathbf{A}_{cl}(\mathbf{x}) = \mathbf{A}(\mathbf{x}) - \mathbf{B}(\mathbf{x})\mathbf{R}^{-1}\mathbf{B}'(\mathbf{x})\mathbf{P}_{ss}(\mathbf{x}). \tag{8.50}$$

4. Calculate the value of $\mathbf{D}$ by solving the algebraic Lyapunov equation [13]:

$$\mathbf{A}_{cl}\mathbf{D} + \mathbf{D}\mathbf{A}'_{cl} - \mathbf{B}\mathbf{R}^{-1}\mathbf{B}' = 0. \tag{8.51}$$

5. Solve the DLE:

$$\dot{\mathbf{K}}(\mathbf{x}, t) = \mathbf{K}(\mathbf{x}, t)\mathbf{A}'_{cl}(\mathbf{x}) + \mathbf{A}_{cl}(\mathbf{x})\mathbf{K}(\mathbf{x}, t) - \mathbf{B}(\mathbf{x})\mathbf{R}^{-1}\mathbf{B}'(\mathbf{x}). \tag{8.52}$$

The solution of (8.52), as shown by [14], is given by:

$$\mathbf{K}(\mathbf{x}, t) = e^{\mathbf{A}_{cl}(t-t_f)}(\mathbf{K}(\mathbf{x}, t_f) - \mathbf{D})e^{\mathbf{A}_{cl}'(t-t_f)} + \mathbf{D}. \tag{8.53}$$

6. Use change-of-variables procedure to calculate the value of $\mathbf{P}(\mathbf{x}, \mathbf{t})$ from (8.49):

$$\mathbf{P}(\mathbf{x}, t) = \mathbf{K}^{-1}(\mathbf{x}, t) + \mathbf{P}_{ss}(t). \tag{8.54}$$

7. Finally, calculate the value of the optimal control $\mathbf{u}(\mathbf{x}, t)$ as:

$$\mathbf{u}(\mathbf{x}, t) = -\mathbf{R}^{-1}\mathbf{B}'(\mathbf{x})\mathbf{P}(\mathbf{x}, t)\mathbf{x}(t). \tag{8.55}$$

Basically, $\mathbf{P}(\mathbf{x}, t)$, instead of solving backward in time from (8.46), we now obtain $\mathbf{P}(\mathbf{x}, t)$ from (8.54) in terms of $\mathbf{K}(\mathbf{x}, t)$, (8.53), the analytical solution of the linear DLE (8.52), which itself requires the solution of the ARE (8.48) and the solution of algebraic Lyapunov equation (8.51).

8.5.3 Finite-Horizon Tracking for Nonlinear Systems

Consider the nonlinear system given by (8.42), which can be re-described in the form (8.43), and let $\mathbf{z}(t)$ be the desired, or reference output. The goal is to find a state feedback, control law that minimizes a cost function given by:

$$\mathbf{J}(\mathbf{x}, \mathbf{u}) = \frac{1}{2}\mathbf{e}'(t_f)\mathbf{F}\mathbf{e}(t_f) + \frac{1}{2}\int_{t_0}^{t_f} \left[\mathbf{e}'(t)\mathbf{Q}(\mathbf{x})\mathbf{e}(t) + \mathbf{u}'(\mathbf{x})\mathbf{R}(\mathbf{x})\mathbf{u}(\mathbf{x})\right] dt, \tag{8.56}$$

where the error $\mathbf{e}(t) = \mathbf{z}(t) - \mathbf{y}(t)$. To minimize the cost function (8.56), a feedback control law is given as:

$$\mathbf{u}(\mathbf{x}) = -\mathbf{R}^{-1}\mathbf{B}'(\mathbf{x})[\mathbf{P}(\mathbf{x})\mathbf{x} - \mathbf{g}(\mathbf{x})], \tag{8.57}$$

where $\mathbf{P}(\mathbf{x})$ is a symmetric, positive-definite solution of the differential SDRE of the form:

$$-\dot{\mathbf{P}}(\mathbf{x}) = \mathbf{P}(\mathbf{x})\mathbf{A}(\mathbf{x}) + \mathbf{A}'(\mathbf{x})\mathbf{P}(\mathbf{x}) - \mathbf{P}(\mathbf{x})\mathbf{B}(\mathbf{x})\mathbf{R}^{-1}\mathbf{B}'(\mathbf{x})\mathbf{P}(\mathbf{x}) + \mathbf{C}'(\mathbf{x})\mathbf{Q}(\mathbf{x})\mathbf{C}(\mathbf{x}), \tag{8.58}$$

with the final condition:

$$\mathbf{P}(\mathbf{x}, t_f) = \mathbf{C}'(t_f)\mathbf{F}\mathbf{C}(t_f), \tag{8.59}$$

and $\mathbf{g}(\mathbf{x})$ is a solution of the state-dependent nonhomogeneous vector differential equation:

$$\dot{\mathbf{g}}(\mathbf{x}) = -[\mathbf{A}(\mathbf{x}) - \mathbf{B}(\mathbf{x})\mathbf{R}^{-1}(\mathbf{x})\mathbf{B}'(\mathbf{x})\mathbf{P}(\mathbf{x})]'\mathbf{g}(\mathbf{x}) - \mathbf{C}'(\mathbf{x})\mathbf{Q}(\mathbf{x})\mathbf{z}(\mathbf{x}), \tag{8.60}$$

with the final condition $\mathbf{g}(\mathbf{x}, t_f) = \mathbf{C}'(t_f)\mathbf{Fz}(t_f)$. The resulting differential SDRE-controlled trajectory becomes the solution of the state-dependent closed-loop dynamics:

$$\dot{\mathbf{x}}(t) = [\mathbf{A}(\mathbf{x}) - \mathbf{B}(\mathbf{x})\mathbf{R}^{-1}(\mathbf{x})\mathbf{B}'(\mathbf{x})\mathbf{P}(\mathbf{x})]\mathbf{x}(t) + \mathbf{B}(\mathbf{x})\mathbf{R}^{-1}(\mathbf{x})\mathbf{B}'(\mathbf{x})\mathbf{g}(\mathbf{x}). \tag{8.61}$$

Similar to Section 8.5.2, an approximate analytical approach is used and the DRE (8.58), and the nonhomogeneous differential equation (8.60), can be solved in the following steps at each time step:

1. Solve for $\mathbf{P}(\mathbf{x}, t)$, similar to the differential SDRE regulator problem in Section 8.5.2, in Steps from 1 to 6.
2. Calculate the steady-state value $\mathbf{g}_{ss}(\mathbf{x})$ from the equation:

$$\mathbf{g}_{ss}(\mathbf{x}) = [\mathbf{A}(\mathbf{x}) - \mathbf{B}(\mathbf{x})\mathbf{R}^{-1}(\mathbf{x})\mathbf{B}'(\mathbf{x})\mathbf{P}_{ss}(\mathbf{x})]'^{-1}\mathbf{C}'(\mathbf{x})\mathbf{Q}(\mathbf{x})\mathbf{z}(\mathbf{x}). \tag{8.62}$$

3. Use change-of-variables technique and assume that:

$$\mathbf{K}_g(\mathbf{x}, t) = [\mathbf{g}(\mathbf{x}, t) - \mathbf{g}_{ss}(\mathbf{x})]. \tag{8.63}$$

4. Solve the differential equation:

$$\mathbf{K}_g(\mathbf{x}, t) = \mathbf{e}^{-(\mathbf{A} - \mathbf{B}\mathbf{R}^{-1}\mathbf{B}'\mathbf{P})'(t - t_f)}[\mathbf{g}(\mathbf{x}, t_f) - \mathbf{g}_{ss}(\mathbf{x})]. \tag{8.64}$$

5. Use changing-of-variables procedure to calculate the value of $\mathbf{g}(\mathbf{x}, t)$:

$$\mathbf{g}(\mathbf{x}, t) = \mathbf{K}_g(\mathbf{x}, t) + \mathbf{g}_{ss}(\mathbf{x}). \tag{8.65}$$

6. Calculate the value of the optimal control $\mathbf{u}(\mathbf{x}, t)$ as:

$$\mathbf{u}(\mathbf{x}, t) = -\mathbf{R}^{-1}(\mathbf{x})\mathbf{B}'(\mathbf{x})[\mathbf{P}(\mathbf{x}, t)\mathbf{x}(t) - \mathbf{g}(\mathbf{x}, t)]. \tag{8.66}$$

8.6 Summary

This chapter presents an overview of the theory and techniques arising in modern control systems such as optimal control of linear and nonlinear and deterministic and stochastic systems. In particular, it presents LQR, linear quadratic tracking (LQT) using matrix differential and AREs, Pontryagin principle and HJB equation. Further, for nonlinear systems, a recent technique called SDRE is presented for both finite-horizon regulation and tracking.

8.7 Thoughtful Questions to Ensure Comprehension

1. Why is optimization important for anyone (arts, humanities, science, engineering and technology)?

2. Most of the optimal control work centers around the contributions by Jacopo Franceso Riccati (1676–1754), an Italian mathematician, Lev Semyonovich Pontraygin (1908–1988), a Russian Mathematician, and Rudolf Emil Kalman (1930–2016), an electrical engineer and mathematician. What is the significance of each contribution?

3. Optimal control can be implemented in either open-loop or closed-loop configuration. Which one is preferred and why?

4. What is the difference between regulation and tracking of optimal systems?

5. In CLOC problem, one is faced with matrix DRE to be solved backward in time. What are the resulting issues involved in this situation?

6. Recently, there is interest in closed-loop, optimal control of nonlinear systems using SDRE? What are the issues involved?

7. Is optimal control poised to penetrate into the areas of cyber, physical, and biosystems (CPBS) to enhance cybersecurity and resilience?

Further Reading

The all-important design is accomplished, besides optimal control, by other various tools such as state feedback or pole placement [15], optimal control [3, 16, 17], robust control [18], adaptive control [19], nonlinear control [20], and intelligent control [21].

References

1 Bacsar, T. (ed.) (2001). *Control Theory: Twenty-Five Seminal Papers (1932–1981)*. New York: IEEE Press.
2 Pontryagin, L.S., Boltyanskii, V.G., Gamkrelidze, R.V., and Mishchenko, E.F. (1962). *The Mathematical Theory of Optimal Processes*. New York: Wiley-Interscience.
3 Naidu, D.S. (2003). *Optimal Control Systems*. Boca Raton, FL and London: CRC Press, a Division of Taylor & Francis. A *Solutions Manual* for this senior graduate level textbook was prepared by the author and is available with the publisher. A *Special Indian Edition* of the original book, *Optimal Control Systems*, was published by Taylor and Francis India, New Delhi, India, 2015. An *International Edition* of the original book, *Optimal Control Systems*, was published by Taylor and Francis India, New Delhi, India, 2016. An expanded and updated *Revised Version* of this book is under preparation.
4 Rieger, C.G. and Naidu, D.S. (2004). Linear Quadratic Regulator and Tracking Control Algorithms Implemented in MATLAB. *Technical report*. Pocatello, ID: Measurements and Controls Engineering Research Center (MCERC), Idaho State University.
5 Jódar, L. and Navarro, E. (1992). Analytic solution of Riccati equations occurring in open-loop Nash multiplayer differential games. *International Journal of Mathematics and Mathematical Sciences* 15 (2): 359–366.
6 Anderson, B.D.O. and Moore, J.B. (1990). *Optimal Control: Linear Quadratic Methods*. Englewood Cliffs, NJ: Prentice Hall. Republished unabridged by Dover Publications, Inc., Mineola, New York, 2007.
7 Bellman, R.E. (1957). *Dynamic Programming*. Princeton, NJ: Princeton University Press.
8 Bellman, R.E. and Kalaba, R.E. (1965). *Dynamic Programming and Modern Control Theory*. New York: Academic Press.

9 Lewis, F.L. (1986). *Optimal Estimation: With an Introduction to Stochastic Control Theory.* New York: Wiley.

10 Çimen, T. (2012). Survey of state-dependent Riccati equation in nonlinear optimal feedback control synthesis. *AIAA Journal of Guidance, Control, and Dynamics* 35 (4): 1025–1047.

11 Nazarzadeh, J., Razzaghi, M., and Nikravesh, K. (1998). Solution of the matrix Riccati equation for the linear quadratic control problems. *Mathematical and Computer Modelling* 27 (7): 51–55.

12 Naidu, A.K.D.S. and Kamel, A.M. (2014). Nonlinear finite-horizon regulation and tracking for systems with incomplete state information using differential state dependent Riccati equation. *International Journal of Aerospace Engineering* 2014: 1–12.

13 Gajic, Z. and Qureshi, M. (2008). *The Lyapunov Matrix Equation in System Stability and Control.* New York: Dover Publications.

14 Barraud, A. (1977). A new numerical solution of xdot=a1*x+x*a2+d, x(0)=c. *IEEE Transaction on Automatic Control* 22 (6): 976–977.

15 Williams, R.L. II and Lawrence, D.A. (2007). *Linear State-Space Control Systems.* Hoboken, NJ: Wiley.

16 Clarke, F. (2013). *Functional Analysis, Calculus of Variations and Optimal Control.* London: Springer-Verlag.

17 Berkovitz, L.D. and Medhin, N.G. (2013). *Nonlinear Optimal Control Theory, Chapman & Hall/CRC Applied Mathematics and Nonlinear Science Series.* Boca Raton, FL: CRC Press, Taylor & Francis Group.

18 Liu, K.-Z. and Yao, Y. (2016). *Robust Control: Theory and Applications.* Singapore: John Wiley & Sons (Asia) Pte Ltd.

19 Tao, G. (ed.) (2003). *Adaptive Control Design and Analysis.* New York: Wiley-Interscience.

20 Khalil, H.K. (2015). *Nonlinear Control.* Upper Saddle River, NJ: Prentice Hall of Pearson.

21 Szuster, M. and Hendzel, Z. (2018). *Intelligent Optimal Adaptive Control for Mechatronic Systems.* Cham, Switzerland: Springer International Publishing AG.

9

Human System Interfaces

Ronald Boring

Center for Advanced Energy Studies, Idaho National Laboratory, Idaho Falls, ID, USA

Objectives

The objectives of this chapter are to guide the design-oriented engineer through planning, prototyping, and evaluating a human-system interface and to treat human-system interfaces in the context of control systems. The reader will gain an understanding of the importance of designing for the end user of the system, learn basic methods for evaluation of a human user of a system, and have an informed basis for selecting planning, prototyping, and evaluation methods as part of a graded development approach.

9.1 Introduction

9.1.1 Control Systems

Chemical plants, oil refineries, electric power plants, and factories involve a carefully controlled process to transform raw materials into a finished product. These are process control industries. Process control extends beyond production. Similar types of control are deployed in air traffic control, grid, logistics and distribution coordination centers, transportation hubs, or even within complex vehicles such as aircraft, ships, and trains. The common thread of these applications is a control system. Where there is a control system, there is invariably a human operator to control and monitor those systems. Even with the advent of highly automated process control systems, there is still a necessary interface to the human user of those systems for activities such as monitoring.

This chapter focuses on the connection between the control system and the human user of that system. In broad terms, this chapter discusses human-system interfaces (HSIs). Of course, the exact technology varies, and terms like "human-computer interface (HCI)" and "human–machine interface (HMI)" are largely synonymous to refer to some form of user interface to a technical system. The human user interacts with a system, and the field of developing interfaces is known as human-system interaction. Human-system interaction involves design and evaluation – two key methods that will be introduced later in this chapter.

Resilient Control Architectures and Power Systems, First Edition.
Edited by Craig Rieger, Ronald Boring, Brian Johnson, and Timothy McJunkin.
© 2022 The Institute of Electrical and Electronics Engineers, Inc. Published 2022 by John Wiley & Sons, Inc.

9.1.2   History of Humans and Control Systems

The purpose of a control system is typically to coordinate distributed processes. Localized control systems for simpler systems also exist. The common characteristics of most process control systems are

- There are multiple components operating at the same time.
- The components are not always located in the same area.
- There is a need to coordinate individual components in sequence.
- There is a consequence to failed coordination.

One of the earliest forms of a control system came in ships. As described in [1], the need to coordinate disparate ship activities, such as navigation, steering, and propulsion, resulted in a central coordination center on the bridge of the ship. The captain of the ship would coordinate activities in remote parts of the ship (i.e. giving commands to the engine room while receiving navigation information). Such coordination was initially accomplished through vocal commands, but later engine order telegraphs allowed communication between the bridge and the engine room. These systems were eventually replaced by remote control, allowing the bridge to control and monitor a remote system. Remote sensors provided indications back to the bridge, thus allowing centralized control and the advent of the control room concept. A control room centralizes instrumentation and controls and coordinates across multiple components or processes. This control room concept has been successfully translated from the ship bridge to be the common model for complex process control across industries. Whether the cockpit of a plane, the launch control center for a rocket, or the control room for a nuclear power plant, the control room model is the basis of human control systems.

9.1.3   Common Elements of Control System HSIs

Distributed control systems (DCS), programmable logic controllers (PLCs), or supervisory control and data acquisition (SCADA) systems are the backbones of modern control systems. Each system features sensors, a computational system, indicators, and controls, and they each play a role in the HSI:

- Sensor information is transmitted to the indicators for monitoring by human operators.
- The computational system may handle routine data transmission such as a databus for transmitting sensor and control signals, to logging data in a data historian, to performing data analysis such as trending or prognostics, to performing automation of processes, all of which serve to aid the human operator.
- Indicators – whether in the form of standalone lights, alarms, trends, or gauges or integrated into a computer display[1] – allow the operator to gather feedback on past, present, or projected future states of the process.
- Controls serve as inputs to the system. These may be as simple as flipping a switch or starting a pump or nuanced activities requiring monitoring and regulation within prescribed bands. Manual actions directly engage the system, while soft controls embedded in a control system may set in motion functions entailing a chain of automated actions. A turbine system, for example, may be operated manually by opening and closing throttle valves. Alternately, the system may be engaged by the operator to have the automated control system perform a series of actions such as

---

1   In this chapter, the physical device is the display and data presented on the display is the screen.

ramping up turbine speed to a predetermined setpoint, increasing speed according to technical specifications. Manual controls must be individually activated and sequenced by the operator, whereas automated controls perform a series of actions toward the operator goal to change the system state.

A simplified representation of the role of the user and the system in an HSI is depicted in Figure 9.1. The system gathers sensor data and outputs this information as display or auditory information, which becomes the perceptual input to the user. The user monitors this information, makes decisions, and then takes control action in the form of keyboard or button press, mouse click, or screen touch, which serves as input to the system. Those system inputs serve as actuations on controls. The system and the human both have other inputs and outputs – the system connects to sensors and controls, while the human interfaces with other humans, job aids such as procedures, and outside sources of information. The HSI is the point at which the system and human uniquely provide inputs and outputs to each other.

9.1.4 Consequences of Poor HSIs in Control Systems

As noted in the previous section, a process that requires a control system is likely complex and layered across multiple systems, with potential consequences for failure. A benign failure would result in stopping the process, which has financial implications as a minimum. Because processes are sequenced, it is possible that a failure of the control system also results in damage to the equipment. Processes may also be safety critical, meaning they have the potential to harm the environment or endanger human users or bystanders if they malfunction. Process control ensures process completion, prevents equipment failures, protects the environment, and prevents harm to humans.

Figure 9.1 The relationship between the system and the user in a human-system interface.

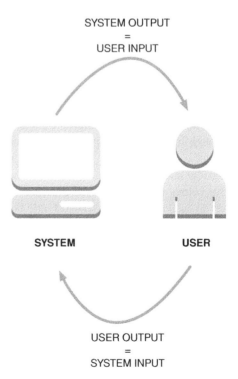

SYSTEM OUTPUT
=
USER INPUT

SYSTEM

USER

USER OUTPUT
=
SYSTEM INPUT

A poorly designed HSI can contribute to any of these consequences, from shutdowns to industrial accidents. Human error has been found to be a significant contributor in up to 85% of industrial accidents [2]. Human error is increasingly identified as the root cause in high-profile events. For example, the Chernobyl nuclear meltdown can be attributed to a failure to follow procedural guidance by shutting off automated safety systems during a test of the reactor [3]. The Deepwater Horizon accident was caused by a failure to monitor and interpret warning signs associated with well kick, resulting in the catastrophic explosion on the offshore drilling rig and a sustained leak of oil into the Gulf of Mexico [4]. The separate crashes of two Boeing 737 Max aircraft were due to a faulty auto-leveling system designed to prevent stalls while ascending. While the system itself was of faulty design due to reliance on a single sensor, the system proved confusing for pilots to override [5].

A human user of a control system must perceive, understand, and project [6] what the process is doing, make timely decisions, and take actions to exert control over the system. Each of these activities presents the opportunity for human error [7]. For example, if the operator does not detect a process deviation such as a high flow rate, this can result in a lack of situation awareness and application of the wrong mental model. *Situation awareness* is the operator's comprehension of what is happening at that time. A *mental model*, in contrast, represents the operator's long-term learning and understanding of the process, including the cause and effect of elements within that process. Failure of situation awareness of the high flow rate means that the operator has not detected the anomaly. As a result, the operator may be applying a long-standing mental model of the process that is attuned to the wrong situation. In this example, the mental model represents a normal flow level, which guides the operator actions. As a result of applying the wrong mental model, the operator may not take appropriate actions to reduce the flow level or otherwise adjust the process in response to the high flow.

Human error is caused by actions or inactions that disrupt the optimal control of the process. These are called errors of commission and errors of omission, respectively [8]. The operator *commits* the wrong action, derailing the process, or the operator *omits* a required action, likewise resulting in a failure in the process.

The quality of the HSI contributes greatly to the error likelihood. A poor HSI may increase the likelihood of human error as much as tenfold [9]. Table 9.1 presents some of the types of human errors that can result from a poorly designed HSI.

It may be tempting to eliminate the opportunity for human error by eliminating the human through increased automation. This is rarely the desired end state. Many key decisions like the rate of a process require the human to integrate information. For example, an operator may consider market demand and anticipate the correct process parameters to meet that demand. Paul Fitts coined the concept HABA-MABA – "Humans are better at-machines are better at" – as a way to catalog those things that should be automated versus left to humans [10, 11].[2] While the capabilities of machines have increased, humans still maintain the edge in terms of generalization and flexibility, judgement and decision-making, response to novel events and degraded conditions, and creativity in problem-solving. Automated process control is ideal for routine, prescribed circumstances. However, when a novel event arises, requiring adaptation of the process, human operators are typically better equipped to handle such situations. While humans are error prone when not given the right HSI, humans are also resilient and adaptive to novel operating situations.

---

2 Fitts' gender-specific terminology has been updated here.

Table 9.1 Human errors in process control attributed to human-system interface.

Type of human activity	Example error of commission	Example error of omission
Perceive what control system is doing	Operator looks at wrong indicator due to its salience over a more important indicator	Operator fails to notice abnormal state due to unclear or poorly salient indicator
Understand what control system is doing	Operator develops incorrect mental model about abnormality in the process because indicators point to wrong fault or they mask the critical fault	Operator does not realize there is a process abnormality because indicators do not point to the fault
Project where control system is going	Operator recognizes a fault but misdiagnoses it due to indicator ambiguity, anticipating a different outcome than the actual one	Operator fails to realize the severity of the emerging problem because of a lack of clear trending indicators
Make decision about control action	Operator decides wrong corrective action to take in response to an abnormal condition because of wrong mental model about process due to inadequate indicators	Operator fails to decide to take corrective action because of improper situation awareness about process through inadequate indicators
Take action to control the process	Operator initiates wrong control because of similar looking and grouped controls	Operator fails to initiate specific control because of multitasking multiple controls simultaneously

9.2 Basic Methods

9.2.1 Introduction to User-Centered Design

User-centered design (UCD) is the process of incorporating end user feedback into the development of a product [12]. In terms of control systems, UCD means optimizing the HSI by evaluating it using representative operators of that system. Most product development, including development of a professional control system, involves several phases of development, corresponding to milestones. An example of software development process for a new digital turbine control system for a nuclear power plant is shown in Figure 9.2 [13]. The development cycle included three phases: planning and analysis, design, and verification and validation (V&V). The design phase was subdivided into three milestones, corresponding to 30%, 70%, and 100% software completion. Below the milestone phases are the UCD counterparts:

- *Planning and analysis phase* – as a part of the software specification, an HSI style guide was developed to guide the design elements such as the specification of screen widgets and the general look and feel of the HSI.
- *Design phase* – during design, three prototypes were built and evaluated. The first was called "static screen," which corresponded to an early mockup of control system screens. The second prototype was called the "dynamic screen," which represented a functional prototype that mimicked the look and feel of the planned system and allowed operators to perform simulated scenarios. The final prototype was denoted as "Pre-ISV." Integrated system validation (ISV) is where the completed and debugged system is deployed and tested. A factory acceptance test (FAT) is commonly performed. ISV represents a FAT of the user – actually testing the operator using the final system to determine any remaining HSI issues. Pre-ISV is a trial run of the ISV.

- *Verification and validation phase* – In the V&V phase, formal FAT and ISV were performed. As the system was ready for deployment and had already undergone quality assurance including debugging, problems encountered at this stage could have delayed deployment and led to project and cost overruns. In some industries, particularly those dealing with safety critical process control, V&V may be overseen by a regulatory authority as a required step to certifying or qualifying the control system.

Each of these development phases presents the opportunity for evaluation. UCD involves an iterative process of design and evaluation, whereby feedback from the evaluation is used to improve the design, and the resulting design is then fed back into evaluation, and so on. Prematurely locking in a specification based on the first round of design prevents the opportunity to refine the design across the system development life cycle and to correct potential areas or tasking where human error is likely.

The next sections detail the design, prototyping, and evaluation processes. While UCD is meant as an iterative process, each project will have different requirements. A qualified control system will likely follow a multiphased approach like outlined in Figure 9.2. A graded approach may also be adopted for other types of projects, which reduces these phases and iterations. For example, an upgrade to an existing system has already undergone all development steps. In such a case, new features may be amended to the deployed system and evaluated. This may be single-pass HSI development. For another example, systems meant for research, including classroom development activities, may select only a representative set of UCD phases in the interest of facilitating a shorter development process. Research and academic development may not seek to progress to actual deployment, and the development process may stop short of V&V.

Design, prototyping, and evaluation are often performed by an integrated and interdisciplinary team of experts with different skills, backgrounds, and experiences. The specification of features may be led by an information architect or architect engineer who understands the process control system. The HSI design may be led by a graphic designer or user interface designer. Prototyping may be led by a software engineer who is skilled at developing software implementations. The evaluation may, in turn, be led by a human factors engineer, who is trained on the psychological aspects of human-system interactions.

Smaller development teams may necessitate each individual taking on multiple roles. For example, the designer and prototype developer might be the same individual. Despite such accommodations to smaller teams, one tenet of UCD remains the independence of the evaluation from the design and development. To gather objective feedback on the system design, the evaluation should not involve the design or development team as users. An often repeated warning to designers and developers is "You are not the user!" It is important to understand that designers and developers of the proposed system understand the system in a way that is fundamentally different than the typical user who was not involved in the development of the system. The designers and developers are therefore unlikely to uncover issues that end users would encounter

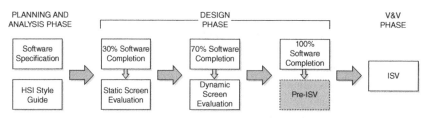

Figure 9.2 Development phases for user-centered design. Source: Boring et al. [13].

in their novel exploration of the system. To be effective, HSI evaluation should reach beyond the design and development team. The old adage – "a second set of eyes" – is crucial when evaluating the ability of an operator to use a process control system.

9.2.2 Design Planning

Design is envisioning what the end product will look like and how its respective features will satisfy the specific requirements. For example, the design of a robotic arm will feature mechanical considerations of what the arm must do and how it will be technically realized. Additionally, that robotic arm must feature some form of control, from fully teleoperated to fully automated. The example robotic control system and the level of human engagement with that system must be carefully considered.

Three documents typically guide the design of an HSI:

- The general approach to how a human will interact with the system is defined in the *design philosophy*. The design philosophy specifies at a high level how the interactions should proceed. For example, the design philosophy might suggest that the human will interface with a control station on a workstation at a desk, that the system will be highly automated, that the human will issue predefined scripts via soft control buttons on a screen, and that the human will monitor the status through graphical visualizations of key indicators.
- The interactivity with human users is documented in the *HSI style guide*. The design philosophy is a high-level conceptual document, whereas the HSI style guide suggests how the system will be implemented and what the "look and feel" of the system should be. The design philosophy generally will not suggest a particular implementation (i.e. it is neutral on topics such as Windows versus Linux, the type of input device, or the color scheme used for on-screen graphics). However, the HSI style guide details how system information is output to the user and how the user inputs information. For the output, it would define the color palette, font sizes, line weights, windowing conventions, preferred modes of graphically presenting sensor information, and alarm conventions. For the input, it would define the acceptable devices the human has for communicating with the system (e.g. keyboard versus touchscreen versus mouse). An HSI style guide can actually be used across different systems and even different vendors to achieve consistency in the user experience with the system. The HSI style guide should be informed by relevant HSI and human factors standards to ensure the safety and accessibility of the system.
- The specific features and requirements are articulated in the *engineering specification*. This document explains which features will be implemented. With an HSI style guide in place, the specification will not need to articulate *how* these features are implemented. This separation of style guide (i.e. the *how*) and specification (i.e. the *what*) also ensures that the feature set – the core function of the system – is not necessarily subject to changes over time. A robust specification can be implemented in many different ways across many different systems. The style guide directs that implementation and can help create a cohesive ecosystem of process control systems that are familiar to the users.

Developing the HSI style guide may benefit from a user needs assessment. This task involves identifying who the different users of the system will be and identifying the features they will be using. A process monitoring system, for example, may feature three types of users: the control operators of the system, engineering staff who monitor overall trends in system performance, and maintenance staff who troubleshoot and fix problems. The operators are interested in maintaining production, and they need indications and controls to help them optimize the process. In contrast, the engineering staff may be monitoring the efficiency of particular components instead of overall production. Their goal is to determine, for example, that a cooling system is operating below optimal levels and may be failing. They hand this information off to the maintenance staff who will use sensor and historic data to pinpoint if the below-capacity performance of the cooling system is due to a failing pump or a slow leak in piping. The same system should support all three user types. If the primary HSI is a workstation with a display, keyboard, and mouse, for example, the control system should include three different screens (i.e. windows on a physical display) to track relevant indicators and provide relevant controls. The operators should not be burdened with historic efficiency trending information that interferes with their ability to monitor and control current production. That information is only useful to engineering. Likewise, engineering should not be able to control current processes because this role is the responsibility of the operator and changing operational parameters could endanger production or safety. Detailed historic information on individual components may not be of particular interest to operations or engineering if that information is primarily used to troubleshoot maintenance activities. The HSI needs to provide the right level of information and the right controls for each type of user. Surveying current users on their tasks and mapping out what different users will do can shape the HSI to ensure it meets all users' needs.

Much of the initial development of a design philosophy, HSI style guide, and engineering specification is first accomplished as part of a planning and analysis phase. However, these documents should reasonably become living documents that can be refined as new insights are gleaned from prototyping and evaluating the system. A prototype is created that then allows the user to interact (at some level) with the designed system. Issues, such as functions that do not fit the mental model of the user, should be addressed at the specification level. Issues of the interactivity with the system implicate the need to revisit the HSI style guide. Incorporation of evaluation feedback to improve the design is crucial to UCD; such an evolution of the design is only possible when the guiding documents like the HSI style guide and engineering specification are not prematurely frozen to changes. Successful design teams agree to maintain this flexibility and understand the fluidity of designing for human users.

9.2.3 Prototyping Process

Prototyping is translating a design into a representation with which a user may interact. A prototype may be as simple as a sketch, storyboard, or wireframe that is used to present the user with the core idea of the HSI. It may also be a functional prototype such as a software mockup that mimics key functionality and interactivity. A functional prototype need not be created in the same development environment as the final product and need not be subject to the same quality assurance standards as the final product. A 3D-printed version of a physical device or an HSI developed in LabView are common forms of the functional prototype.

A final form of the prototype is a predeployment version of the system. This is the beta version of the final product prior to factory acceptance testing. For example, this might be the control system implemented on the actual DCS platform interfaced with a simulator or physical test loop. Prototypes are inherently exploratory and are therefore used to evaluate specific aspects of a design or the design in totality.

Each phase of prototyping presents a representation of the design that is getting closer in fidelity to the final product. Therefore, prototypes range from conceptual design artifacts, to proof-of-concept demonstrations, to predeployment runtime systems. At each phase of prototyping, there is the opportunity to evaluate the design, as discussed in the next section. Evaluating the system provides valuable information on the effectiveness of the HSI and provides opportunities to refine the design. Therefore, it is anticipated that as the fidelity of the prototypes increases, so will the quality of the design. The development process must be flexible enough to allow improvement of the design as the HSI is tested with users.

Figure 9.2 provided an illustration of the evolution of the design across successive design phases. Initially, a static mockup was developed. The static mockup consisted of a realistic representation of system screens according to the features in the specification and HSI style guide. These static screens allowed prospective users (i.e. operators of the turbine control system) to provide feedback on the proposed design prior to implementation. Such an evaluation would identify significant mismatches between the system and the users' mental models of the system. It would also identify concerns with any stylistic elements from the HSI style guide. Next, a partially functional prototype of the system was created. During an evaluation, operators would provide not only their impressions but also demonstrate their interactivity with the system through defined scenarios. The functional prototype would reveal any issues with operator performance, such as inability to find key information or controls in a timely manner. Finally, as the system design draws to a close, it must be tested to ensure it comports with the specification and meets any further quality requirements needed for implementation.

9.2.4 Evaluation Process

The purpose of evaluation is to gather data about the operator interactions with the HSI to ensure the usability of the system. Specifically, evaluation will answer questions like

- Were operators able to complete the task with the HSI as designed?
- What were the problems with the HSI?
- Did the operators like the HSI?
- What changes or additional features would the operators like?

Evaluation will most commonly take the form of a usability test, in which a prototype version of the HSI is presented to a representative group of users. The usability test may take several forms, including

- *Discoverability of the HSI* in which a first-time user is asked to navigate through the HSI. Often, this is guided by a script with simple tasks like "Show me which screen you would use to start the circulation water pump." The steps taken by the user are documented and any unusual outcomes are noted (e.g. if the user cannot find the screen to start the pump).
- *Scenario walkthroughs* which are more extensive than discoverability in that the user will actually interact with the system rather than just navigate within it. Instead of finding a particular component, the user would actually operate it, often as part of a more complex scenario involving multiple functions of the control system.
- *Benchmarking* where two or more HSI variants are compared to determine which one works best. Benchmarking can be a comparison against a competitor product or an internal test of different implementations of the control system HSI.
- *Stress testing* in which complex what-if scenarios are performed. Often, these are performed as part of certification and entail simulated failure of the system such as during an accident to test user responsiveness and recovery of the HSI.

The purpose of a usability test is to detect any issues a user may have with the HSI. Each user will provide new insights. According to [14], testing five users will identify approximately 85% of the possible usability issues. Additional testing with more users will account for more issues but in a diminishing capacity. While the optimal number of users to test is therefore often considered around five, even testing just a few users can greatly improve the design by identifying the most pressing issues with a design.

As part of usability testing, human performance measures are gathered. The types of performance measures are almost limitless [15], but several common measures are explained briefly below.

- *Operator comments* – during walkthroughs for scenarios, unless timing is a key measure, the user should be encouraged to think aloud and express what their thoughts as they are exploring and using the HSI. Additionally, there should be a chance for a debrief following the walkthrough during which the evaluator will ask questions related to the user experience and the user will offer insights into what worked or did not work for them while interacting with the HSI. Such comments will often provide invaluable insights into how to improve the HSI. These comments may be recorded but will typically be transcribed by the evaluator after the usability test and translated into design issues or design recommendations.
- *Operator surveys* – following walkthroughs, the evaluator may wish to have the user complete surveys. Common surveys will use a numeric scale that the operator assigns to gauge factors like user satisfaction [16], situation awareness [17], or workload [18]. These surveys will provide numeric data on the user's subjective experience with the HSI. For example, when performing a benchmark study, such data are particularly useful to quantify the differences between HSIs.
- *Operator performance* – these are usually objective measures that can be recorded in an unobtrusive manner while the user performs tasks. Common measures of task performance include time to complete a task, number of actions taken to complete a task, or number of task errors. Often, these metrics will be compared against a set of baselines or optimal measures.
- *Operator logs* – a prototype HSI may feature a data historian that logs parameters from the process control system and the various inputs and outputs of the system. These are particularly useful for measuring time between an indicator such as an alarm and the resulting control action taken by the user. They may also be used to measure the quality of the control (e.g. the amount of electricity produced in a power plant control system simulator [19]).
- *Physiological measures* – these are measures that are instrumented to the user to gather data about their physiological states. Common forms include eye tracking, which may be used to determine the visual scan paths when searching on a screen; heart rate, which may be used to determine stress or excitement; and galvanic skin response, which measures skin conductivity and is a good measure of stress as manifest in perspiration.

In most engineering disciplines, there is a clear preference for objective quantitative measures. These are measures that can be directly observed and objectively assigned a level. For example, reaction time is directly observable (i.e. the time between starting and finishing a task) and offers a number (e.g. seconds, minutes, or hours) that can be readily compared against others. Within engineering fields, the value of subjective or qualitative measures may at first glance seem less useful. Subjective refers to someone estimating the performance such as when the person performing the task rates how easy or difficult a task was or when a subject matter expert rates the quality of performance for someone else doing the task. However, subjective measures are often the only and best way to gather certain insights. For example, satisfaction and perceived ease of use are mental phenomena that may not readily be observed outside the performer's head. This is a long-standing quandary within psychology – how to look inside the head to determine thoughts and feelings. While some strong emotional responses like fear or disgust may manifest physiologically, much of

Table 9.2 Typical subjective agreement Likert scale used in user research.

1	2	3	4	5
Strongly disagree		Neutral – Neither agree/disagree		Strongly agree

mental life is not readily observable to the outsider. Subjective measures become the lifeline into these important insights that have a bearing on the system's design. A common way to gather subjective measures is to ask the user to rate their experience on a Likert scale such as the five-point scale seen in Table 9.2.

One of the oldest methods for gathering subjective user feedback with a five-point Likert scale is the system usability scale (SUS) developed by Brooke [20]. The SUS features ten questions that are asked after a user performs a task with a system:

1. I think that I would like to use this system frequently.
2. I found the system unnecessarily complex.
3. I thought the system was easy to use.
4. I think that I would need the support of a technical person to be able to use this system.
5. I found the various functions in this system were well integrated.
6. I thought there was too much inconsistency in this system.
7. I would imagine that most people would learn to use this system very quickly.
8. I found the system very cumbersome to use.
9. I felt very confident using the system.
10. I needed to learn a lot of things before I could get going with this system.

To score the SUS, the scale is standardized to the same polarity, since half of the questions are positively worded, and half are negatively worded. For each of the odd-numbered questions, subtract one, and for the even-numbered questions, subtract the user score from five. This process brings all items to a positively weighted score from 0 to 4, where 4 is the most positive rating. These corrected scores are then summed and multiplied by 2.5 to produce a total scale from 0 to 100.

There are many variants of the SUS, and it may not be necessary to ask all ten questions for each usability evaluation. Other questions may be more appropriate for certain types of applications.

Two other subjective measures commonly used in process control are workload and situation awareness. Workload refers to the mental effort required to perform a task. The most common measure of workload is the National Aeronautics and Space Administration (NASA) task-load index (TLX), which provides a series of subjective measures to assess different dimensions of mental workload [18]. The NASA TLX and other workload measures tell the difficulty in performing a task. Situation awareness is the user's awareness of what is happening. For more complex tasks, the cause and effect may not be readily transparent to the user, resulting in increased complexity. Higher situation awareness is associated with better performance. Situation awareness is also typically assessed by providing users with a subjective questionnaire based on Likert or similar scales. Lau and Boring [21] provide a simplified guide to understanding when to use various situation awareness measures. Workload and situation awareness are popular measures to use in human-in-the-loop process control studies, but they are not always necessary for basic usability evaluations of systems [22].

Additional subjective insights may be gathered from open-ended questions in which the user provides their thoughts and ideas about the system they have used. Example questions include

1. What did you like best about the system?
2. What did you like least about the system?
3. How would you improve the system?

These questions do not provide quantitative responses unless they are systematically categorized and tabulated by an expert analyst later. However, these types of questions do provide direct feedback on the system and the user's experience with it. The qualitative insights are especially useful during the design phase of a system, when features are still being finalized. A qualitative insight that a feature did not work the way a user wanted and an explanation of what the user expected instead may prove more informative to refining the design than an objective measure that shows they performed the task slowly or a subjective measure that shows they did not like the interface. Qualitative responses will answer *why something didn't work* and often *how to fix it* in a way that quantitative data may not reveal [23]. However, such feedback must also be weighed against the requirements of the system. A user's wish list of design features may not be readily implementable. Such qualitative data should generally be used to drive the refinement of existing features rather than become a focus group for new features beyond the original scope of the design.

9.2.5 Validation versus Verification

The type of evaluation described up to this point is actually *validation*, which is the observation of a user interacting with the HSI. Another form of evaluation is *verification*, which is where the HSI is compared to an external guideline like a standard or style guide [24]. Validation entails operator-in-the-loop studies, while verification entails a subject matter expert like a human factors engineer scrutinizing the HSI according to an established standard for HSIs.

V&V can occur at different phases of the system design lifecycle. *Formative evaluation* refers to inputs that help shape (i.e. form) the design, while *summative evaluation* refers to establishing that the overall (i.e. summed) design and implementation meet the requirements for the user. Thus, formative evaluation is done earlier in the design process and may feature more informal evaluations, while summative evaluation is done later in the design process and features more formal evaluations (see Figure 9.3).

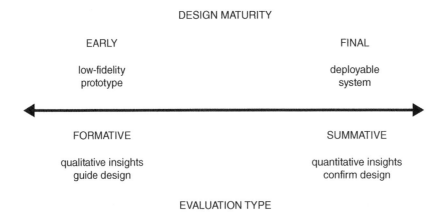

Figure 9.3 The relationship between design maturity and evaluation type.

- Formative validation tends to encompass usability tests of different aspects of the system design. These tests are designed to refine the design of the system. Such tests benefit from qualitative insights that shape design recommendations.
- Summative validation tends to encompass formal operator-in-the-loop studies of a completed design to determine it meets design requirements. As noted earlier, an FAT for users of the system is called an ISV. The entire range of functions of the system should be evaluated with representative users in typical use scenarios. Summative validation tends to make more use of quantitative and objective findings.
- Formative verification tends to involve evaluating a system against requirements. In the early stages of design, this may involve a simplified checklist, such as [25], to ensure good practices for design are met. This may also involve ensuring internal requirements like an HSI style guide are met.
- Summative verification tends to entail comparing the system against applicable standards and specifications. Three widely used and easily accessible standards for process control are
  o MIL-STD-1472G (CHANGE-1), *Department of Defense Design Criteria Standard: Human Engineering* [26]
  o NASA/SP-2010-3407, *NASA Human Integration Design Handbook (HIDH)* [27]
  o NUREG-0700, Rev. 3, *US Nuclear Regulatory Commission's Human-System Interface Design Review Guidelines* [28].

The *Guideline for Operational Nuclear Usability and Knowledge Elicitation* (GONUKE) [24] further delineates evaluation phases and types (see Table 9.3). It provides guidance for the pre-formative planning and analysis phase of design through the postsummative implementation and operation phase. Additionally, it considers an additional type of evaluation called *epistemiation* in which information is elicited from expert users of a system [23]. Epistemiation is particularly useful for upgrades to an existing system with an established user base.

All phases of evaluation are not expected nor required for all system development activities. Instead, a graded approach may be exercised in which only select evaluations are performed to help improve the system design. The degree to which all levels of evaluation are performed is a product of regulatory thoroughness; the technology readiness level [29] of the system; the budget

Table 9.3 Evaluation phases and types for GONUKE.

		Evaluation phase			
		Preformative (planning and analysis[a])	Formative (design[a])	Summative (verification and validation[a])	Postsummative (implementation and operation[a])
Evaluation type	Expert review (verification)	Design requirements review	Heuristic evaluation	System verification	Requalification against new standards
	User study (validation)	Baseline evaluation	Usability testing	Integrated system validation	Operator training
	Knowledge elicitation (epistemiation)	Cognitive walkthrough (task analysis)	Operator feedback on design	Operator feedback on performance	Operator experience reviews

a) Corresponding stages in NUREG-0711.
Source: Boring et al. [23].

available to support the design effort; and the potential financial, safety, or security consequences of system failure. Almost any system will benefit from some design input such as formative V&V.

9.3 Summary

The process for UCD is presented as a checklist in Table 9.4. This checklist presents each of the key elements from this chapter. Figure 9.4 accompanies the table and reminds the design team of types

Table 9.4 Short checklist of HSI design and evaluation activities.

Design planning			
□ Include?	Design philosophy		
□ Include?	HSI style guide	→	User needs assessment
□ Include?	Engineering specification		
Design prototype			
□ Include?	Static mockup	♻	□ Include iteration?
□ Include?	Dynamic (functional) prototype		
Design evaluation			
□ Include?	Formative validation	→	Usability testing (see Figure 9.3)
□ Include?	Formative verification	→	Expert usability checklist
□ Include?	Summative validation	→	Integrated system validation
□ Include?	Summative verification	→	Standards compliance

Note: Items may be included or excluded as part of a graded approach.

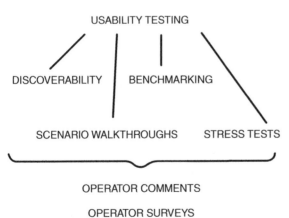

Figure 9.4 Types of usability testing and related human performance measures.

USABILITY TESTING

DISCOVERABILITY   BENCHMARKING

SCENARIO WALKTHROUGHS   STRESS TESTS

OPERATOR COMMENTS

OPERATOR SURVEYS

OPERATOR PERFORMANCE

OPERATOR LOGS

of evaluation and performance measures that may be gathered. The checklist is set up according to a graded approach, meaning each element includes a checkbox for inclusion. Best practices would entail including all elements, but individual design circumstances may dictate other requirements. For example, a student design project might include a brief engineering specification, a single functional prototype, and a usability test with fellow students as part of formative validation. Such a project would not be iterated but would include suggestions from the evaluation on how to improve the system if it was iterated.

Following the guidance in this chapter can ensure a usable design. While the examples from this chapter mainly point to process control systems, the guidance is largely universal and can be used for a variety of products, from web design to development of complex software. A thoughtful consideration of the HSI and vetting with users will minimize errors in the use of the system and increase efficiency and satisfaction [16]. Ultimately, these advantages combine to create a more resilient HSI in which the user is more likely to be successful at completing tasks with the system and more able to recover from challenging situations. In a process control setting, this resilience translates into increased production and safety.

Thoughtful Questions to Ensure Comprehension

1 Why is it advantageous to have a multidisciplinary team involved in developing an HSI? What functions of UCD are accomplished by different disciplines?

2 How can you design an HSI to reduce human error?

3 What human errors can result from a poorly designed HSI?

4 Why is evaluation of the HSI such an important activity?

5 What is the difference between a design philosophy, HSI style guide, and engineering specification? How would you gather information from users to help develop the HSI style guide?

6 Is it feasible simply to automate a system to eliminate the human users? What would be a disadvantage of that approach?

7 Why does the approach for gathering information on user performance tend to be different for formative versus summative evaluation? When would qualitative measures be advantageous? When would quantitative measures be appropriate? Why?

8 Is it necessary to have a high-fidelity prototype to evaluate a system? What types of information about users can be gathered for a low versus high-fidelity prototype?

9 What are the advantages and disadvantages of objective versus subjective measures of assessing users?

10 When is the graded approach to design and evaluation appropriate?

Further Reading

Albert, W. and Tullis, T. (2013). *Measuring the User Experience: Collecting, Analyzing, and Presenting Usability Metrics*, 2e. Morgan Kaufmann Publishers.

Crampin, T. (2017). *Human Factors in Control Room Design. A Practical Guide for Project Managers and Senior Engineers*. Wiley.

Hollifield, B.R., Oliver, D., Nimmo, I., and Habibi, E. (2008). *The High Performance HMI Handbook. A Comprehensive Guide to Designing, Implementing and Maintaining Effective HMIs for Industrial Plant Operations*. PAS.

Nielsen, J. (1993). *Usability Engineering*. Morgan Kaufmann Publishers.

Norman, D. (2013). *The Design of Everyday Things: Revised and Expanded Edition*. Basic Books.

Pannafino, J. and McNeil, P. (2017). *UX Methods A Quick Guide to User Experience Research Methods*. CDUXP LLC.

Rosenberg, D. (2019). *UX Magic*. Interactive Design Foundation.

References

1 Boring, R., Ulrich, T., and Lew, R. (2016). RevealFlow: a process control visualization framework. *Lecture Notes in Artificial Intelligence* 9744: 145–156.

2 Kletz, T. (2001). *An Engineer's View of Human Error*, 3e. CRC Press.

3 International Nuclear Safety Advisory Group (1992). *The Chernobyl Accident: Updating of INSAG-1, Safety Series 75-INSAG-7*. International Atomic Energy Agency.

4 Transocean (2011). Macondo Well Incident. Transocean Investigation Report, Volume 1.

5 Spielman, Z. and Le Blanc, K. (2020). Boeing 737 Max: expectation of human capability in highly automated systems. *Advances in Intelligent Systems and Computing* 1210: 64–70.

6 Endsley, M.R. (2000). Theoretical underpinnings of situation awareness: a critical review. In: *Situation Awareness Analysis and Measurement* (eds. M.R. Endsley and D.J. Garland). Lawrence Erlbaum Associates.

7 Whaley, A.M., Xing, J., Boring, R.L. et al. (2016). *Cognitive Basis for Human Reliability Analysis, NUREG-2114*. US Nuclear Regulatory Commission.

8 Swain, A.D. and Guttmann, H.E. (1983). *Handbook of Human Reliability Analysis with Emphasis on Nuclear Power Plant Applications, Final Report, NUREG/CR-1278*. US Nuclear Regulatory Commission.

9 Gertman, D., Blackman, H., Marble, J. et al. (2005). *The SPAR-H Human Reliability Analysis Method, NUREG/CR-6883*. US Nuclear Regulatory Commission.

10 Fitts, P.M. (1951). *Human Engineering for an Effective Air-Navigation and Traffic-Control System*. National Research Council.

11 Swain, A.D. (1980). *Design Techniques for Improving Human Performance in Production*, Revisede. Industrial and Commercial Techniques, Ltd.

12 Norman, D.A. and Draper, S.W. (1986). *User-Centered System Design: New Perspectives on Human-Computer Interaction*. Erlbaum.

13 Boring, R.L., Joe, J.C., Ulrich, T.A., and Lew, R.T. (2014). Early-stage design and evaluation for nuclear power plant control room upgrades. *Proceedings of the Human Factors and Ergonomics Society Annual Meeting* 58: 1909–1913.

14 Nielsen, J. and Landauer, T.K. (1993). A mathematical model of the finding of usability problems. Conference on Human Factors in Computing Systems: CHI 1993 Proceedings, 206–213.

15 Boring, R.L., Joe, J.C., Ulrich, T.A., and Lew, R. (2015). *Operator Performance Metrics for Control Room Modernization: A Practical Guide for Early Design Evaluation, INL/EXT-14-31511*, Rev. 1. Idaho National Laboratory.

16 Bailey, J.E. and Pearson, S.W. (1983). Development of a tool for measuring and analysing computer user satisfaction. *Management Science* 29: 530–545.

17 Hart, S.G. and Staveland, L.E. (1988). Development of NASA-TLX (Task Load Index): results of empirical and theoretical research. *Advances in Psychology* 52: 139–183.

18 Taylor, R.M. (1990). Situation awareness rating technique (SART): the development of a tool for aircrew systems design. In: *Situational Awareness in Aerospace Operations* (Chapter 3. France: Neuilly sur-Seine, NATO-AGARD-CP-478.

19 Ulrich, T.A., Lew, R., Werner, S., and Boring, R.L. (2017). Rancor: a gamified microworld nuclear power plant simulation for engineering psychology research and process control applications. *Proceedings of the Human Factors and Ergonomics Society Annual Meeting* 61: 398–402.

20 Brooke, J. (1986). SUS: a 'quick and dirty' usability scale. In: *Usability Evaluation in Industry* (eds. P.W. Jordan, B. Thomas, B.A. Weerdmeester and A.L. McClelland). Taylor and Francis.

21 Lau, N. and Boring, R.L. (2017). Situation awareness in sociotechnical systems: concepts and applications. In: *Human Factors in Practice: Concepts and Applications* (eds. H.M. Cuevas, J. Velázquez and A.R. Dattel), 55–72. CRC Press.

22 Boring, R.L. (2015). Envy in V&V: an opinion piece on new directions for verification and validation in nuclear power plants. *Proceedings of the Annual Meeting of the Human Factors and Ergonomics Society* 59: 1746–1750.

23 Boring, R.L., Lew, R., and Ulrich, T.A. (2016). Epistemiation: an approach for knowledge elicitation of expert users during product design. *Proceedings of the Annual Meeting of the Human Factors and Ergonomics Society* 60: 1699–1703.

24 Boring, R.L., Ulrich, T.A., Joe, J.C., and Lew, R.T. (2015). Guideline for operational nuclear usability and knowledge elicitation (GONUKE). *Procedia Manufacturing* 3: 1327–1334.

25 Mohlich, R. and Nielsen, J. (1990). Improving a human-computer dialogue. *Communications of the ACM* 33: 338–348.

26 US Department of Defense (2019). Department of Defense Design Criteria Standard: Human Engineering. MIL-STD-1472G (CHANGE-1).

27 National Aeronautics and Space Administration (2010). *Human Integration Design Handbook (HIDH)*NASA/SP-2010-3407. NASA.

28 US Nuclear Regulatory Commission (2000). *Human-System Interface Design Review Guidelines, NUREG-0700*, Rev. 3. NRC.

29 Ramirez-Marques, J.E. and Sauser, B. (2009). System development planning via system maturity optimization. *IEEE Transactions on Engineering Management* 56: 33–548.

Part IV

Metrics Fundamentals

Resilient Control Architectures and Power Systems, First Edition.
Edited by Craig Rieger, Ronald Boring, Brian Johnson, and Timothy McJunkin.
© 2022 The Institute of Electrical and Electronics Engineers, Inc. Published 2022 by John Wiley & Sons, Inc.

10

Differentiating Resilience

Jeffrey D. Taft

Energy and Environment, Pacific Northwest National Laboratory, Richland, WA, USA

Objectives

The objective of this chapter is to acquaint the reader with the discipline of grid architecture as the basis for introducing a structural approach to the definition, characterization, and analysis of electric power system resilience. The use of structural concepts will provide the reader with an appreciation for the concept that interconnection of grid elements is as important in the consideration of resilience as are the characteristics of individual grid components and systems. A structural-based definition of resilience leads to an understanding of why simply hardening components is not a sufficient strategy for resilience improvement.

10.1 Introduction

In 2014, Pacific Northwest National Laboratory (PNNL) introduced the concept of grid architecture to the US Department of Energy (DOE) as a discipline that could address the grid as a whole and provide the means to manage the inherent complexity of grid modernization. Grid architecture has its antecedents in system architecture, but we combined that discipline with additional elements from network engineering, software engineering, control theory, and various forms of mathematics, including optimization methods and graph theory, and applied these to electric power systems. Specifically, PNNL defined grid architecture this way:

> **Grid architecture** is the application of system architecture, network theory, and control theory to the electric power grid. A grid architecture is the highest-level description of the complete grid and is a key tool to help understand and define the many complex interactions that exist in present and future grids.

Roughly speaking, an architecture is an abstract model of a complex system that we use to reason about its structure and behavior and to predict its characteristics. A grid architecture is the same thing for electric power systems. We refer to any particular representation of this type as a grid architecture and to the discipline itself as grid architecture.

Resilient Control Architectures and Power Systems, First Edition.
Edited by Craig Rieger, Ronald Boring, Brian Johnson, and Timothy McJunkin.
© 2022 The Institute of Electrical and Electronics Engineers, Inc. Published 2022 by John Wiley & Sons, Inc.

In general, a system architecture is comprised of three kinds of elements:

- Black box components – elements whose internal workings are not of concern (hence, "black box"); components may be of any scale from a device like a transformer to an entire system such as a communications network
- Structure – how the components are connected or related
- Externally visible characteristics – these include characteristics of the components, the structures, and the whole system.

Components with their characteristics combine though structure (with its characteristics) to yield systems. While we treat components as black boxes, this is not an excuse for the architect to specify magic boxes that cannot exist, so we use various methods to ensure that the architectures we specify are grounded in reality.

The uses of grid architecture are many and include

- Identifying legacy constraints
- Removing structural barriers and redefining essential limits
- Managing system complexity (and therefore risk)
- Setting essential bounds on system behavior
- Identifying gaps in theory, technology, organization, etc.
- Facilitating communication among stakeholders
- Defining platforms and interfaces and informing interoperability

That said, an architecture is not a design. An architecture admits many possible implementations, any of which must fit inside the bounds or constraints defined by the architect, whereas a design is a specific expression of an architecture that admits exactly and only one implementation. One role of the architect is to define the smallest set of constraints needed to ensure the resulting system will have the right characteristics and behavior while minimizing limitations on the grid developers, designers, and engineers as much as possible. For the grid, complexity makes this a challenging proposition.

An important aspect of grid architecture methodology is the extensive focus on structure. This is because structure determines the essential limits or bounds on what a complex system can and cannot do. Most design and implementation processes explicitly or implicitly assume a structure, whereas in the grid architecture work, we are primarily concerned with *specifying* structure. There are two reasons for this

- Get the structure right early on and the system pieces fit into place neatly, the downstream decisions are simplified, and investments are future-proofed.
- Get the structure wrong and integration is costly and inefficient, investments are likely stranded, and benefits realization is limited.

Grid architecture treats the grid as an interconnected set of structures in a paradigm called network of structures. While the system of systems paradigm concentrates on IT systems as components, grid architecture views the grid as comprised of seven classes of structures:

1. Electric infrastructure – the physical means of electricity delivery
2. Industry structure, including electricity market structure (not market rules)
3. Regulatory structure (relationships, not regulatory rules)
4. Digital superstructure (information systems and communications networks)
5. Control, treated as a structure rather than an application

6. Convergent networks (electric, gas, water, traffic, etc.) as might be found in a smart city environment
7. Coordination framework – the often-intangible structure that determines how disparate elements of the grid cooperate to deliver electricity

These structure classes have complex relationships whose details vary by region, industry segment, regulatory regime, and other factors.

We have inherited a massive amount of structure from the twentieth century grid and with it many structural constraints that inhibit grid modernization. Consequently, we define the core problem of grid architecture as follows:

Determine the appropriate new structures or minimal structural changes to the grid that

- Relieve crucial constraints on new capabilities
- Limit propagation of undesired change effects
- Strengthen desirable grid characteristics
- Simplify design and implementation decisions.

The complexity of the grid goes well beyond the levels we associate with many sophisticated systems. In fact, we designate it *ultra-large-scale (ULS) complexity*, using a concept developed at Carnegie Mellon University. This level of complexity arises from the fact that the grid is comprised of many already complex structures, and these structures are interconnected and interact in complex ways. Such systems have characteristics that pose special challenges, including

- Inherently conflicting diverse requirements
- Decentralized data, control, and development
- Continuous (or at least long time scales) evolution and deployment
- Heterogeneous, inconsistent, and changing elements
- Operation involving wide time scales
- Operation involving wide geographic scales
- Normal failures (failures are common and frequent, not exception events).

The consequences of these ULS properties are many and are sometimes surprising. One consequence is that is it not possible to define a comprehensive set of use cases for grid modernization, and so architectural methods that depend on use case sets as inputs fail to yield strong results. Instead, we use a combination of emerging trends, systemic issues, and legacy constraints as key inputs to the process of developing new architectural views for the grid (see Figure 10.1).

Additional information about the discipline of grid architecture can be found online.[1]

One of the most important systemic issues driving grid modernization is the rising emphasis on improving grid resilience. Grid modernization activities cite resilience (sometimes called resiliency) as a key electric power grid characteristic to be improved or maximized, and so it is crucial for the development of resilient grid architectures that the concept of grid resilience be clear and quantifiable. However, attempts to define and quantify a concept of resilience for electric power grids have mostly relied upon ad hoc definitions that do not have much underlying rigor and are often closely tied to reliability, sometimes only differing in terms of scale and frequency of events. While grid reliability would seem to have strong definition given the IEEE

---

1 https://gridarchitecture.pnnl.gov.

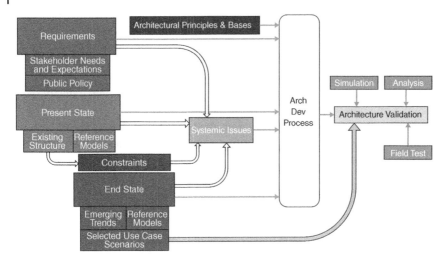

Figure 10.1 Grid architecture development process.

1366 standard, grid resilience is a term that is less clear. For the purposes of grid architecture, a clear definition based on foundational principles yields not only clarity of concepts but provides a set of architectural principles for grid resilience improvement and points the way toward rational quantification and valuation of resilience.

10.2 Conventional Views of Grid Resilience

The Electric Power Research Institute (EPRI) definition of resilience specifies it in terms of three factors: prevention, recovery, and survivability.[2] These terms are described rather than being defined and do not have metrics associated with them. Also, as will become clear later, we will argue that one of them (recovery) does not belong in the definition of resilience as applied to grid architecture.

The industrial control systems cyber emergency response team (ICS-CERT) has produced a discussion that has some similarities to the EPRI definition [1]. It uses the National Infrastructure Advisory Council 2009 definition: "the ability to reduce the magnitude and/or duration of disruptive events." This is somewhat useful in that it represents resilience as a grid characteristic, independent of external or internal events. However, no metrics are provided. The same work defines reliability for grids as "the ability of the power system to deliver electricity in the quantity and with the quality demanded by users." This definition conflates power quality and resource adequacy with service interruption event reaction, and so it creates a difficult situation in terms of defining metrics: nonorthogonality with other metrics. Some models of power quality include reliability as a subset [2].

Presidential policy directive 21 defines resilience for the grid as "the ability to prepare for and adapt to changing conditions and withstand and recover rapidly from disruptions. Resilience includes the ability to withstand and recover from deliberate attacks, accidents, or naturally occurring threats or incidents." This definition also conflates resilience with reliability by including recovery.

---

2 EPRI, "Grid Resiliency," available online: https://www.epri.com/research/products/000000003002006784.

The smart grid dictionary defines resilience as "the ability to resist failure and rapidly recover from breakdown" [3]. This definition starts well but then merges utility response with a grid characteristic. The article does suggest that resilience impacts reliability, and thus suggests (perhaps inadvertently) that resilience is a grid characteristic, not an event response. This is essentially the starting point for the resilience discussion in the recently released National Academy of Sciences (NAS) study on electricity system resilience [4], which also combines grid characteristics with utility response to external events. Many definitions or descriptions of resilience include some aspect of ease of recovery, but do not show how to measure it as an intrinsic grid characteristic and for good reason. This approach depends on how the utility handles events, which conflates utility processes with grid characteristics and still ends up being dependent on specific events. The same NAS study indicates that there are no generally agreed-upon metrics in wide use [5].

Some definitions of resilience attempt to position it as a characteristic that applies only after some damage has been done. For example, the resilience analysis process (RAP) developed by Sandia National Laboratories [6] lists as a basic principle that "resilience metrics should be based on the performance of power systems, as opposed to relying on attributes of power systems." Some discussions attempt to use terms like security or availability to address what should be thought of as resilience (or perhaps they should be thought of as component elements of resilience) and then to push resilience into the domain of reliability. A result is that many proposed resilience metrics are in fact reliability measures of various kinds.

The foregoing definitions are not especially useful when it comes to grid architecture. This is due to several problems:

1. They conflate externalities with grid internals
2. They lead to metrics that are backward looking
3. They do not provide a means to understand clearly what to change about grid structure to improve resilience
4. They do not lead to rational means to quantify and value resilience measures

Given these issues, it is appropriate to consider how to define grid resilience in a manner that is useful to grid architecture. To do that we must first be clear about grid characteristics generally.

10.3 Grid Characteristics

Figure 10.2 illustrates the relationship of architectural elements to grid characteristics. Note that grid characteristics are divided into two kinds: arbitrarily labeled qualities and properties. Qualities are characteristics seen from the outside by users of the grid and represent how they wish to be served by it; properties are characteristics seen internally and represent how the grid delivers the services it offers. Qualities are a user's point of view; properties are a developer/builder/operator's view. Figure 10.3 gives a sense of how user needs and public policy are translated into grid architectures.

A cornerstone of the grid architecture discipline is complexity management and that includes clarity of definitions and recognition of definition relationships and, therefore, definition structure. The discussion about grid resilience necessarily includes several related concepts, most notably reliability. Figure 10.4 shows how electric reliability decomposes into subcategories of properties. Most discussions of electric reliability focus on availability, but to get to a clear understanding of resilience, one must also consider usability. Several other characteristics come into play when considering resilience. Figure 10.5 shows a structural relationship diagram for these characteristics.

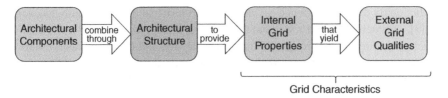

Figure 10.2 Relationship of grid architecture to grid characteristics.

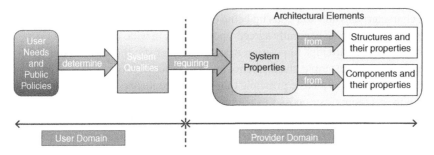

Figure 10.3 Synthesis of grid architectures.

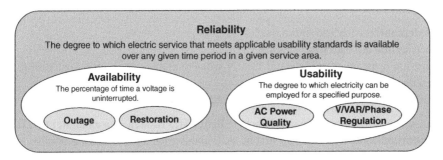

Figure 10.4 Electric reliability elements.

Each has a definition that has been carefully selected to isolate one particular aspect of grid behavior or capability. The definitions are available online[3] and, combined with the structural relationships, collectively provide a powerful means to reason about grid characteristics, including resilience. In the remainder of this chapter, we focus on electric grid resilience.

10.4 Grid Resilience and the Relationship to Electric Reliability

Figure 10.6 provides a definition of grid resilience and illustrates some key issues, such as the difference between resilience and reliability, and how to determine which definition applies in the various phases of grid operation.

---

3 https://gridarchitecture.pnnl.gov/media/methods/Grid_Charactcristics_Definitions_and_Structure.pdf.

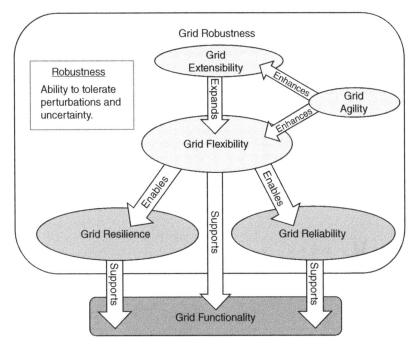

Figure 10.5 Grid resilience in context.

Grid resilience is the ability to avoid or withstand grid stress events without suffering operational compromise or to adapt to and compensate for the resultant strains to minimize compromise via graceful degradation. It is in large part about what does **not** happen to the grid or electricity consumers.

A key concept here is that resilience is *an intrinsic characteristic of a grid* or portion of a grid. A perfectly resilient grid would not experience outages and so any definition or metric that is based on measuring outage frequencies, times, extents, or impacts on customers or systems does not get at the essence of resilience. Resilience applies to the grid under stress: how the grid resists losing capabilities or gracefully degrades is the essence of resilience. This explains why reliability measures are not useful for quantifying resilience. Resilience is in large part about what does **not** happen.

This definition includes the ability to withstand operating excursions outside the normal operating envelope with an inherent tendency to return to operations within the normal envelope. Electric reliability, on the other hand, is a measure of *behavior once resilience has broken*. Standard electric reliability metrics fall into two categories: frequency indices (CAIFI, SAIFI, etc.) and duration indices (CAIDI, SAIDI, etc.). Frequency indices are very roughly related to resilience in the sense that they reflect to some degree how often resilience is broken (but in a nonnormalized fashion, making them unusable as resilience measures). Duration indices measure how well a utility responds to broken resilience (also in a nonnormalized fashion). Therefore recovery, as mentioned in the EPRI resilience definition, actually belongs in the reliability domain.

The dividing line is clear: for electricity delivery, the start of a sustained outage is the transition point from the domain of resilience to the domain of reliability. An understanding of this concept is necessary for the development of resilient grid architectures.

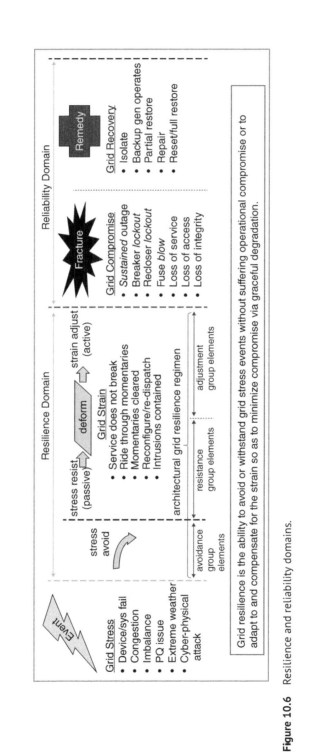

Figure 10.6 Resilience and reliability domains.

For some time, there has been a view in parts of the electric power industry that momentary interruptions should be included in reliability metrics. The impact of momentaries on smart inverters has heightened this issue recently. In the definition and model provided here, momentaries are *power quality* issues that stress the grid. When smart inverters respond to momentaries by pulling off the grid for a period of time, this reflects a lack of resilience in the inverters (by design) that can create a transition to a reliability domain issue.[4] Hence, changing the ride-through behavior of smart inverters can be a resilience improvement with a potential impact on reliability.

It is worth noting that FERC uses a statutory definition of reliability for bulk power systems derived from 18 CFR Part 39[5] that pertains to some of what we here include in resilience *("Reliable Operations means operating the elements of the Bulk-Power System within equipment and electric system thermal, voltage, and stability limits so that instability, uncontrolled separation, or cascading failures of such system will not occur")*. Given the source of this definition, it is not likely that industry definitions will change, nor is it the purpose of this document to suggest that they should.

Resilience and reliability do have a relationship, but it is not simple. The Smart Grid Dictionary discussion referenced above suggests that resilience "impacts" reliability. In fact, increasing resilience *may* improve reliability, but this is not guaranteed, since reliability as measured by standard metrics depends on some set of events that result in outage and resilience is intended to avoid such outcomes. It can be the case that grid events and stress do not intersect with the resilience change so that reliability metrics may improve, not show a change, or degrade. This is one of the reasons that the definition and metrics for resilience are so important – to be useful in the development of resilient grid architectures and the design of resilience tools, device, and systems, it must be possible to determine resilience directly from grid descriptions, not from reliability indices.

Note that the definitions of both resilience and reliability can be applied at various scales. This means it is possible to discuss the resilience of a whole power system, a transmission or distribution system, a single circuit, or portion of a circuit, even a single device. The choice of scale will depend on the nature of the problem being solved and is up to the engineer, operator, regulator, or other stakeholder to determine appropriately. Scale should be specified so that the extent and scope of the analysis is clear. Scale includes not just physical system extent but also time scale. This is important for dividing certain operations into resilience versus reliability functions. For example, recloser operation and fault location, isolation, and service restoration (FLISR) activity can depend on time settings, which may vary from utility to utility.

In the 2014/2015 work on grid modernization, the DOE energy policy and systems analysis (EPSA) group defined cybersecurity as an element of resilience. While cybersecurity involves more than what is covered by resilience, cybersecurity is clearly an aspect of resilience for electric grids. This means that the definition of resilience and the identification of transitions from resilience domain to reliability domain must be sufficiently general to include cyber-defense failures as well as ordinary electricity delivery failures.

10.5 Characterization of Resilience

We think about grid resilience via rough analogy to how materials respond to external mechanical forces and so we adapt some of the terminology of metal stress and strain to fit the grid resilience problem.

---

4 This can be taken advantage of in an IoT environment to become a cybersecurity vulnerability.
5 Rules Concerning Certification of the Electric Reliability Organization; And procedures for the establishment, approval, and enforcement of electric reliability standards.

10.5.1   Stress and Stressors

Stress refers to how a system internally deals with external duress applied to it in terms of something akin to elastic deformation (i.e. the system bends a bit but returns to its original shape when the external disturbance is removed). Alternately, we may view it as the buildup of internal forces that resist the external duress. For example, a feedback control system that performs regulation will exert internal control force to maintain its control variable at the regulation set point when a disturbance occurs, applying more control force as needed (up to its limit), and then automatically returning the internal corrective control action (internal force) to its nominal value when the external disturbance abates. A transmission tower will experience the buildup of internal mechanical force in response to wind loading and will return to proper shape if its stress limit is not surpassed. Therefore, a **stressor** is an external (to the systemic scope being considered – more on that later) source of duress that disturbs system operation, potentially to the point of causing degradation of performance or outright failure.

Some examples of standard grid stressors are

- Exogenous impacts such as tree taps or vehicle damage
- Device/subsystem/system failure or fatigue
- Transmission congestion
- System imbalance
- Power quality disturbances
- Extreme weather, including geomagnetic-induced currents
- Cyber-physical attack, including electromagnetic pulse (EMP).

In addition, we consider unconventional stressors such as

- Distribution circuit constraint violations (thermal, voltage, and protection)
- Communication network congestion and performance constraints versus increasing data flow
- Accidental damage (cutting optical fiber during construction, etc.)
- Changes in fuel availability for generation
- Changes in water availability for cooling

We also include stressors related to operational issues and noncircuit exogenous forces

- Integration of new systems or capabilities and/or third-party energy services organizations
- Impact of software upgrades
- Operator errors and errors in configurations
- Changing requirements due to new regulations and/or social issues
- Impact of unevenly distributed and increasing connection of nonutility responsive/interactive devices and systems to the grid.

10.5.2   Physical Scale

Resilience and reliability both must be considered in terms of systemic scale. This means that the concepts of resilience domain and reliability domain can and should be applied at various scales, from a single device to whole-grid interconnections and at all levels in between. For example, if a single device fails, it enters the reliability domain. The circuit or system to which it is attached may stay in the resilience domain or may enter the reliability domain depending on the consequences of the device failure and the grid reaction to it.

Example: a fault occurs in a section of a partially meshed distribution feeder. A FLISR[6] system isolates the faulted section and performs line switching to quickly (definition of "quickly" depends on the utility involved) restore power to all the feeder sections but the faulted section. In such a case, the faulted section has entered the reliability domain, but the rest of the feeder has remained in the resilience domain.

This concept also resolves a potential ambiguity in the definition of a stressor. When a component fails, it become a stressor on the subsystem to which it belongs. The concept may be viewed as a reverse recursion in that failure of a subsystem becomes a stressor on the whole system. However, when considering the failed component, some stressor (current overload, ice buildup, degradation through aging, etc.) must have caused the component failure. In the case of a cascading failure (component fails → subsystem fails → system fails), we may view the stressor that caused the original component failure to be the *root stressor*.

The systemic scope concept must be applied in this manner generally; otherwise, it will not be possible to separate resilience from reliability, which would defeat the entire purpose of creating the new definitions. We also need the concept of systemic scope to properly define resilience measures.

10.5.3 Temporal Scale

For electricity delivery, the start of a sustained outage is the transition point from the domain of resilience to the domain of reliability. Thus, momentaries are *power quality* issues that stress the grid; therefore, they are part of the resilience domain. However, the definition of the time span of a momentary can vary from utility to utility. This is perfectly fine; it means that the boundary between resilience domain and reliability domain is determined by each utility in an appropriate manner, thus requiring no changes to reliability metrics.

> The combination of the concepts of systemic scope and temporal scope provide the means to resolve classification issues without the need to resort to non-deterministic methods.

10.5.4 Strain

Continuing the materials analogy, strain is the change in a system that occurs in response to externally applied force or disturbance. Strain may manifest in a variety of ways, such as pole flexing, device overheating, or exertion of control action in a feedback loop.

10.5.5 Resilience Domains

To support practical application of the resilience definition, we define the three principal regimes of the resilience domain, along with certain subdomains, as shown in Figure 10.7.

10.5.5.1 Stress Avoidance
Action on the stressor or the system so that stress on the system does not happen (is avoided). Stress avoidance refers to measures taken to ensure that potential stressors never impact the grid in the first place.

---

6 Fault location, isolation, and service restoration.

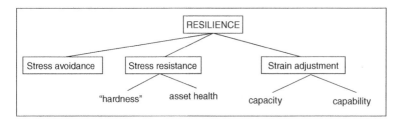

Figure 10.7 Resilience domains and subdomains.

Examples:

- Vegetation management – removes the stress from vegetation physically contacting the system
- Flood wall – placement around a nuclear reactor containment building so that tsunami waves do not impact the building
- Animal guards – prevents animals from contacting energized conductors and thereby causing a short circuit
- Preventive maintenance – avoids conditions leading to equipment failure
- Software containers – allows e-mail and attachments to be opened only in a safe container (sandbox) so that viruses cannot be accidentally unleashed in an information system
- Equipment placement – avoids contact with a stressor, for example, by placing equipment above high-water mark or below ground, depending on what is appropriate

10.5.5.2 Stress Resistance

A strengthening (hardening) of the system so that stressors have minimal or no impact on the system and so the system operation is not degraded. The stress is effectively absorbed or rebuffed. Stress resistance is a limited *mitigation strategy* that can be quantified by the amount of "spare" internal capacity for compensation that is available to deal with disturbances. Stress resistance occurs within the bounds of a nominal operating paradigm and with nominal operating parameters and settings, so we may think of it as limited deformation within the elastic bounds of the component or system; removal of the stressor allows the component or system to spring back into shape automatically. In other words, stress resistance is intrinsic to grid architecture and normal system function.

Examples:

- Protective enclosures – the system or equipment is shielded from the stressor. The enclosure absorbs the energy of the stress.
- Hardening of electronics – adding shielding, thermal management subsystems, and electrical filtering suitable for substations (as opposed to data centers) to substation servers, routers, and remote terminal units.
- Wind resistance – increasing wind resistance hardens against tower collapse in severe weather.
- EMP/geomagnetic induced current (GIC) shielding and decoupling – electromagnetic shielding and capacitive decoupling of low-frequency/ direct current (DC) currents in transmission systems harden against EMP and GIC-induced stresses.
- Encryption – increasing the difficulty in accessing the protected information.
- Closed loop regulation of an operating parameter such as voltage of system frequency.

Component hardening can be understood in terms of withstand ratings, such as wind load limits or overcurrent ratings for electrical components, but hardening may also be understood in other

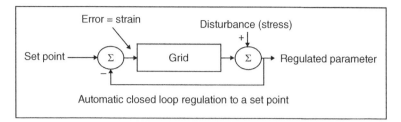

Figure 10.8 Closed-loop control stress and strain.

terms. For closed-loop control systems, hardness refers to how well the controlled variable can be regulated in the presence of disturbances. Consider Figure 10.8, which can represent a closed-loop automatic control for feeder voltage regulation, for example.

In this subsystem, a disturbance that tends to cause a change in regulated voltage is the stress. The resultant strain is the error signal (difference between setpoint and actual voltage). When the error exceeds the limits of available control action, then resilience is broken. Therefore, the range of available adjustment to maintain voltage within regulation limits represents the hardness of the control system. Any stress that stays within the strain limit will be handled and when the stress is removed, the system automatically reverts to nominal conditions as the strain is relieved.

10.5.5.3 Strain Adjustment

Flexibility of the system to adapt to stress. The nature of the change may be parametric, structural, or modal. A parametric change would be a new value for a set point, threshold, or other system operating value intended to compensate for a disturbance to limit system degradation. A structural change would be a reconfiguration of a device, circuit, or subsystem (example: isolation of a circuit fault and switching of circuits to reroute power as in FLISR). A modal change would be a switching from one operating paradigm or algorithm to another, again to limit system degradation.

Examples:

- Structural reconfiguration – the distribution system under stress shifts loads from one feeder to another via circuit switching, thus minimizing the impact of the stress on the first feeder
- Load management – stress on the system is reduced by reducing demands via responsive loads
- Rerouting communications – a communication network changes routing tables to alter data flow paths to reduce performance degradation due to congestion
- Use of reserves – a bulk energy system uses spinning reserves to compensate for load and generation changes within a given level of operational flexibility
- Modal change – switching the use of storage from augmenting system inertia to shaving peak load in a contingency situation

Note that for strain adjustment to work it is necessary to be able to measure the strain on the system [7].

All resilience domains have the characteristic of mitigating one or more vulnerabilities. Table 10.1 clarifies the nature of the mitigations for the three primary resilience groups.

Table 10.2 extends the classifications to subgroups shown in Figure 10.3 for avoidance, resistance, and adjustment in a manner that is helpful in developing ways to quantify the impact of resilience measures.

Unlike traditional measures of resilience that focus on the amount of damage that is done by external events or the time to recover, these characterizations suggest measures that focus on intrinsic grid characteristics derived from component properties and grid structure.

Table 10.1 Resilience group characteristics.

Stress avoidance	Stress resistance	Strain adjustment
Preventative action	Supports normal operation mode	Planned response outside of normal mode of operation
Ancillary to operational components and structures	Operational components and structures unaltered	Exercises alternative operational structure or configuration
	Normal coordination-control-decision process	Corrective action coordination-control-decision process
	Sustains economic or other operational objectives-based decisions	Impacts economic or other operational objectives/decisions

Table 10.2 Resilience groups and subgroups.

Resilience group and subgroup	Comments
Stress avoidance	Prevention of stress events from occurring in the first place
Stress resistance Asset/subsystem/system hardness	Inherent strength; how much stress can be applied before a component or system begins to yield (degrade performance) (analogous to yield point: amount of stress that can be applied before transition from elastic to plastic deformation)
Stress resistance Asset/subsystem/system health	Ability to accept rated load (power, data flow, computational burden, etc.) without degradation of operation or excessive loss of life
Strain adjustment Adjustment capacity	Reserve ability to handle stress that may arise (e.g. generation flexibility)
Strain adjustment Adjustment capability	Ability of a grid to use available compensation capacities – this implies functioning mechanisms to invoke and control whatever capacities are present and useable

Note that strain adjustment capacity and strain adjustment capability are treated separately. This is because the mechanisms to make use of available capacity may subject them to stresses.

10.5.6 Foundational Support

Several capabilities or measures apply to all the resilience domains and subdomains. Table 10.3 lists these key foundational elements.

10.6 Architectural Principles and Concepts for Resilience

Based on the definitions described above, we have developed a set of architectural principles to guide grid modernization. These include

1. Grid resilience is the ability of the grid to avoid or withstand grid stresses without suffering operational compromise or to adapt to and compensate for the resultant strains to minimize compromise via graceful degradation. Grid resilience should be understood in terms of grid vulnerability, not in terms of hypothetical large rare events. Resilience elements comprise countermeasures to grid vulnerabilities.

Table 10.3 Resilience foundational elements.

Foundational resilience element	Description/comments
Situational awareness	All resilience-improving measures require some form of situational awareness, whether in the form of real-time sensing and measurement or knowledge of system vulnerabilities and stressor impacts (either forecasted or determined after the fact)
Planning/design for resilience	None of the resilience-improving measures happens by accident or can be relied upon to emerge spontaneously from other aspects of grid architecture and so these measures must be planned and designed in. Planning includes development of a resilience strategy; design includes allocation of resilience measures
Interoperability	While possibly viewable as a resilience measure for IT systems, in the larger sense this applies to all sorts of interconnection issues, including mechanical, electrical, communication, control, coordination, and data/information exchange interconnections, and applies not just to IT systems but to general grid codes/interconnection agreement issues

2. The resilience domain has three regimes: stress avoidance, stress resistance, and strain adjustment. The scope of resilience includes more than just device/system failures and large exogenous events; it includes stressors related to operational issues and noncircuit exogenous forces.

3. Grid resilience architectures use the following as foundational elements:

    3.1 Situational awareness (monitoring, forecasting, and extended grid state)

    3.2 Planning/design for resilience

    3.3 Interoperability and interface standards.

4. Grids are ULS systems, subject to normal failures [8]. Consequently, architecture should not treat failures as exception conditions but as part of normal operating modes.

5. The architecture views should employ redundancy/critical component backup to eliminate single points of failure and to provide options for strain adjustment.

6. The architecture views should incorporate buffering for the purpose of decoupling volatilities that may be induced by stochastic generation sources and the behaviors of controllable loads.

7. Architectures should make use of modularity principles, namely module strength and module (de)coupling, to limit propagation of stresses and strains and therefore cascading failures.

8. Dependencies should be minimized to limit system brittleness.[7] This may involve decoupling or may involve structures that provide some form of compensation for dependency, such as resilience loops in gas–electric systems [9].

9. The architecture views should use structures that enable agility, adaptation, and configurability (including energy resource flexibility) to support strain adjustments. This is related to redundancy and modularity.

10. Architectures should provide structures that can support or enable graceful degradation[8] in the event of grid strains.

---

7 A brittle system is characterized by a sudden and steep decline in performance as the system state changes. A system is brittle if it is unable to effectively absorb stress-induced shock.

8 Graceful degradation is the ability of a system or network to maintain limited functionality even when a portion of it has been rendered inoperative. In graceful degradation, the operating efficiency or speed declines gradually as an increasing number of components fail. It is somewhat analogous to ductile failure in materials.

11. Architectures should provide structures that can support or enable fault tolerance[9] in the event of stress-induced failures.
12. Known resilient structures (distributed systems, layering, platforms, etc.) are to be preferred where appropriate.

In addition to the guiding architectural principles, a set of core concepts PNNL employs in architecture development inform resilient grid architecture specification decisions.

10.6.1   All Hazards Approach

Resilience is not limited to selected large, rare events or cyberattacks. All forms of stress on the grid and its component elements are considered, including those arising from processes, such as software upgrades, operator/configuration error, and device miscalibration.

10.6.2   Situational Awareness

The need for situational awareness is taken as fundamental. This involves more than just grid power state. This work uses the concept of Extended Grid State, as defined in the work done in the US DOE Grid Modernization Initiative Sensing and Measurement Strategy project [10].

10.6.3   ULS Normal Failures Approach

Applying ULS system theory to the grid leads to the view that faults and faults must be treated as normal events rather than exceptions. Architectural structure should be selected to mitigate the impact of such failures in the same way as with other grid stresses. In other words, the ULS complexity of the grid is a source of grid stress. Examples include intermittent communication link failures, bad software maintenance and upgrades, noisy and/or missing measurements, and any number of other failures that plague large complex systems.

10.6.4   System Hardness

Hardness for grid resilience purposes usually associated with components and this is addressed below in this reference architecture. In addition, it is useful to consider the "hardness" of subsystems and whole systems to identify structural weaknesses and determine structural improvements that affect structural hardness (the ability of a grid structure to withstand stress).

10.6.5   Flexibility

The ability of a system (in this case the grid) to make (usually automatic) adjustments to grid stresses. This can include traditional energy resource flexibility, and increasingly, energy resource flexibility afforded using distributed energy resources (DER), including net load flexibility. It also includes structural flexibility, such as automated capability for fault isolation and circuit reconfiguration, and the ability to change or add functions and behaviors as new circumstances warrant.

———
9 Fault tolerance is the property that enables a system to continue operating properly in the event of the failure of (or one or more faults within) some of its components. A highly fault-tolerant system may continue at the same level of performance even though one or more components have failed.

10.6.6 Extensibility

The ability to add functionality without major restructuring is a key to future-proofing investments and is a legitimate focus for grid architecture. The driving forces in the extensibility case come from at least one of the following:

- Changing customer expectations
- Availability of new technology
- New public policy mandates

Structural extensibility is a means to address such trends, by providing the foundation for agility in the implementation of new capabilities. The platform structure mentioned earlier is an example of such a foundation.

10.6.7 Agility

The speed with which either flexibility or extensibility can be exercised.

10.6.8 Distributed Versus Centralized Systems

A decentralized system is one in which the elements are separate (usually geographically dispersed but not always) and act independently, with perhaps some small amount of supervision to provide set points, etc. A distributed system is a decentralized system in which the elements cooperate to solve a common problem. This implies some form of communication among the decentralized elements. A centralized system is one in which all the computing, logic, control, data analysis, etc., is performed at a single element. A centralized system is a degenerate form of distributed architecture and distributed systems may have a central element that participates in the overall processes, the latter being sometimes characterized as a hybrid of central and distributed architectures.

The distributed architecture concept is applied to the grid in several ways. The most obvious not only applies for coordination and control but also applies to generation and storage and to data management and analytics (intelligence). Note that markets are always distributed systems, even when the market-clearing mechanism is centralized, as is the case with organized wholesale electricity markets.

10.6.9 Buffering

Most complex systems have buffers of one kind or another. Communications systems use jitter buffers to even out the flow of data bits from uneven sources. Logistics systems use buffers – called warehouses. Water and gas systems use buffers – called storage tanks. In each case, the buffer serves to even out flow variations. By acting as a kind of shock absorber and providing springiness or sponginess, they cushion a system against stresses due to volatility of sources. Such springiness gives the system resilience. Lack of buffering is a vulnerability.

Power grids (especially at the distribution level) lack buffering due in part to limitations on technology and for the distribution case due to lack of a need for it in the twentieth century. Buffering addresses resilience vulnerability in several ways, including even some forms of cyber-attack.

10.6.10 Structural Resilience

Architectural structures may be more or less resilient due to intrinsic characteristics. Identification of structural resilience vulnerabilities in legacy grid structures and determination of minimal

Table 10.4   Grid structures and issues.

Structures	Key issues
Industry and market structure	Tier bypassing, scaling, and cybersecurity
Electrical (circuit) infrastructure	Coupling and hidden coupling
Controls	Hidden coupling, scaling, latency, fault tolerance, and cybersecurity
Information systems and communication networks	Brittleness via coupling and cybersecurity
Converged networks (such as gas/electric systems)	Dependency
Coordination frameworks and structures	Fault tolerance, extensibility/flexibility, and scaling

changes to structure to address those vulnerabilities is a key aspect of grid architecture work. Structures and structural principles whose resilience properties are known to be strong include

- Layering and platforms, including structures derived from layered decomposition
- Core/edge network structure
- Modularity and (de) coupling

Modularity bears a strong relationship to interoperability in that modularity requires clear interface definitions and those interface locations and definitions derive from architectural structure and from basic module characteristics such as module strength and module coupling.

The grid structures to be treated in terms of structural resilience are detailed in Table 10.4.

10.6.11   Redundancy

Avoidance of single points of failure is a structural approach to system resilience. In this regard, a "point" may be a single component, or may be a whole subsystem, such as a communication network. Redundancy may involve the use of multiple components that can serve the same function but may not be implemented identically. An example would be the use of two communication systems that employ different physical media (optical fiber and microwave for example) to carry the same data traffic between the same endpoints. It may involve backup components that are technologically different from primary components, such as the case where primary relays are digital but backup relays are electromechanical. Structurally, it may involve redundant pathing, such as for electric circuits with more than one available path configuration or communications systems with physically separate paths (separate optical fiber physical layout routes, for example).

See the grid architecture website for application of these principles in reference grid architectures.[10]

10.7   Structural Resilience Quantification and Valuation

For grid resilience and the architectural principles and reference architectures presented above to be of full use, they must be connected to planning processes. This requires means to express or determine how much resilience is present, how much can be added by various changes to the grid, and how to trade off various alternatives to achieving a resilience-improvement objective.

---

10 https://gridarchitecture.pnnl.gov/library.aspx.

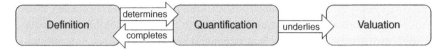

Figure 10.9 Characteristics definition, quantification, and valuation.

A problem here is that the various conventional definitions have not led to rational means to quantify the amount of resilience at stake or the comparative worth of resilience alternatives. A logical sequence aids in addressing these issues, shown in Figure 10.9. We require definitions of grid characteristics before we can determine how to quantify them, but definitions are not complete until quantification can be established. Valuation cannot be specified until the means for quantification are available and can be applied to specific situations. This is true for grid characteristics of all kinds, and holds for system characteristics in general, not just for the grid. Skipping to valuation without properly addressing definition and quantification is problematic, but as has been pointed out, existing definitions are weak, and quantification is almost unknown for grid resilience.

To develop methods of quantification for grid resilience, three principles are needed:

1. Resilience is a *system characteristic* and as such, depends on both component properties and structure.
2. Resilience depends on the entire power delivery chain, as seen from the point of view of a delivery end point or set of points.
3. Resilience must have a time horizon associated with it.

Individual devices and subsystems can contribute to system resilience, but the way they make their contributions depends strongly on the structure into which they fit. Without taking structure into account, it is not possible to be accurate about how much contribution to grid resilience would be made by a change (or even if the contribution is positive or negative).

The second principle above (the "As Seen By" principle) is analogous to the Thevenin Theorem in circuit analysis, in that the whole power delivery system can be viewed as having an equivalent system resilience, as seen by the user of power at a single point such as a hospital, residence, or service area. Consequently, there is no single value for grid resilience, but rather a possibly unique specific equivalent resilience for any particular endpoint, which may differ from that for another endpoint (because different power delivery paths and components of the grid may be involved).

The third principle requires that any quantification of resilience has a time horizon associated with it. This is a pragmatic measure since withstand ratings for electrical components will frequently have time limits and where storage is used as a resilience improving buffer, upper limits on storage capacity place effective time limits on functions such as outage ride-through. This concept is congruent with the "mission time" approach used in calculating system reliability in electronics and aerospace applications.

We describe the contributions that components or subsystems may make to grid resilience as resilience *determinants*. Such determinants combine with the grid and each other through structure to affect grid resilience. Given this approach, the concept of a resilience determinant must be a multiscale property, that is, it may be applied to a single device, to a subsystem, or to a larger portion of a power delivery system, up to a whole regional grid.

Relating device physical characteristics (e.g. maximum energy storage capacity of a grid storage device) depends on each device or subsystem type. For example, hardness of a utility pole may be measured in terms of wind-withstand rating, while hardness of an electrical component may be described in terms of maximum temperature or overcurrent. Combining or comparing disparate determinants in terms of physical component characteristics is not possible, so it is necessary to use the characteristics to compute normalized dimensionless functions that can be combined via reduction rules that can apply structure to determine how to convert complex resilience determinants into equivalent simple determinants and finally system resilience. This approach is commonly used in control theory, reliability analysis of complex (nonelectric utility) systems, and even electric circuit analysis.

From the point where resilience and modifications intended to improve resilience can be quantified, valuation can then be determined on a rational basis.

10.8   Summary

The improvement of resilience in electric power systems has been of growing importance in the United States for several years. Progress has been made in various areas, but much remains to be done in terms of the basic architecture of the power grid. A limiting factor has been the lack of a connection between foundational grid architecture principles and methods on the one hand and clearly defined relationships between resilience improvement objectives and actual means for assessing, planning, and implementing resilience measures on the other. At the core of this limitation is the need for principled definition, quantification, and valuation of the resilience impacts of grid architectures and architecture changes. The use of structural concepts provides a framework for these issues and provides a new means to obtain insight into how resilience may be analyzed and improved.

Thoughtful Questions to Ensure Comprehension

1   How does resilience differ from reliability?

2   Why is resilience not just reliability for large, rare events such as hurricanes and why is it not useful to define resilience in terms of external events and resultant consequences (such as outages)?

3   What role does system structure play in determining system resilience?

4   Is there a single resilience quantity for a power delivery system?

5   Why is it helpful to divide resilience into three categories (stress avoidance, resistance, and strain adjustment)?

6   How is the concept of system scale applied to resilience as defined here?

7   How does buffering improve grid resilience?

8   In the context of grid resilience, what is the difference between a stress and a strain?

9   Under what circumstances might a strain become a stress?

10 Why does grid architecture focus so strongly on structure?

11 What is the difference between system qualities and system properties?

12 What is the difference between system architecture and system design?

Further Reading

Taft, J.D. (2018). *Electric Grid Resilience and Reliability for Grid Architecture*. PNNL https:// gridarchitecture.pnnl.gov/media/advanced/Electric_Grid_Resilience_and_Reliability_v4.pdf.
Widergren, S., Shankar, A., Kelley, B. et al. (2018). *Toward a Practical Theory of Grid Resilience*. GMLC https://gridarchitecture.pnnl.gov/media/advanced/Theory_of_Grid_Resilience_final_GMLC.pdf.

References

1 Clark-Ginsberg, A. *What's the Difference Between Reliability and Resilience?* Stanford University https://www.researchgate.net/profile/Aaron-Clark-Ginsberg/publication/320456274_What%27s_ the_Difference_between_Reliability_and_Resilience/links/59e651230f7e9b13aca3c2ba/Whats-the-Difference-between-Reliability-and-Resilience.pdf.

2 Brown, R.E. (2002). *Electric Power Distribution Reliability*, 40. Marcel Dekker, Inc.

3 Herzog, C. (2013). *Smart Grid Dictionary*, 5e. GreenSpring Marketing LLC http://www .smartgridlibrary.com/2012/03/26/grid-resiliency-is-required-for-improved-grid-reliability.

4 National Academy of Sciences (2017). *Enhancing the Resilience of the Nation's Electric System*, 1–4. The National Academies Press https://www.nap.edu/catalog/24836/enhancing-the-resilience-of-the-nations-electricity-system.

5 National Academy of Sciences (2017). *Enhancing the Resilience of the Nation's Electric System*, 2–28. The National Academies Press.

6 Vugrin, E., Castillo, A., and Silva-Monroy, C. (2017). Resilience Metrics for the Electric Power System: A Performance-Based Approach. Sandia Report SAND2017-1493.

7 Rieger, C. (2014). *Resilient Control Systems Practical Metrics Basis for Defining Mission Impact*. Idaho National Laboratory INL/CON-14-31971, August 2014. https://inldigitallibrary.inl.gov/ sites/sti/sti/6269308.pdf.

8 Feiler, P., Gabriel, R.P., Goodenough, J. et al. (2006). *Ultra-Large-Scale Systems,*" Software Engineering Institute. Carnegie-Mellon University.

9 Taft, J.D. and Becker-Dippmann, A. (2015). *Grid Architecture*. Pacific Northwest National Laboratory Section 6.2, Electricity/Gas Interaction. https://gridarchitecture.pnnl.gov/media/white-papers/Grid%20Architecture%20%20-%20DOE%20QER.pdf.

10 Taft, J.D., Li, Z., and Stewart, E. *GMLC Sensing and Measurement Extended Grid State Task Team, Extended Grid State Definition Document*. Pacific Northwest National Laboratory PNNL-SA-141027. https://gridarchitecture.pnnl.gov/media/white-papers/Extended_Grid_State_Definition_v3.3_GMLCFormat_final.pdf.

11

Cross-architecture Metrics

Timothy McJunkin

Energy and Environment Science and Technology, Idaho National Laboratory, Idaho Falls, ID, USA

Objectives

The objective of this chapter is to produce useable metrics based on the definition of resilience. The chapter also introduces elements of transdisciplinary areas that are needed to create a resilient system. It is important to obtain "buy-in" from stakeholders in different disciplines so that clarity is critical in the defining stages. The chapter also points to resources that explore the space of resilience for the readers' own research. No magic, 100% satisfactory metric or even definition of resilience has been achieved. A student in this area is encouraged to consider many perspectives and develop new ways to attack this open area of research.

11.1 Definition of Resilience

We can begin this quest with the dictionary definition of resilience from Merriam-Webster:

(a) *Capable of withstanding shock without permanent damage or rupture*
(b) *Tending to recover from or adjust easily to misfortune or change.*

A capable definition is a characteristic of humans, animals, or ecosystems describing a desirable response to an adverse experience. It could be used for a man-made system. But it may not be in a form that we can measure. We would need to define permanent damage and create a way to measure it. At any rate, we gather some important words that do have actions or potential measures. What is the size of the shock? How long does it take to recover?

We should explore some other references that have been offered. In the US federal government, the 12 February 2013, Presidential Policy Directive 21: Critical Infrastructure Security and Resilience, used resilience 44 times. Although a definition is not coined, many ideas are implied in the statement: "all Federal department and agency heads are responsible for the identification, prioritization, assessment, remediation, and security of their respective internal critical infrastructure that supports primary mission essential functions." We know from here that we as a society need to focus on those systems that deliver a productive and comfortable

Resilient Control Architectures and Power Systems, First Edition.
Edited by Craig Rieger, Ronald Boring, Brian Johnson, and Timothy McJunkin.
© 2022 The Institute of Electrical and Electronics Engineers, Inc. Published 2022 by John Wiley & Sons, Inc.

environment to enable well-being in a modern society, focusing on critical infrastructure. The key words from the Directive are the following:

1. Identification – inventory of what is in a system
2. Prioritization – consideration for what is most important
3. Assessment – what is the status of the health of the system and what could go wrong
4. Security – how is it protected.

In the electric grid, the domain we consider the most in this book, the Electric Power Research Institute (EPRI) focuses on three areas in which to "harden" the grid: prevention, recovery, and survivability. Prevention focuses on positioning the system to mitigate the effects on the system. Recovery in the EPRI view is about deploying crews in natural events. Survivability encompasses keeping key elements of a community up and running for public safety. Common themes to the dictionary version of resilience and the governmental view can be established though they are binned in a different manner.

Resilience and reliability are related topics. One summary that was published in the *Department of Homeland Security's Industrial Control Systems Cyber Emergency Response Team Industrial Control Systems Joint Working Group* March 2016 newsletter, (https://www.researchgate.net/profile/Aaron-Clark-Ginsberg/publication/320456274_What%27s_the_Difference_between_Reliability_and_Resilience/links/59e651230f7e9b13aca3c2ba/Whats-the-Difference-between-Reliability-and-Resilience.pdf), describes some commonality to these two R words with some contrasts. Learning and improving from the past can be related to either but is a focus of resilience. The author points out that, some actions that a utility may take to improve a resilience posture like rolling blackout or preparation for cyberattacks may adversely impact reliability statistics in the short run.

The National Academies of Science have a published work considering the improvement to resilience of the United States power grid [1]. A key recommendation to the US Department of Energy (DOE) is given as follows:

"**Recommendation 1 to DOE**: Improve understanding of customer and societal value associated with increased resilience and review and operationalize metrics for resilience."

DOE responded by initiating efforts in the Grid Modernization Laboratory consortium (GMLC) efforts in 2016 [2]. Metrics considered in the foundational metrics trend toward capturing large outage events that are not always captured in the traditional outage events. The metrics identified separated customers from critical customers due to larger events:

- Cumulative (critical) customer-hours outage
- Cumulative (critical) customer energy demand not served
- Average number (or percentage) of (critical) customers experience an outage during a specified time
- Time to recover
- Cost to recover
- Loss of utility revenue
- Cost of grid damages
- Avoided outage
- Critical services without power (e.g. hospitals, emergency services)
- Critical services without power for more than N hours

These metrics tend to look back in time at the result of events rather than provide an assessment of how prepared the system is. To some degree, the authors of this book seek a transformation to forward looking metrics rather than the rearview mirror. Many recommendations on infrastructure and architecture improvements are mentioned. Running preparedness drills is also a key factor. For the topic of metrics, this source is important for considering the restore phases of resilience.

Another DOE effort in the GMLC program considered metrics derived from the Resilient Control Systems community. The Wikipedia page for Resilient Control Systems Contains information and many additional resources (https://en.wikipedia.org/wiki/Resilient_control_systems).

We now consider a definition that has been developed by a researcher in the Resilient Control domain that puts at least a notionally actionable construction to the context of critical infrastructure. From [3, 4], **A resilient control system is one that maintains state awareness and an accepted level of operational normalcy in response to disturbances, including threats of an unexpected and malicious nature.**

The key elements of this definition are the following:

1. *Maintain state awareness* – a declaration that throughout a disturbance the operators will be able to "see" the state of the system accurately to a level of fidelity that can allow decisions to be made and control actions can be executed.
2. *Acceptable level of normalcy* – the system will perform in a minimal but satisfactory manner.
3. *Disturbance* – sets the understanding that disturbances can occur from many sources and will not all be anticipated or predictable.

A notional manner of depicting the performance of the system has been ubiquitously captured in the Disturbance, Impact, Resilience, and Evaluation (DIRE) curve shown in Figure 11.1. In Figure 11.1, important time frames in the evolution of the effects of a disturbance on the system are indicated as Reconnaissance, Resistance, Response, Recovery, and Restoration. Different abilities of the system play out at different times. Assets with properties of a system that are intrinsic, meaning oppose the disturbance without the need for sensors and control action, are put into the *resist* category. For example, in the bulk electric grid resistance to frequency droop is provided by the accumulative spinning mass referred to as inertia of the large synchronous machines that are generators and large machines of industry. Assets that act based on feedback from sensors but work to react to a disturbance in near real time are able to *respond* and further arrest the drop in performance. Response epic could kick in as early as milliseconds depending on a particular

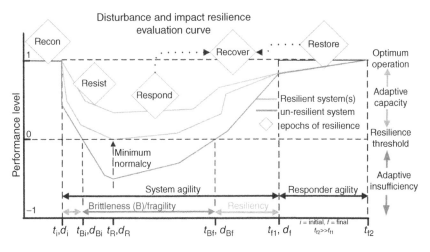

Figure 11.1 The notional form of the measure of performance of a system under a disturbance.

sensor and control system response or could be in minutes or hours if a human response requiring analysis is required. In the electric grid example, generators may have automatic generation control, which measures the frequency of the grid, compares the measurement to the desired frequency, and then applies a controller to a governor that controls the amount of heat or other source of torque to the generator shaft. An example that requires more processing and likely human analysis and decision-making before control would be executed is re-dispatching of generation. Depending on the state of the generator, there would be at least a decision latency and possibly some other startup latency to mobilize the asset. In the *recover* phase, assets that were not needed or held in reserve would now be deployed to move the system back toward optimal or normal mode of operation. This could be recharging a storage asset from remote generators or in the case of hydroelectric power with storage available waiting for the reservoir to naturally fill or drain, in the case where the response was to reduce volume through the turbine and the natural flow into the reservoir increase above the desired level. The final postevent epic is the *restore* time frame. Restoration will typically begin after the system recovers to the highest point possible given the resources that were in place prior to the disturbance. Maintenance or replacement of assets that were utilized in the response needs to occur to ensure any damage or degradation to components or systems is remedied. A failure to restore properly will put the system into a more fragile state for a future event or could even contribute to a failure that would be a disturbance. Restoration may include a decision to make longer-term investments to add or improve the resilience of the system based on lessons learned, if the system failed to respond to the event in the expect manner or if the event indicates that a larger disturbance may need to be accounted for. An example of an event that required major restoration after a failure to maintain functionality would be the tsunami that flooded Fukushima Diaiichi Nuclear Power Plant in 2011, disabling the backup generators and causing a nuclear catastrophe on top of the devastation of the flooding itself. In that case, the disturbance was greater than the design basis for the floodwalls, and more to the point, the generators were placed at a low elevation rather than being placed at a higher elevation. Near misses can drive the needed investment for a more resilient posture. Another "R" word to consider for the postlessons learned period is revamp for longer time frame design adaptation.

We initially skipped over the *recon* phase, which is a continuous process during the operation to assess the state and health of the system. The situational awareness allows for operation in a way that the system is also biased in a position to withstand a disturbance per design. When the system does not conform to the desired margins, the operators and managers of the system would have that information and take actions to put the system in the appropriate state. While operating understanding major contingencies to further act to mitigate when operating margins would be low if one or more components in the system failed. A metric that could be used in design may possibly be operationalized by applying the actual state of the system to calculate the resilience as designed and per the actual conditions.

Who is important? Who are the stakeholders?

Figure 11.2 gives a start of the word "salad" for disciplines that are required and/or have an interest in making a system resilient. Control systems, cybersecurity, and human factors are considered the base legs of a resilient system for modern technology. However, many other elements are required. Ultimately, the designers, operators, security experts, and implementors impact the safety and availability of the critical infrastructures. Regulators working from policy are charged with protecting the interests of the public. Effective policy should be based on the objectives of a system and the risk of failure that impacts the safety and well-being of the public. Vendors and utilities are interested in revenue and efficiency in implementing the resource. How we measure and ultimately value resiliency is influenced and impacted by economics. The point here is that

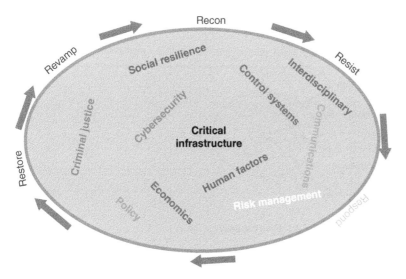

Figure 11.2 Some of the disciplines that are required for design and operation of a resilient system.

there are a lot of areas to be influenced positively through a solid and understandable definition of and measure of resilience.

11.2 Notional Capture of Resilience Adaptive Capacity

A recasting of the resilient control community definition above gives the ability to provide a path to a straightforward measure of resilience in terms of preparation for disturbances. If we ask: how can we measure the magnitude and duration of disturbance that a system can withstand and not impact the system critical functions, or in terms of the definition minimal normalcy. To do this we must do the following:

1. Define objectives of the system in normal conditions and situations where the system is under duress (i.e. minimum normalcy).
2. Make a quantitative assessment of the assets of the system or the portion or portions of the system for which we are interested, possibly an area with a critical load (e.g. hospitals, fire departments, law enforcement).

In a power system, we can consider the availability to adjust real and reactive power. Generally, real power is associated with matching load to generation to maintain frequency. Reactive power is likewise used for voltage regulation. Figure 11.3 shows the concept of controllable assets in the yellow circle and uncontrollable elements as a cloud. Critical loads are represented in orange. The cartoon is used to represent portions of an electrical system (e.g. a microgrid, a distribution feeder, a bus on a transmission system, or even at the scale of a balancing authority). A group of assets that have some commonality with respect to electrical conductivity, ownership, and goals are referred to as an economic unit. An economic unit that has adaptive capacity can be self-sufficient for some period of time and may be able to provide support to connected units. Also, an economic unit where the cloud exceeds the magnitude of the controllable assets needs outside help to maintain functionality. Types of elements are listed in the figure as examples. Note, loads actually can be controlled but at the expense of shedding a paying customer; however, to maintain system stability

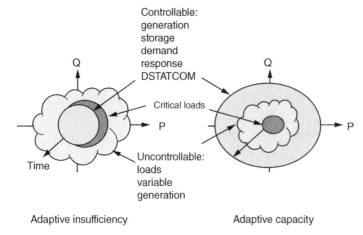

Adaptive insufficiency   Adaptive capacity

Figure 11.3 A cartoon view of adaptive capacity versus uncontrollable aspects. Source: Rieger et al. [5]. ©
2019, Springer Nature.

and supporting critical loads, this would be part of the plan. On DIRE curve, this would be a loss
in performance that is necessary to remain above the minimum normalcy line.

Economic unit grouping examples for a distribution system are shown in Figure 11.4. There is
an aggregation of economic units into larger groups to arrive at scales that have an appreciable
impact at the next level up. To understand how one distribution system affects subtransmission or
transmission, the grouping would be the contribution of all the assets in the distribution system.
The figure serves the purpose of notionally describing the groupings capabilities. However, there
may be restrictions on how much of the capability can be exported based on the available capacity
and health of the connective distribution or transmission system lines.

11.3 Response Epoch: Adaptive Capacity on an Asset-Level Development

The Rs of resilience will be discussed out of order addressing the response epoch first and then
resist. The reason is that we will need to know the size of the disturbance and the latencies for the
elements of the system to know how much time the resist epoch will need to hold the system up
before the assets that rely on those measurements. We can then test the amount of inertia available
to slow the effects of the disturbance to allow the system to withstand the impact prior to the control
response.

We now turn to consider an efficient manner to describe the capabilities of an asset. An asset
could be a generator, battery storage, flexible loads (i.e. demand response loads), etc. One useful
example to show an asset's capability to respond to a disturbance is an energy-limited device, such
as a battery storage system. A battery storage system has power electronics with a reactive component such that it can supply or absorb both real and reactive power. There is a maximum amount
of each described as Pmax (real power in units of Watts) and Qmax (reactive power in units of
VARs). Additionally, in the real-power area, there are energy constraints on a battery that affect
the amount of time the asset will be available if utilized at full capacity and the energy limit is the
power used multiplied by the applicable unit of time. The scientific unit would be Joules, but typically, the power system terminology is watt-hour or megawatt-hour. The important factor here is

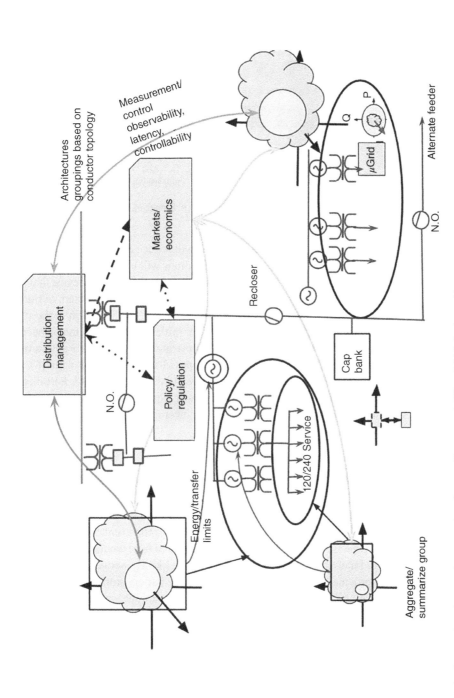

Figure 11.4 Economic unit grouping examples. Source: Rieger et al. [5]. © 2019, Springer Nature.

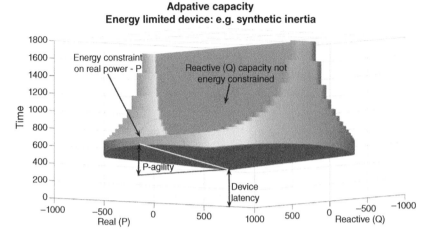

Figure 11.5 Illustration of one asset with energy constraints. Source: Rieger et al. [5]. © 2019, Springer Nature.

that once the energy capacity is used, the real-power output will reduce to ZERO. Reactive power may still be available after that point in time. Other considerations will be the limits on the assets with respect to time – one is device latency. This is the time it takes to activate the device that may be due to a communication latency in a control system, or due to the time of the system or human operator to make a decision to use the capability. A second temporal consideration is the ramp rate of the device. A time constant, rate, or time to move from initial to maximum rate is needed. That ability or limitation will be referred to as agility. The illustration in Figure 11.5 shows these considerations. The specifics of the calculation of general shapes of assets can be found in [6]. The entire shape is important to consider the response to frequency and voltage deviations. Forming the shapes in cylindrical coordinates with the angle related to power factor and the radius as the magnitude if the device is commanded to respond at a particular power factor allows multiple assets to be easily added together to form an aggregated capability.

11.4 Adaptive Capacity on an Aggregated-Level Development

The resulting shape describes a combination of assets that produce a shape that allows for the analysis of the response through time after beginning to respond to the event. Any narrowing of the shape indicates that a response to a disturbance of that length of time would not be able to be supported. The shape would include assets that could add power, like a generator; subtract a load, like demand response; or provide for up and down response, like battery storage. An example of a complex shape that is a combination of spinning inertia, battery storage, spinning reserve generation assets are given in Figure 11.6. The narrow waste early in time is the limit of the response to a disturbance. This is referred to as the sustainable disturbance magnitude (SDM). If and when the maximum response drops below that level, further out in time, the limit is known as the sustainable disturbance duration (SDD). This can be used as a design analysis by asking the question, "what type of asset would the system need to increase the magnitude and/or the duration of sustainability?" The cost of that asset then could be evaluated against the desire to increase resilience. The graphic shows additional information in regard to the resources outside the SDM and SDD that would be available for recovering phase of the event. In all likelihood, recovery will be assisted

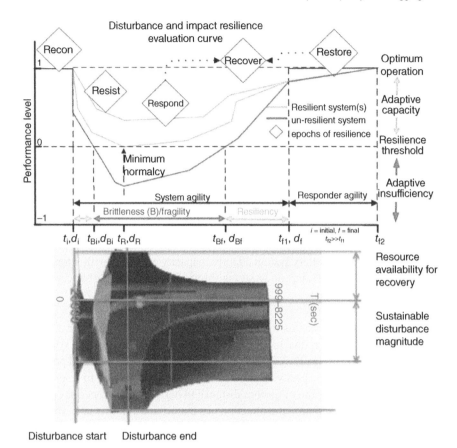

Disturbance start   Disturbance end

Figure 11.6 Juxtaposition of the manifold with corresponding epochs of resilience depicted in the DIRE curve. Source: Rieger et al. [5]. © 2019, Springer Nature.

by longer-term assets or external support to recharge batteries and allow the system to return to the pre-event position.

Example 11.1

A battery system has the capability to produce or absorb power from the grid. If the storage capacity of the batter has a capacity of 20 MJoules (MW-second) and a peak power output of 100 KW in either absorbing or supplying power, then assume that the battery is biased at 50% capacity or 10 MJ and the battery system is currently providing/absorbing 0 W. Assume that the full power can be achieved with a ramp over 10 seconds, and there is a latency of 5 seconds to engage this device. Draw the two-dimensional shape of the adaptive capacity of this device in the real power versus time plane. The shape would be a piecewise linear graph equation:

$$P(t) = \begin{cases} 0\,\text{W for } t < 5 \\ \pm 100\,\text{KW}\dfrac{t-5}{10} \text{ for } 5 < t \leq 15 \\ \pm 100\,\text{KW for } 15 \leq t < \dfrac{10\,\text{MJ} - 100\,\text{KW} * 10\frac{\text{s}}{2}}{100\,\text{KW}} + 5\,\text{s or } 100\,\text{s} \\ 0 \text{ for } t \geq 100\,\text{s} \end{cases}$$

This first time period describes 0 W adaptive capacity during the initial latency waiting for measurement, decision, and transmittal of a command. The second time period describes the ramping. The third part describes the period that the system can run at maximum capacity, which runs from the time the system completes the ramp to the time when the battery is either completely full or depleted. Finally, the fourth period describes the time after which the system can no longer produce due to the energy constraint.

Exercises

1 Draw a graph of the adaptive capacity for a battery system that is initially at 0 W output that has a latency of 2 seconds, a ramp time of 0 seconds to maximum output, a max input/output of 200 KW, and a usable full capacity for storage of 2 MJ with the battery currently at 50% usable capacity.

2 Express the general equation for a battery system adaptive capacity over time with latency Tl, max power +/− Pmax, ramp time of Tramp, and a storage limit of Emax assuming battery starts at 0 output and storage at 50% of Emax.

Note that battery systems or storage systems in general will not respond the same for a given state of charge or energy stored. However, this does serve as a starting point for what the system's contribution to respond to a disturbance is. If we consider a system that is not at a natural point of symmetry (e.g. 0 output and 50% energy capacity) and we consider the asset's ability to respond to a disturbance, we really need to consider the minimum distance to maximum in power or energy, since we do not have the luxury of anticipating the direction of a disturbance. In this case, you can consider the limiting condition in reactive and real power constraints. If the operating point for the asset is not neutral, there may be instances where considering an asymmetric form of the adaptive capacity may be useful.

When there are multiple assets that can be grouped, then the capacity can be a simple addition of the capabilities over time. Sampling the value of the assets over time and simply adding that capability is the means of doing this in two dimensions. The full s-plane metrics perform this aggregation by considering the radial, power factor angle, and magnitude on an asset over time. For this textbook, we will simply consider the real power plane. The summation of individual assets sampled over time can provide that aggregation with the following:

$$P_{tot}(nT) = \sum_i P_i(nT)$$

where T is the sampling time.

Exercises

3 Combine result of Example 1 and Exercise 1 to form an aggregated adaptive capacity, using the aggregation method above. Hint in Figure 11.7 – positive side of adaptive capacity shown.

4 If we neglect the first four seconds of the disturbance as being covered by the inertia of the system, how long could the system withstand a disturbance of magnitude 200 kW?

5 In the graphical hint for Exercise 3, there is a gap both at the frontend due to latencies (hopefully covered by sufficient inertia) and when the asset from Exercise 1 is exhausted. If we experienced

Capacity shown:

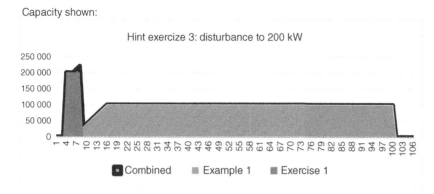

Figure 11.7   Graph of adaptive capacity combining assets from Example 1 and Exercise 1.

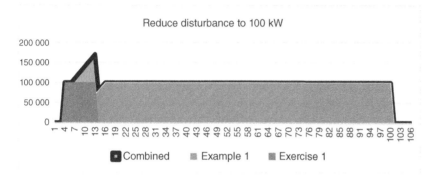

Figure 11.8   Graph showing the effect on the adaptive capacity if disturbance is reduced to 100 kW.

a disturbance of only 100 kW, the first asset would use the energy less aggressively and would be available for a longer duration shown below. How long could the disturbance of that magnitude be supported? Once again neglected the initial gap. See Figure 11.8.

What could be done to "fill the gap" to allow a disturbance of 100 kW magnitude to be supported?

11.5   Cybersecurity Considerations

The study of resilient control systems involves cyber-physical security due to the fact that many if not most control systems in the modern world have at least begun to include digital components that are networked together to provide operators, managers, and other stakeholders up-to-date information about the system. Thus, we need to at least consider some of the possible effects of malfunctions or compromise of those digital components that are tied directly or indirectly to physical impacts. Figure 11.9 looks at two possible impacts. The first shown with the cartoon on the left illustrates the effect of taking a controllable asset and compromising it such that a portion or the entirety of the asset is turned from a controllable asset to an uncontrollable component.

One may speculate, "what's the worst thing an adversary could do?" That may be to disable the device, deleting the contribution to the system. Another might be to make the device do the exact opposite of the intended function (i.e. changing the sign from a negative to positive on a feedback loop). For those familiar with control systems, this would be clear (i.e. turn a finely tuned compensator into an oscillator hell-bent on tearing the system apart).

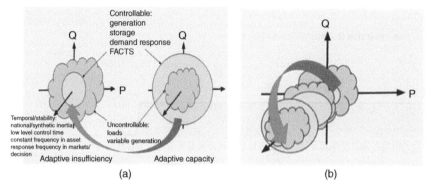

Figure 11.9 Disabling the control of a device will decrease the controllable assets and potentially increase the uncontrollable region (a). Delaying a control signal may push the assets response out later in time (b). Source: Rieger et al. [5]. © 2019, Springer Nature.

For those starting their exploration, imagine your car steering wheel produces a turn to the left when you are attempting to turn to the right to avoid an oncoming truck. The second instance is more subtle. If commands to use the asset are given through a network and those commands are delayed via a "denial-of-service" attack (see the chapters on cybersecurity). In the case on the right, an asset that would normally be available early after an event now has a delayed response. The ramifications of either of these scenarios can be severe. A pure latency can alter stability posture due to a shift in the phase margin. A system that had a gain of less than one when the phase of a Bode plot analysis crossed 180° is stable. However, a delay such as one from a networked control system will advance the phase as a function of frequency. Enough delay introduced by network traffic could cause the phase to be 180° when the gain is greater than one causing the system to oscillate out of control [7]. Other possible control issues could be introduced through cybersecurity channels that would produce unpredicted results, such as modification of gains and set points. Those combined with other attacks that make the human operators unaware of the malfunction or spoof the human such that they make a mistake are among the concerns that will cause assets to act in unexpected and possibly damaging ways. Examples, such as STUXNET and the Ukraine Grid attacks, demonstrate the concern. Analysis of what could go wrong followed by consideration of the physical effects on the tools that are used to respond to disturbances have now become a critical consideration. The metrics shown here provide a method to objectively consider possible ramifications and put sufficient efforts into protecting the system from the potential misuse of system functions.

The illustration of a possible distribution system control network is given in Figure 11.10, as an example of a distributed system where control actions can be disturbed by a network. Communication problems such as disconnects and inconsistent latencies could cause issues in performance and even the stability of the system. Those issues may have other causes than malicious cybersecurity attacks but anything that relies on digital devices could have a cause related to intentional human actions. Data integrity attacks (e.g. man in the middle) can also be a matter of concern causing the system or the humans operating the system to lose or be provided incorrect information about the system. Information in any of the control or feedback channels could be a target of an adversary. Consideration of these issues should logically drive designers to architectures that would push low-level time-sensitive control loops down to levels that do not require complex communication. Higher levels of supervisory control and optimization still require such communications but can be constructed in a manner that does not put the stability of the system at risk due to the needed network (see Chapter 17 for more information).

A final cyber security example, in Figure 11.11, shows a representation of an aggregation of assets that provide enough adaptive capacity in the left volume. In the center image, the control of an

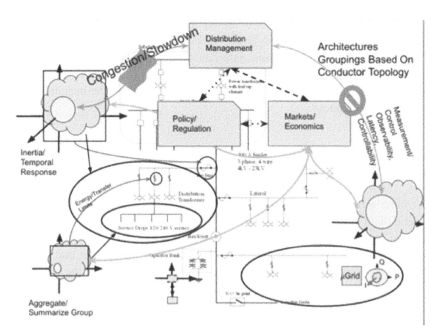

Figure 11.10 Illustration of a possible communication and control architecture and the potential effects of communication issues with a cybersecurity connection. Source: Rieger et al. [5]. © 2019, Springer Nature.

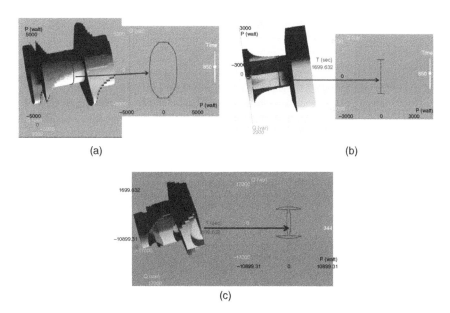

Figure 11.11 The illustration in (a) is the baseline system with sufficient adaptive capacity, in (b) a set of assets with a common vulnerability is disabled, and in (c) an example where an increased latency pushes the response of a significant asset out in time. Both examples reduce the duration of disturbance the system can withstand.

important asset is removed due to failure or attack of an important asset or set of assets, such as battery storage systems. Here the adaptive capacity in the real power dimension drops to zero very soon after the disturbance is initiated. The right side of the figure shows a delay of those assets that push the availability of those assets later in time. The impact is similar in the reduction of adaptive capacity an early time in the progression of the disturbance. This shows that the magnitude of disturbance would be markedly reduced.

11.6   Consideration of Resist Epoch (Inertia)

Now we consider the initial gap mentioned above. The following hands-on examples evaluate the resilience of a microgrid that is initially islanded, meaning it is not supported by an external grid. The resist phase provided by inertia must be able to support the system until the primary frequency response can take effect. The first and last step will be how we consider the support that inertia gives in the resist phase of an event. The first step is to calculate the kinetic energy of the spinning machine and then evaluate the amount that can be used. The kinetic energy of a synchronous generator is a function of the moment of inertia (J) of the turbine, shaft and generator core, and the rotational frequency (ω):

$$K = \omega^2 J$$

We can relate that to the frequency of the alternating current of the grid in the typical units of frequency in hertz and the number of poles of the generator as follows:

$$K = \left(\frac{2\pi f}{Np/2} \right)^2 J$$

Now considering the allowed frequency range before generators and other sources of energy or loads may trip as $f_o \mp f_l$, where f_o is the nominal frequency (60 Hz for the North American electric grid) and f_l is allowed variation in frequency before the system gets to trip settings that will disable generators to protect them from damage and further exacerbate the problem. Two values can be found for the minimum and maximum kinetic energy. From there we can show the amount of energy the spinning machines can support and determine an approximation for the amount of time given the magnitude of a disturbance:

$$K_m = [2\pi(f_o - f_l)/Np/2]^2 J$$

$$K_M = [2\pi(f_l + f_o)/Np/2]^2 J$$

where K_m and K_M are the minimum and maximum kinetic energy of the synchronous machines. Given the microgrid is operating at the nominal frequency, the amount of kinetic energy available K_a to support the resistance phase is dependent on the direction of the power disturbance, Pd

$$K_a = \begin{cases} K(f_o) - K_m & \text{for } Pd < 0 \\ K_M - K(f_o) & \text{for } Pd > 0 \end{cases}$$

An approximation of the time duration of the support for inertia, T_a is

$$T_a = \frac{K_a}{|P_D|}.$$

The industry often describes the moment of inertia as an inertial constant, H, that is the ratio of the kinetic energy of the machine at the nominal rotational speed and the rating of the machine in million volt-amps (MVA).

$$H = \frac{J\omega_0^2}{2} / \text{Rating}_{\text{MVA}}.$$

Example

A plant is rated at 100 KVA, has a rotational speed of 20.9 rad/sec, and an H given as four seconds. What is the moment of inertia J of the plant?

$$J = \frac{2H\text{Rating}_{\text{MVA}}}{\omega_0^2} = 2 * 4 * \frac{100(10^3)}{20.9^2} = 0.23(10^3)$$

If the nominal grid frequency of 60 Hz can vary by up to 2 Hz in the system before tripping and the gap to other responses is three seconds that can be supported by the inertia of this asset, and if this is a 36-pole generator, what is the disturbance magnitude that can be supported?

$$K(f_o) = \left(\frac{2\pi(60)}{18}\right)^{20.23 = 100.9 \text{ kW}}$$

$$K_m = \left(\frac{2\pi(58)}{18}\right)^{20.23 = 94.28 \text{ kW}}$$

$$K_M = \left(\frac{2\pi(62)}{18}\right)^{20.23 = 107.73 \text{ kW}}$$

$$K_a = \begin{cases} 100.9 - 94.28 = 6.72 \text{ kW} & \text{for } Pd < 0 \\ 107.73 - 100.9 = 6.83 \text{ kW} & \text{for } Pd > 0 \end{cases}.$$

Taking the minimum, which is in the droop, negative frequency direction:

$$|Pd| = \frac{6.72 \text{ kWs}}{3 \text{ s}} = 2.24 \text{ kW}.$$

For the examples and exercises in this chapter, we would have an insufficient amount of inertia for the assets to respond to the system since $Pd < 200$ kW. The system would need approximately 100 times more inertia or a response from the battery system that had a much shorter latency to activation. The student should consider which might be more realistic?

11.7 Consideration of Recover and Restore Epochs

This chapter did not delve deeply into the recovery epoch. This epoch is really the phase where the system is restored to a bias that where the system can be in position to take another impact. Once the system has ridden through the disturbance, energy is required to bring several things back up to normal. For example, we may have the grid frequency that has diverged away from normal. This requires energy to push the frequency back to normal. We also need to bring storage back to a neutral bias. This can only happen once the cause of the disturbance is resolved and additional generation could be dispatched to adjust the posture of the system. The amount of energy used in the response can be calculated by dividing the power available for recovery by the amount of time required to recover.

Restoration involves the assessment maintenance and repair of equipment that may have been used in an unusual manner. Also, in the event that the system was unable to maintain a predefined minimum normalcy, this epoch would cover the steps needed to storage the system to that stage. Examples of contributions may be black start abilities of some assets that would be required to restart the critical infrastructure during a wind scale failure of the grid.

An open source version of the metrics tool is now available [8].

Thoughtful Questions to Ensure Comprehension

To check your comprehension of the chapter, challenge yourself to answer the following.

1 What are several sources for definitions of resilience? How do they differ and what do they have in common?

2 What are the epochs of resilience? Describe in your own words what occurs during each epoch.

3 What are the potential physical world effects of cybersecurity to a control system? Would there be enough inertia in the system if the latency of the battery system in the exercises were increased to 10 seconds? What is the maximum amount of latency that the inertia in the system could support?

References

1 National Academies of Sciences, Engineering, and Medicine (2017). *Enhancing the Resilience of the Nation's Electricity System*. Washington, DC: The National Academies Press https://doi.org/10.17226/24836.

2 GMLC (2016). Foundational metrics analysis. Grid Modernization Laboratory Consortium. https://gridmod.labworks.org/projects/foundationalmetrics-analysis (accessed 27 February 2017).

3 Rieger, C.G. (2014). Resilient control systems Practical metrics basis for defining mission impact. *2014 7th International Symposium on Resilient Control Systems (ISRCS)* (August 2014), 1–10.

4 Rieger, C.G., Gertman, D.I., and McQueen, M.A. (2009). Resilient control systems: Next generation design research. *2009 2nd Conference on Human System Interactions, Catania*, 632–636. doi: https://doi.org/10.1109/HSI.2009.5091051. http://ieeexplore.ieee.org/stamp/stamp.jsp?tp=&arnumber=5091051&isnumber=5090940.

5 Rieger, C., Ray, I., Haney, M.A., and Zhu, Q. (2019). *Industrial Control Systems Security and Resiliency*. Springer Nature.

6 McJunkin, T.R. and Rieger, C.G. (2017). Electricity distribution system resilient control system metrics. *2017 Resilience Week (RWS)*, Wilmington, DE, 103–112. doi: https://doi.org/10.1109/RWEEK.2017.8088656. http://ieeexplore.ieee.org/stamp/stamp.jsp?tp=&arnumber=8088656&isnumber=8088637.

7 Ramakrishnan, K. and Swarnalakshmi, N. (2018). Impact of gain and phase margins on stability of networked micro-grid frequency control system. *2018 4th International Conference on Electrical Energy Systems (ICEES)*, Chennai, 126–132. doi: https://doi.org/10.1109/ICEES.2018.8442412.

8 Open source PowDDeR Tool on Git Hub (2020). https://github.com/IdahoLabCuttingBoard/PowDDeR (accessed 31 August 2021).

Part V

Resilience Application

Part V, "**Resilience Application,**" provides a resilient control system perspective for application of disciplinary contributions, with the intent to evolve from multidisciplinary to interdisciplinary. A system application gaming environment provides a thoughtful means for students to apply these considerations.

Chapter 12: What is with a game in a resilience class? Well, there are many reasons to use a game for education. One, it is something for you to look forward to as part of a class. An event where you compete and cooperate with your fellow students can reveal more about resilience than reading papers, doing homework, or studying for an examination. The only thing that you might learn more from is creating your own game or really using a project development to dig into the understanding of a subject. The Grid Game has evolved from a simple swing equation simulation of the real power aspects of a microgrid to a multiplayer game that enables players to experience the impacts of unexpected events. As resilience is multidisciplined as you have been learning in this textbook, a simulated game gives you a chance to think about strategy and improvements to the human interaction with a system.

Chapter 13: Modern power grids rely heavily on TCP/IP networks to monitor and control physical processes. This reliance opens the door to potentially new and powerful cyberattacks against them. In this chapter, we introduce technologies that are used to operate the power grid and security challenges facing the power grid, present previous attacks, and discuss research efforts to improve the security and resiliency of the grid.

Chapter 14: In this chapter, we introduce methods to address resiliency issues for control systems. The main challenge for control systems is its cyber-physical system nature that strongly couples the cyber systems with physical layer dynamics. Hence, the resiliency issues for control systems need to be addressed by integrating cyber resiliency with physical layer resiliency. We introduce frameworks utilizing a games-in-games paradigm that can provide a holistic view of the control system resiliency and enable an optimal cross-layer and cross-stage design at the planning, operation, and recovery stage of control systems. The control systems are often large-scale systems in industrial application and critical infrastructures. Decentralized control of such systems is indispensable. We extended the resiliency framework to address distributed and collaborative resiliency among decentralized control agents.

Chapter 15: Technological advancements have resulted in highly critical infrastructure, which has increased the infrastructure's attack surface and made them more vulnerable to cyberattacks.

Resilient Control Architectures and Power Systems, First Edition.
Edited by Craig Rieger, Ronald Boring, Brian Johnson, and Timothy McJunkin.
© 2022 The Institute of Electrical and Electronics Engineers, Inc. Published 2022 by John Wiley & Sons, Inc.

The constantly evolving threat landscape and sophisticated attack vectors boasts intelligent and adaptive threat actors that can surpass traditional engineered and deployed defenses. A skilled cybersecurity workforce is essential; furthermore, there is an immediate need for anticipatory defense measures that reflect the adaptive and dynamic nature of the threat actors. Developing anticipatory cyber strategies requires understanding the human aspects of cyberattacks: how adversaries organize, strategize, adapt, and function effectively, and how defenders secure grids and make effective decisions in cyber defense and system operation when experiencing cyber-attacks. One effective mechanism to train the future workforce in this space is by gamifying cybersecurity.

12

Introducing the Grid Game

Timothy McJunkin

Energy and Environment Science and Technology, Idaho National Laboratory, Idaho Falls, ID, USA

Objectives

The objectives of this chapter are to explain the installation, instructions, and objectives of the Grid Game. The chapter also discusses some of the fundamental concepts that underlie the game. Players are encouraged to offer feedback to help improve the Grid Game.

12.1 Introduction

You may or may not be aware how the electrons are pushed from the coal, nuclear, hydro, solar, or wind driven generators to your light switch. The Grid Game is an entertaining way to experience part of the process of controlling a small electric grid, sometime referred to as a microgrid, demonstrating the considerations of a resilient control system (Figure 12.1). Maintaining the desired frequency of the alternating current of the grid is the focal point of this game. As generators are spun by water, steam, or wind, the magnetic poles of the generator induce the electric currents that change direction at a rate proportional to speed of the machine. The player's goal is to sell as much power as possible by growing their customer base. The Grid Game allows the player to "recruit" more customers and grow their business. If they grow too many customers without building up the appropriate generation, they will likely not be able to maintain their microgrids stability. The Grid Game has many interesting features, including the potential for a market where players can buy and sell power to each other. It also has the ability to show some possible hazards of our reliance on computer systems for control, namely the potential for so-called "cyber-attacks" to have negative impacts on systems. Players are given the opportunity to protect themselves from or react to remedy the attacks during the game. You can find out more details about the workings and past uses of the Grid Game in the 2015 conference paper by McJunkin et al. [1].

12.2 Download/Install the Game

The installer is a standard Microsoft Windows installer (http://www.gridgame.org). It has been tested on Windows 7 and 10 and earlier. Source code is also available. Contact Timmcj@gmail.com for more information.

Resilient Control Architectures and Power Systems, First Edition.
Edited by Craig Rieger, Ronald Boring, Brian Johnson, and Timothy McJunkin.
© 2022 The Institute of Electrical and Electronics Engineers, Inc. Published 2022 by John Wiley & Sons, Inc.

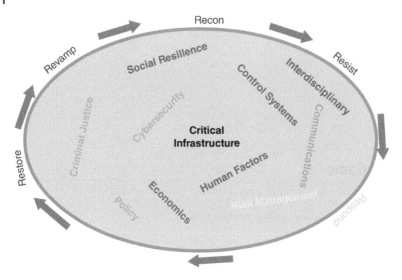

Figure 12.1 The ecosystem of a resilient control system.

12.3 Play the Grid Game

The following are screen captures from the tutorial imbedded within the game for your reference. Figures 12.2–12.15 provide the instructions to operate the game.

First instruction: The primary goal of the game is to stabilize the frequency of your microgrid (see Figure 12.2). This is accomplished by matching the power generated to the amount consumed by the loads.

The grid operator has several tools at their disposal to stabilize the grid in this game. The first is a storage source (e.g. battery, pumped storage) that can be used to balance the net power level by injecting power into or absorbing power from the grid into/out of the storage (Figure 12.3).

The players control the power put onto the grid with this (see knob in Figure 12.4) input to the power inverter connected to the battery. Positive values will add power to the grid, taking it

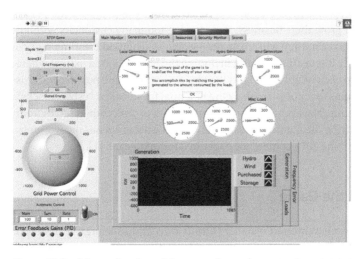

Figure 12.2 Primary function of the controller and operator is maintaining frequency within acceptable limits. Source: Grid Game (www.gridgame.org).

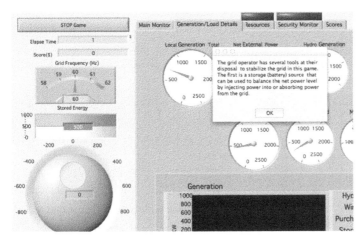

Figure 12.3 Tools for the operator. Source: Grid Game (www.gridgame.org).

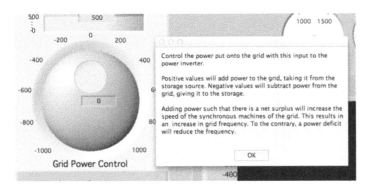

Figure 12.4 The battery system control knob. Source: Grid Game (www.gridgame.org).

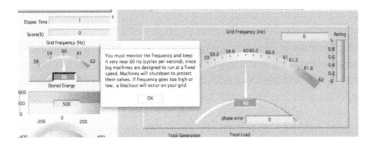

Figure 12.5 Frequency monitor. Source: Grid Game (www.gridgame.org).

from the storage source. Negative values will subtract power from the grid, giving it to the storage. Adding power such that there is a net surplus will increase the speed of the synchronous/spinning machines of the grid. This results in an acceleration of the machines' speed of rotation and an increase in grid frequency. To the contrary, a power deficit will reduce the frequency.

Monitor the frequency, shown in Figure 12.5, and keep it very near 60 Hz (cycles per second), since big machines are designed to run at a fixed speed. Machines will shut down to protect themselves. If frequency goes too high or low, a blackout will occur on your grid and your customers will not be happy.

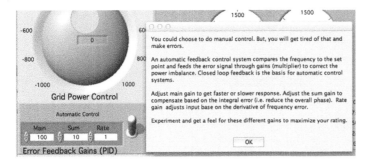

Figure 12.6 Choose your gains. Source: Grid Game (www.gridgame.org).

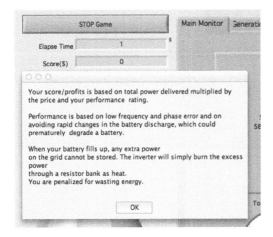

Figure 12.7 Bank some points. Source: Grid Game (www.gridgame.org).

You could choose to do manual control, but you will get tired of that and make errors. An automatic feedback control system compares the frequency to the set point and feeds the error signal through gains (multiplier) to correct the power imbalance. Closed loop feedback is the basis for automatic control systems. Adjust main gain to get faster or slower response. Adjust the sum gain to compensate based on the integral error (i.e. reduce the overall phase), see (Figure 12.6). Rate gain adjusts the input based on the derivative of frequency error. Experiment and get a feel for how these different gains change the response of the control system and try to maximize your rating.

Your score/profit is based on total power delivered multiplied by the price and your performance rating (Figure 12.7). Performance is based on low frequency and phase error and on avoiding rapid changes in the battery discharge, which could prematurely degrade a battery. When your battery fills up, any extra power on the grid cannot be stored. The inverter will simply burn the excess power through a resistor bank as heat. You are penalized financially for wasting energy.

Monitor the detailed history versus time in the generation/load details tab, when you have time to analyze things (Figure 12.8). You may find patterns that inform your strategy by understanding how loads and generation vary.

In the screen shown in Figure 12.9, you have the opportunity to manage (buy) grid assets. As your score improves, add more standard (residential) customers, and more generation units to grow your grid. The more generation and customers you have, the more energy you sell and the more cash you earn. With increase in storage, you may be in position to sell power to other grids.

When other microgrids put power on the market, this section, Figure 12.10, will provide an opportunity to purchase power for an offered price and duration. The amount of power will be

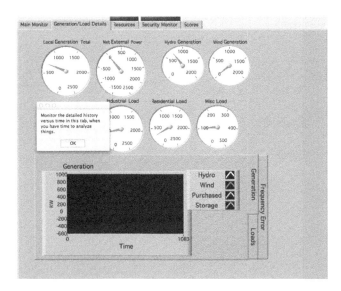

Figure 12.8 Details of load and generation. Source: Grid Game (www.gridgame.org).

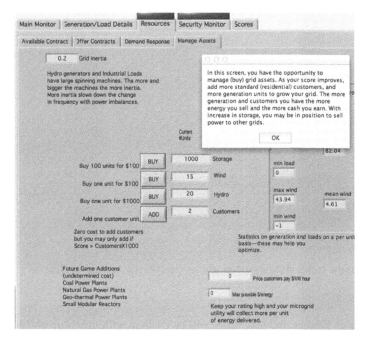

Figure 12.9 Purchase assets and grow your customers. Source: Grid Game (www.gridgame.org).

added to your power balance for the duration listed. If their grids blackout during the transaction, your grid will lose that power source but will be compensated with a refund of the full contract. The details of purchased contracts are shown on this screen as well (see Figure 12.11).

The offer contracts page (Figure 12.12) provides the interface to sell contracts to other grids for net income to your grid. Once another grid purchases your contract, the specified amount of power

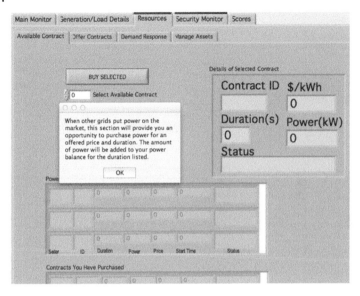

Figure 12.10 Buy contracts. Source: Grid Game (www.gridgame.org).

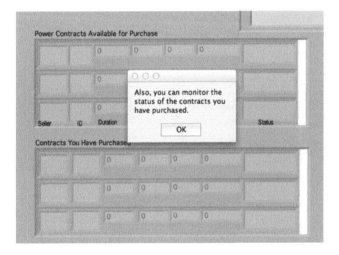

Figure 12.11 Also, you can monitor the status of the contracts you have purchased. Source: Grid Game (www.gridgame.org).

will be subtracted from your power balance and your score will be credited with the value of the contract.

Industrial loads are customers that consume a fixed amount of power, making them easier to account for since their load is constant. These customers have made an agreement to allow you to turn off their power for a specified duration in return for compensation, Figure 12.13. Recruiting industrial customers is another way to increase your score by selling more power. Negotiate the terms of your contract and recruitment cost.

Your grid is a cyber-physical system. Software and firmware have a possibility of containing vulnerabilities. This screen shows you activities that may be abnormal and allows you to purchase

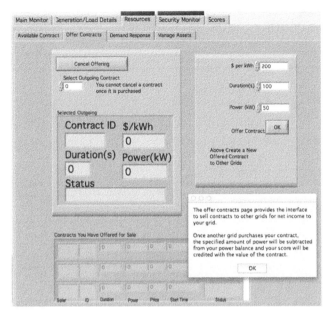

Figure 12.12 Try to sell excess power. Source: Grid Game (www.gridgame.org).

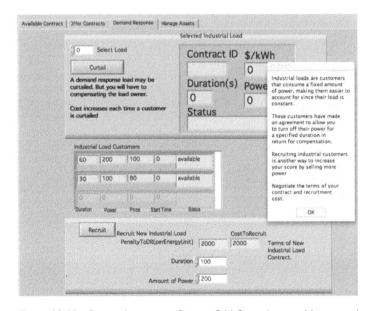

Figure 12.13 Demand response. Source: Grid Game (www.gridgame.org).

services to prevent some and remedy others, Figure 12.14. Beware the Blue Frog…it may make you go open loop!

It is now time to play. Click OK as in Figure 12.15. Grow your grid and earn more revenue. If your friends or foes join the game on another grid client, you can see who can best manage the grid. Decide whether to compete or cooperate, or a little bit of both.

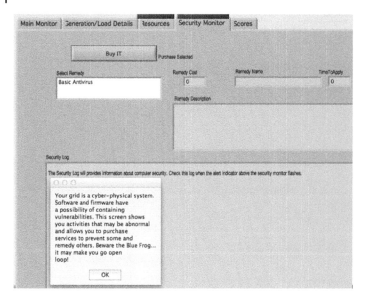

Figure 12.14 Protect yourself from malicious hackers. Source: Grid Game (www.gridgame.org).

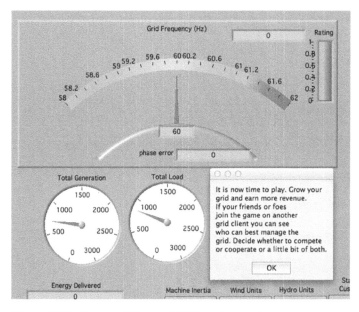

Figure 12.15 Let's play!. Source: Grid Game (www.gridgame.org).

12.4 Fundamentals

The grid is more than the electrical delivery mechanism of these abstract things called generators. The grid can be seen as a distributed electromechanical system, made up of large spinning machines like those in the hydroelectric generators of Hoover Dam (see Figure 12.16). Those machines are turned by the force of the water pushed by gravity through turbines. The turbine torque turns a generator, which rotates wires through an electromagnetic field to induce voltage

Figure 12.16 The electric grid is made up of generators that are large spinning machines tied synchronously to the grid such that their rotational speed is proportional to the alternating current wave voltage and current wave forms. Source: Jon Sullivan, https://commons.wikimedia.org/wiki/File:Hoover_Dam's_generators2.jpg, Public Domain.

and current into the grid that the loads of the system utilize. The machines are designed to run in a narrow band of rotational speeds, which produce an equivalent narrow band of frequencies in the grid, in North America this is 60 Hz. The change speed of the large machines is determined by the balance of generation with loads and losses in the network. If there is more generation, the frequency goes up. If there is more load, the frequency goes down. The frequency must be controlled to keep it in a safe operating range. A detailed examination of this can be found in Grainger's 1994 book [2], but is simplified here as the swing equation for a single pair of magnetic pole generator (most have many more, which means the machines turn slower than 60 Hz):

$$Pg - Pl = J\,\omega\left(\frac{d\omega}{dt}\right)$$

where, J is the moment of inertia and ω is the rotational velocity of the machine, which is the same as the electricity waveforms. An important factor in the equation is that the amount of inertia in the system affects the rate that the system changes. If J is big, the speed changes more slowly with a difference in Pg and Pl.

The grid is made up of different loads and generation sources as shown in Figure 12.17. As loads and generation vary, so will the frequency. For the purposes of the game, you have a battery system that you can control either manually or with an automatic control system.

A control system compares a desired set point to a measured value of that parameter. A gain is applied to amplify the error to control the amount of response. In this case, frequency is measured compared to 60 Hz and a control signal based on the error, multiplied by a set of gains, is sent to the battery.

A simple control system loop is shown in Figure 12.18. Again, Kp, is applied to the error signal indicating the amount of power to take from the battery or put into the battery from the grid. A large Kp would lead to a large response for a small error. You could think of this in terms of a drive that is said to have a lead foot. If you set the gain too large, the system will overshoot, have oscillations, and

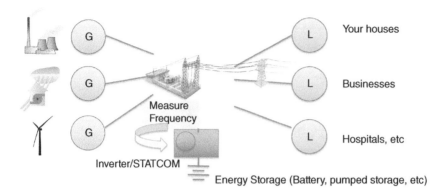

Figure 12.17 This illustration depicts the generators and loads on a system that are connected to the transmission and distribution networks of the grid. A battery system is supplied in the game to allow a controllable device that can be used to regulate frequency.

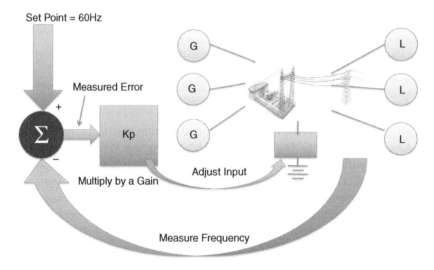

Figure 12.18 The parts of a control system.

possibly go unstable such that protection circuits shut off your generators and make your grid go dark. Too small of a gain and the system may hit a frequency limit before the response is effective.

Kp is the proportional response gain, named because it is applied to the proportional error of the frequency directly. Other types of compensation can be designed into the controller. In the game you are given two other gains: one that responds to the integrated error, referred to as the integral gain, and one that is applied to the derivative rate of change of the error, called the derivative gain. Together they form a common proportional, integral, derivative controller, or PID. We will not discuss theory in this chapter, but the game is here for you to play with those gains to see how they improve or hurt the stability of the grid (Figure 12.19).

12.5 Evaluate the Grid Game and Players (Yourself and Others)

Let us think about the operator for a minute. For details in that topic, please refer to the Human Factors chapters. We have taken the manual control tedium away from the operator. Do we now

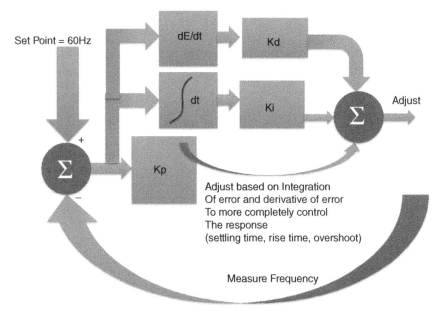

Figure 12.19   A PID control system. Source: McJunkin et al. [1]. ©2015, American Society for Engineering Education.

save some money – **Fire the Operator**? Well that is one option, but the answer is probably NO! The low-level controls that need to respond quickly and persistently – a very monotonous job for a person – should be done by a machine, but allow people to make the higher-level decisions, and monitor, predict, and respond to events. What did you have to do in the game? The answer is monitor energy storage because low-level automatic control only works if controls are not to their limits (saturation – out of stored energy). Humans can also consider larger factors to continually improve the design a more resilient system! The rest of this book helps you consider how that might be done in this course. What did you do in playing it to improve its resilience? What could you do if you changed the design of the system, from human-machine interface to more sophisticated control? What other tasks would you like to automate if you could? Are there negative potential consequences in doing so?

What are the potential problems with digitizing the control system? You might experience some of those if the Red Team attacks your:

– Finances – hijack accounts
– Data integrity from sensors
– The control system itself

A game server provides a communication channel between microgrids to enable energy trades, but opens up vulnerabilities.

Connecting microgrids would

– Provide flexibility and the possibility to support each other, but
– Open up physical vulnerabilities when connections are needed and can be disrupted.

12.6 Play Together

An interactive scoreboard is available via a server. If you want to set a server up for your class or your group of friends, contact the author of this book.

12.7 Improve the Game

There are lots to add to the game and lots of uses for the game. Some uses of the game from the past are included in references [3–8]. The source code can also be made available to change or advance the game. It was written in National Instruments LabVIEW.

Thoughtful Questions to Ensure Comprehension

1 What items in the ecosystem of resilient systems (Figure 12.1) did you or could you explore with the game?

2 What gain settings seemed to make the most difference to the response of the frequency to variations in load and generation? Could you make the system go unstable? Try again for fun!

3 Was the amount of inertia in the hydroelectric assets important to the stability of the system? How? Why?

4 What would you do to improve the game? This could be a nice project. If you are serious about doing this, you can contact the author to discuss getting code, etc.

References

1 McJunkin, T.R., Rieger, C.G., Johnson, B.K. et al. (2015). Interdisciplinary education through "Edu-tainment" electric grid resilient control systems course. 2015 ASEE Annual Conference and Exposition. Seattle, Washington, ASEE Conferences (June 2015). https://peer.asee.org/24349 (accessed 1 September 2021).

2 Grainger, J.J. (1994). *Power System Analysis*. McGraw Hill.

3 Rege, A., Parker, E., and McJunkin, T. (2017). Using a critical infrastructure game to provide realistic observation of the human in the loop by criminal justice students. 2017 Resilience Week (RWS) (September 2017), 154–160. https://doi.org/10.1109/RWEEK.2017.8088665

4 Rege, A., Biswas, S., Bai, L. et al. (2017). Using simulators to assess knowledge and behavior of "novice" operators of critical infrastructure under cyberattack events. 2017 Resilience Week (RWS), (September 2017), 50–56. https://doi.org/10.1109/ RWEEK.2017.8088647.

5 Chiou, F., Fry, R., Gentle, J.P. et al. (2017). 3D model of dispatchable renewable energy for smart microgrid power system. 2017 IEEE Conference on Technologies for Sustainability (SusTech), (November 2017), 1–7. https://doi.org/10.1109/SusTech.2017.8333537

6 Chiou, F., Gentle, J.P., and McJunkin, T.R. (2016). Dispatchable renewable energy model for micro- grid power system. IEEE Conference on Technologies for Sustainability (October 2016).

7 Davis, N., Johnson, B.K., McJunkin, T.R. et al. (2016). Dispatch control with PEV charging and renewables for multiplayer game application. IEEE Conference on Technologies for Sustainability, Phoenix, Arizona (October 2016).

8 McJunkin, T.R., Rieger, C.G., Rege, A. et al. (2016). Multidisciplinary game-based approach for generating student enthusiasm for addressing critical infrastructure challenges. 2016 ASEE Annual Conference and Exposition, New Orleans, Louisiana: ASEE Conferences (June 2016). https://peer.asee.org/25763 (accessed 1 September 2021).

13

Cybersecurity and Resilience for the Power Grid

Xi Qin[1], Kelvin Mai[2], Neil Ortiz[1], Keerthi Koneru, and Alvaro A. Cardenas[1]

[1]University of California, Computer Science and Engineering, Santa Cruz, CA, USA
[2]University of Texas, Computer Science, Dallas, TX, USA

Objectives

The objectives of this chapter are to help the reader identify the main components and the operation technologies used to monitor and control the power grid infrastructure and learn the scope of reliability and resiliency of its operation, to enumerate prominent attacks that recently threatened power grid operation, causing different levels of interruption, and demonstrate how adversaries could exploit its legitimate vulnerabilities, and to review state-of-art research results and techniques to mitigate security and privacy problems in both the classic power grid and the modern smart grid.

13.1 Introduction

Advances in embedded systems and communication networks have allowed us to modernize and automate the operation of the power grid; however, this increased reliance in computing systems also opens the door to potentially new and powerful cyberattacks against the grid. In this chapter, we introduce technologies that are used to operate the power grid and security challenges facing the power grid and discuss research efforts to improve the security and resiliency of the grid.

This chapter is organized as follows. In Section 13.2, we describe some of the technologies used by power grids with the focus on how adversaries could use these same protocols as a means to carry out their attacks against the power grid. In Section 13.3, we present the security challenges of the grid and discuss details of previous attacks. Finally, in Section 13.4, we present research efforts at the bulk and distribution systems and the state-of-the-art solutions.

13.2 Operation Technologies in the Power Grid

As stated in previous chapters, the power grid is generally divided in three major parts as illustrated by Figure 13.1: (i) generation systems that can include hydro, coal, wind, or nuclear power generators; (ii) transmission systems that carry electricity at high voltages (to minimize currents and, therefore, minimize energy losses in the transmission wires) over large geographical areas;

Resilient Control Architectures and Power Systems, First Edition.
Edited by Craig Rieger, Ronald Boring, Brian Johnson, and Timothy McJunkin.
© 2022 The Institute of Electrical and Electronics Engineers, Inc. Published 2022 by John Wiley & Sons, Inc.

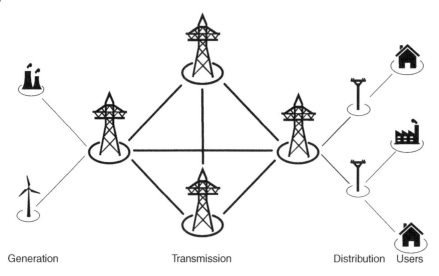

Figure 13.1 Power grid anatomy.

and (iii) distribution systems that carry electricity to local residential and business customers. The first two components combined are commonly referred to as the "bulk power system."

In power grids, the focus of its operation is to ensure the reliability of the system. The operation depends on the information that is gathered about the state of the grid. The operator needs to be aware of the system's conditions to make correct control decisions. That awareness is not possible without proper information.

In that sense, the management of the power grid is now reliant on the communication infrastructure to collect and share data. Ideally, all sections of the network must be able to communicate with one another as a unique system to ensure that the right information gets to the right place at the right time.

Modern power systems have several industrial network control protocols adapted to their specific use cases. For instance, data exchanged between devices in substations needs to be transmitted reliably and with minimal latency. To achieve this, substations can use the network protocol IEC 61850. In contrast, large-scale monitoring of conditions that do not require immediate responses can be done over transmission control protocol/internet protocol (TCP/IP) network protocols, for example a popular industrial network protocol used for supervisory control and data acquisition (SCADA) system between substations and a central control server is IEC 104; while communications between control centers can be done using the Inter-Control Center Communications Protocol (ICCP/TASE.2), an industrial protocol. Figure 13.2 illustrates an example of this ecosystem of industrial control protocols and their uses.

The core of the power grid operation relies on the SCADA, which is responsible for ensuring the reliability of the system. It gathers information from the system and then takes control actions to maintain reliability of the system and optimize its operation and market prices. In the rest of this section, we briefly introduce popular protocols used by SCADA systems to monitor and control a power grid. In Section 13.3, we will focus on IEC 104 and IEC 61850 due their importance in attacking the power grid in Ukraine.

The **Distributed Network Protocol 3 (DNP3)** is a standard communications protocol for data acquisition and control between master stations (i.e. SCADA), and outstations such as remote

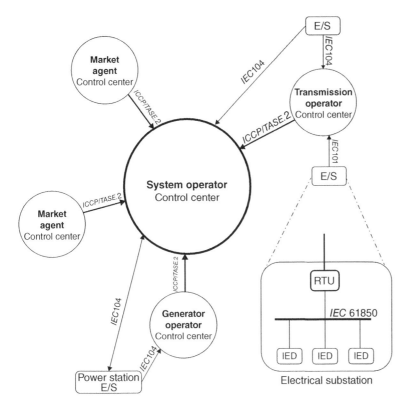

Figure 13.2 Communications protocols used in a power grid.

telemetry units (RTUs) or other intelligent electronic devices (IEDs). DNP3 has gained significant acceptance in North America and competes with IEC 104, which is popular in Europe. The purpose of DNP3 is to transmit relatively small packets reliably, and with messages arriving in a deterministic sequence. DNP3 breaks messages into multiple frames that provide error control and fast communications [1]. DNP3 allows master-outstation, multiple outstations, as well as multiple masters topology. On the other hand, it supports three types of communications: peer-to-peer, broadcast, and unsolicited responses, which means the outstation provides event data to masters without a request [2]. DNP3 also provides time synchronization and time-stamped events that support time-stamped messages for sequence of event (SOE) recording.

IEC 60870-5-101 IEC 101 was originally developed by the International Electrotechnical Commission (IEC) in 1995 [3] to provide a standard that transports basic telecontrol messages (monitoring and controlling commands) between *control stations* and *substations* over a permanently connected serial communication link (i.e. a modem circuit). Then in 2000, as an extension to IEC 101, IEC 104 was defined by IEC to encapsulate IEC 101 telecontrol messages into Application Protocol Data Units (APDUs) and transport IEC 101 telecontrol messages over TCP/IP using Port 2404. This protocol is popular in Europe and Asia, and it can be used to monitor and control large-geographical areas. Figures 13.3 and 13.4 show examples of the format of IEC 104 messages, and a specific example of an I-format packet transmission.

The ICCP allows a control center to exchange data with another control center in a communication based on the client/server model (instead of the master/slave model of previous protocols). A control center can be both a client and a server, depending on the case. ICCP is an object-oriented

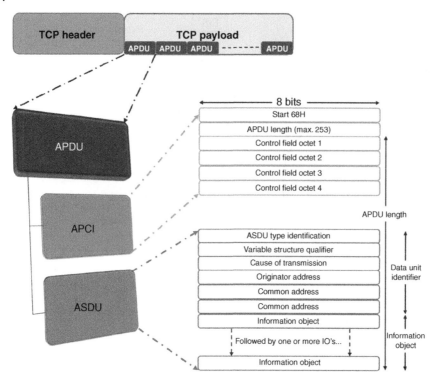

Figure 13.3 IEC 104 APDU as part of TCP payload and APDU structure. Note, APDU can either consist both APCI and ASDU, or just APCI only.

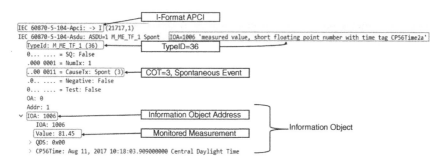

Figure 13.4 I-format APCI with typeID="Measured value, short floating point number with time tag CP56Time2a" typeID, COT="spontaneous," and carried monitored measurement from device address 1006.

protocol that uses objects to describe elements and behaviors, and it uses bilateral tables to control access. Data transfer is initiated with a request by a control center (client) to the control center that owns and manages the data (server). There are three types of requests: single, periodic, and report by exception. The data exchange encompasses real-time measurement and control data that can operate over a point-to-point link or a router-based wide area network (WAN). ICCP uses manufacturing message specification (MMS) for message services that define the nomenclature, catalog, and variable addressing.

The Common Information Model (CIM [IEC 61968 & 61970]) is a communication protocol developed for electrical utility organizations and used for data exchanges between systems, including energy management systems (EMS), distribution management systems (DMS), planning, energy markets, and metering systems. It supports operation and control data, telemetry, data acquisition, and alarm status. Similar to ICCP, CIM is an object-oriented protocol. It not only provides a standard way of representing power grid resources as object classes and attributes but also determines how the classes relate to each other. CIM defines a common language that facilitates the integration of network applications used in electrical energy planning, management, operation, and business. Also, in SCADA systems, it supports power system simulations and inter-control center communication. With a CIM implementation of a power system, objects can communicate directly with each other, reducing the requirement for a single, centralized program or procedure to perform all the processing of data.

IEC 61850 was first issued in 2005 to standardize the exchange of information among multi-vendor IEDs that perform protection, automation, metering, monitoring, and control within an automated substation. For any device, the IEC 61850 model begins with an abstract view of the device and its objects. This abstract model is mapped to the open systems interconnection (OSI) – seven-layer stack for communication divided into three categories as MMS (ISO 9506) [4], TCP/IP, and Ethernet, as shown in Figure 13.5.

The abstract model consists of models that define objects, services, and the behavior of IED devices, known as virtual manufacturing device (VMD) models [5]. Information about configuration and mapping of abstract objects and abstract services to instantiated services is provided in the specific communication service mapping (SCSM) document. The information model is the abstract representation of the devices in the substation [6].

Any of these protocols can potentially be attacked and used by the attacker to deliver malicious payloads. As we will see in Section 13.3 attackers in the 2016 Ukraine power grid attack used two specific control commands from IEC 104 to cause the power outage [7]. These commands had typeID 45 and 46, which are "Single Command" and "Double Command," respectively. According to IEC 101/IEC 104 protocols, "Single Command" is a 1-bit command that represents: 0=OFF and 1=ON. Similarly, "Double Command" is a 2-bit command that represents: 1=OFF, 2=ON, while 0 and 3 are not permitted. Therefore, by utilizing typeIDs 45 and 46, it appeared that one of the attackers' goals was to try to control any field devices that have ON/OFF states such as circuit

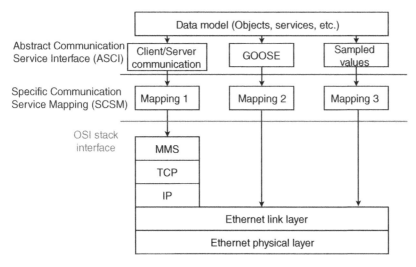

Figure 13.5 Mapping of IEC 61850 model protocols to OSI model.

breakers, relays, which control the flow of electricity. The same attack also targeted substations with IEC 61850; after reading a configuration file with IP addresses using MMS, the attack attempted to interact with them and change their status. The next section provides additional details of these attacks.

13.3   Cyberattacks to the Power Grid

Electricity and its underlying infrastructures, the power grid, have become integral and essential parts of life in the twenty-first century. Electricity provides convenience, comfort, economic growth, and most importantly safety. Therefore, in the United States, a concerted effort between the Department of Energy (DOE), the Department of Homeland Security (DHS), and the Federal Energy Regulatory Commission (FERC) have implemented 27 power grid reliability and resiliency efforts, and spent roughly US$ 240 million between 2013 and 2015 on these efforts, which were reported by the United States Government Accountability Office [8]. These efforts addressed a wide range of threats and hazards, including but not limited to cyberattacks, physical attacks, and natural disasters, all of which can have major impacts on the reliability and resiliency of our electric power system.

Reliability and resiliency of the electric power system or power grid are two critical requirements that ensure: (i) minimal disruptions and (ii) the ability to prepare, adapt, and quickly recover from an outage, whether by natural disasters, or equipment failures, or accidental or deliberated criminal acts.

For security purposes, it is important to differentiate between two types of blackouts. The first type of blackout can be caused by bringing down the bulk power system. This can be potentially catastrophic as bulk power systems are responsible for delivering electricity in large geographical areas (countries or several states in the United States). Bringing the bulk power system back online is a process that takes several days. The second type of blackout can be caused by targeting a distribution system, and while these blackouts can have several negative consequences, they are smaller in terms of scale and can be restored quicker.

Fortunately, attacks that bring down the bulk of the power grid have not happened before against real systems; however, two recent attacks against the power systems of Ukraine in 2015 and 2016 illustrate the threats and risks to these systems. In particular, the attack in 2015 targeted distribution systems, and the attack in 2016 targeted a transmission substation [7]. In addition, the attacks in 2016 were automated with malware (Industroyer) that knew how to automatically target control systems, while the attacks in 2015 required a remote human opening circuit breakers. This difference is illustrated in Figure 13.6.

13.3.1   Attacks in Ukraine

The first attack against Ukraine happened in December 2015. In it, three power grid companies were attacked by a Windows-based malware called BlackEnergy, an all-purpose Trojan for access and reconnaissance, and then another malware called KillDisk for overwriting boot-records.

In spring 2015, employees of these power companies received a phishing email disguised as a message from Ukrainian parliament with a Microsoft Word attachment. When employees opened the document, it prompted a request for macro enabling and activating BlackEnergy. With BlackEnergy, the attackers successfully opened a backdoor in the corporate network, pivoted to the industrial network to find the SCADA server, and then used this server to send control commands to physical equipment. In particular, attackers opened circuit breakers and caused a

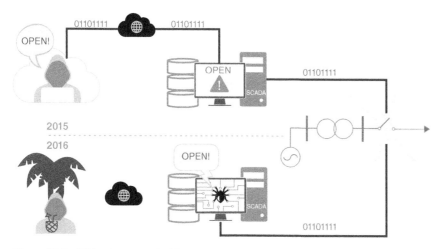

Figure 13.6 While the attack in 2015 required remote hackers operating the system, the attacks in 2016 were automated with malware.

blackout. An engineer from the power grid company took a short video of what he observed at that moment. The video was uploaded by C&M-News [9].

After opening circuit breakers, the attackers damaged the substations' serial-to-Ethernet converters (devices used to take control commands from a network, and change it to a serial line) by overwriting the firmware. The attackers designed malicious updates for the firmware in each company since each company used its own version of DMS. Thus after the attack, the grid operators were not able to control or recover from the attack remotely. To fix the problem, the operators had to physically go to the site affected and manually perform operations.

In addition, the attackers damaged the uninterruptible power supply in some of the power companies. This step left the control stations themselves in the dark while in their own power crisis. The blog [10] named it as "a blackout within a blackout."

The next year the attackers targeted a transmission substation (instead of distribution substations) and in addition, the attackers showed their improved capabilities with fully automated malware to carry out the attacks on its own. This malware was named Industroyer by ESET [11], the first company to report details of the malware (this malware is also known as Crashoverride [12]).

According to the ESET report, the version of malware that was used in the 2016 attack was written specifically to attack four different ICS protocols: IEC 60870-5-101 (IEC 101), IEC 60870-5-104 (IEC 104), IEC 61850, and OLE for Process Control Data Access (OPC DA). Similar to traditional PC trojans, Industroyer has backdoor, launcher, and wiper modules, but that is where the similarities end. Unlike traditional PC trojans, Industroyer includes a payload module that targets the four industrial control protocols mentioned above, thereby allowing Industroyer the capability to carry out multi-pronged attacks against different parts of the power grids. Another noteworthy difference is the wiper module that, according to Dragos [12], could have caused denial-of-service attack against Siemens SIPROTECT protective systems by placing these systems into "firmware update" mode.

To illustrate an example of the malware, we focus on the IEC 104 malicious payloads, Industroyer first reads a configuration INI file that contains configurable instructions, which tells the malware which IP addresses and TCP port to use for IEC 104, and then tells the initial action type (ON/OFF) to carry out and whether or not to flip each subsequent actions between ON and OFF states. These ON/OFF actions will be crafted into specific IEC 104 commands that are designed to switch a circuit breaker between ON and OFF positions. The two specific commands used by Industroyer for the 2016 attack were "Single command: C_SC_NA_1" and "Double command: C_DC_NA_1."

While attackers were able to target a substation of the bulk power system in Ukraine, the attacks did not bring down the overall bulk system; therefore, reconnecting the attacked network to the power system was straightforward. However, future attacks can be potentially more damaging by attempting to bring down the bulk power system, or by targeting and damaging expensive equipment like generators and large transformers.

13.3.2   Other Potential Attacks

Late in 2007, CNN reported on a project performed by the U.S. Department of Energy and Idaho National Laboratory (INL). The project, called the "Aurora generator test," demonstrated how attackers could use cyberattacks to permanently damage physical equipment in the power grid – specifically to rotating equipment such as motors and generators.

The test setup included a 3.8-MVA diesel power generator that provided electricity to a 13.8-kV distribution substation. Electricity generated flowed through circuit breakers that were controlled by a computer program. To carry out the test, the program opened and closed circuit breakers in rapid succession to cause "out-of-phase" conditions to the generator. By opening and closing circuit breakers, the rotation speed of the generator could not synchronize with the remaining circuit and the generator was destroyed [13]. While this was a government test, attackers can potentially launch similar attacks to other generators or big transformers; this equipment is hard to replace as it takes several months to build and costs are in excess of millions of dollars. A coordinated attack taking out multiple pieces of equipment in the power grid can potentially do long-lasting damage to the system and create blackouts lasting several orders of magnitude longer than the attacks in Ukraine.

13.4   Research Efforts

To complete this tutorial on power grid security, we will outline some research efforts for designing more secure and resilient power systems. By looking at security research efforts for the "classical" power grid, and reviewing the security efforts in the so-called "smart grid," and especially how technologies were meant to make this grid smarter, this demonstrated how it created opportunities for adversaries to attack the grid.

13.4.1   Classical Power Grid Systems

Analyzing physical measurements such as voltage, current, active and reactive power, etc., is part of a system monitoring process that ensures equilibrium of the overall electricity generation in the power grid. These physical measurements are often taken and transmitted from remote generation and transmission substations to designated control centers. Based on these measurements, control centers can then perform state estimation analysis of the power grid by using power flow models, which are equations that depict the energy flow on each transmission line. The result of the analysis is then used to determine if certain part of the power grid would need to increase or decrease power generation. Therefore, bad measurements (by faulty equipment) can affect state estimation analysis, which could potentially lead to unacceptable consequences. The literature on detecting bad measurements is extensive [17–21]; however, Liu et al. [14] showed that adversaries could potentially bypass such bad measurement detection algorithms.

They called their attacks *false data injection* attacks (FDI). These researchers pointed out that current state-of-the-art bad measurements algorithms often assume that the square of differences between observed and estimated measurements are significant when bad measurements occur. By exploiting this knowledge, adversaries could hypothetically inject malicious measurements to violate the above assumption, thereby, bypassing bad measurements detection. FDI attacks have been studied in several other parts of the grid, including nuclear generation [22].

Another novel idea to protect power systems is to detect attacks by modeling the physical behavior of equipment or the physical (radio-frequency) emanations of electric systems. Shekari et al. [23] presents a noninvasive IDS based on radio frequency for monitoring substations activities. The system uses, as a side-channel, magnetic field measurements at low frequencies ($< 100\,kHz$) produced by the AC circuits. These measurements (such as current harmonic, impulsive emissions by switching actions, and lightning sferics) are the input to estimate the operating status in a substation and also, as a redundant mechanism to authenticate the collected data. The purpose of this IDS is to detect cyberattacks related to false data and malicious command injection in substations, such as the attacks that occurred in Ukraine, where its SCADA became an untrusted entity. The author proposed the use of lightning sferics as a signature in the measurement data to validate the integrity and authenticity of the receiver's signal. On the one hand, a substation can detect malicious commands (such as circuit breaker malicious switching, transformer malicious tap changing, false data injection to protective relays) coming from a compromised SCADA system. On the other hand, a SCADA system can identify false data injection coming from a communications network with a substation.

Following the physics as anomaly detection, another idea is to learn the physical response of devices operating the power grid via fingerprinting [24]. In this work, the authors develop new fingerprinting methods specifically targeting industrial control systems (ICS). One feature is the cross-layer response time (CLRT) and another feature is the physical operation time. The authors evaluated the results from fingerprint generation with a real-world dataset from both a large-scale power grid and a small lab setting environment.

Another novel area is the combination of cyber- and physical-anomaly detection [25]. In this work, the authors characterize physical behavior as well as traditional cyber behavior in the DNP3 industrial control protocol.

Finally, another new research direction for protecting the power grid is the use of moving-target defense (MTD) for control systems [26]. MTD can be used to obfuscate the network of the system by confusing the adversary with extra devices added to the system sending chaff traffic [26], but MTD can also be used to dynamically perturb the physical system [27]; in this latter work, the goal was to invalidate the attacker's knowledge about the transmission line reactances by dynamically perturbing the power grid line reactances using *distributed flexible AC transmission system* (FACTS) devices.

13.4.2 Smart Grid Systems

The smart grid refers to the modernization of the power grid to provide new functionalities like real-time pricing (RTP), demand response, and advanced metering infrastructures (AMI). AMI systems can be used to detect electricity theft [28]. In their model, they assume an attacker can either get access to the meter or can impersonate it. Then, the attacker can send the fake time series back to the utility vendor to lower their utility bill. The authors obtained real AMI data for six months in a 15-minute reading interval for analysis using adversarial machine learning and the evaluation of classification results. They designed several detectors to catch the invariant relationship between the falsified bill and the true bill within a certain period of time.

In a follow-up paper, Bhattacharjee et al. [16] outlined four potential AMI data falsification attacks: (i) *Fraction of Compromised Meters* in which adversaries could compromise M meters out of N smart meters to perform data falsification attack; (ii) *Attack types* where adversaries simultaneously falsify data from multiple compromised meters using one of three falsification attacks: "deductive, additive, and camouflage" by modifying the actual power consumption value P_t^i by some δ_t value; (iii) *Average Falsification Margin*; and (iv) *Falsification Distribution* both of which are meant to strategically modify P_t^i by some δ_t within $\delta_{min}, \delta_{max}$ such that to avoid being detected by the three mentioned detection methods. To mitigate AMI data falsification attacks, the authors propose a framework that consists of two parts: (i) anomaly-driven consensus with correction and (ii) trust scoring. According to the authors, the consensus correction model provides robust consensus as inputs to the proposed trust scoring model; therefore, it improves the overall classification.

Another potential new attack has been referred to as *load-altering attacks*. Load-altering attacks have been previously studied in demand-response systems by [15, 29–33]. Demand-response programs provide a new mechanism for controlling the demand of electricity to improve power grid stability and energy efficiency. In their basic form, demand-response programs provide incentives (e.g. via dynamic pricing) for consumers to reduce electricity consumption during peak hours. Currently, these programs are mostly used by large commercial consumers and government agencies managing large campuses and buildings, and their operation is based on informal incentive signals via as phone calls by the utility or by the demand-response provider (e.g. a company such as Enel X) asking the consumer to lower their energy consumption during the peak times. As these programs become more widespread (targeting residential consumers) and automated (giving utilities or demand-response companies the ability to directly control the load of their customers remotely), the attack surface for load altering attacks will increase. The attacks proposed consider that the adversary has gotten access to the company controlling remote loads and can change a large amount of the load to affect the power system and cause either inefficiencies to the system, economic profits for the attacker, or potentially cause enough load changes to alter the frequency of the power grid and take down parts of the system.

More recently, Soltan et al. [34] studied the same type of load altering attacks but when the attacker creates a large-scale botnet with hundreds of thousands of high-energy IoT devices (such as water heaters and air conditioners). With such a big botnet the attacker can cause (i) frequency instabilities, (ii) line failures, and (iii) increased operating costs. A follow-up work by Huang et al. [35] showed that while the most powerful Botnet-based load-altering attacks (assuming the attacker compromises more than 8 million air conditioners) might cause the power system to partition and operate as separate islands, or can also cause some controlled load shedding, creating a system blackout – which would require a black start period of several days to restart the grid – or even a blackout of a large percentage of the bulk power grid can be very difficult.

Finally, having well-defined resilient metrics can help critical infrastructure, such as power grid, to continue to operate at an acceptable (safe) measures in the face of adversity or threats. Although modern power grids are considered to be fairly reliable, the new functionalities added by the smart grid development or modernization process also adds further complexity, but more importantly, brings about unintended cyber and physical vulnerabilities [36]. Therefore, the authors believe, during abnormal events, future grids should be able to not only perform real-time analysis on many new and complex data in real time but must also identify new risks while setting forth new system operating limits (SOL) such that it renders the desirable operations despite diverse disturbances that include physical, cyber, and cognitive disturbances. In short, to maintain power system resiliency, the authors outlined four power system time and/or data-dependent attributes from which four measurable metrics were derived and described. These four metrics, according to the authors, allow power system operators to measure the system performance against lower operational boundaries as set forth by the SOL.

13.4.3 Grid Simulator

Grid Game is a simulation game for the players taking the role-play of an operator for a small power grid, i.e. microgrid. The operation goal is to maintain the balance between generation and consumption by adjusting the generator's rotation frequency. The economic goal is to enlarge the customer market and to sell electricity as much as possible. This game can benefit the readers of this chapter mainly in these perspectives: it is a well-designed simulation environment to test attack scenarios and to observe how the power grid reacts. It provokes an interactive environment for the players to play either the adversaries or the operators. As a result, the player can gain hands-on experience to understand the grid resilience and reflect on the defense strategies.

This game has many exciting features for the players to study the resilience and the security problems in a microgrid. The players can also experience the operator's intense defense moments at some level of fidelity when planning out the best strategies to react to the attacks. Moreover, the game allows for a unique contract interaction among microgrid companies to trade electricity whenever the local load and response is imbalanced. The game designers have developed five attack scenarios in three layers, infrastructure security (e.g. Denial of Service [DoS]), information security (e.g. financial attacks), and control system security (e.g. automatic control disruptions) [37, 38]. In the DoS experiment, the adversaries intend to cut off the connections between the operators and the consumers. There are two virus financial attacks, the Little Guy Virus, and the Big Guy Virus, which aim to hurt the revenue of the grid by causing minor damages or interruptions to the operators. There is a group of advanced malware. The first Gluxnet mimics the situation with Stuxnet, pressing stress on the generator's machinery and inducing damages to the physical components. The second and the most sophisticated attack scenario Blue Frog is the analog case of the Ukrainian power grid attack by the Blackenergy malware. To sum up, the uniqueness of this simulation game can facilitate the players to learn the complex relationship among the entities of the cyber system, physical system, the defenders, and the attackers.

13.5 Summary

This tutorial introduced the general technologies that are used to monitor and control the power grid, summarized recent attacks to power systems and new potential attacks, and showed novel research directions that attempt to protect power systems using new unique properties from cyber–physical systems. As governments continue to consider attacks to cyber–physical systems as part of their future military and political conflicts, the importance of securing the power systems will increase, and new research technologies will need to be tested and deployed in these systems to not only prevent attacks but also detect them in a timely manner and respond to these attacks.

This material is based on research sponsored by the Air Force Research Laboratory under agreement number FA8750-19-2-0010. The U.S. Government is authorized to reproduce and distribute reprints for Governmental purposes notwithstanding any copyright notation thereon.

13.6 Thoughtful Questions to Ensure Comprehension

Here are some of the questions suggested:

1. What are the differences between the protocols IEC 61850, IEC 104, and ICCP/TASE.2?
2. What different types of blackouts are there? Describe one type that can take the power grid of a country down for days vs other attacks that can be fixed within hours.

3. What are the main differences between the two attacks to Ukraine's power grid? How did the attacking strategies of the attackers evolve?
4. What are the industrial protocols targeted by the malware in the 2016 attack to the Ukrainian power grid?
5. The Aurora generator test was a proof of concept of how a cyber attack could potentially cause physical damages to equipment in the power grid. Which type of equipment was destroyed in this test?
6. What kind of vulnerabilities are introduced by new functions in the smart grid? Provide one mitigation method for each vulnerability.
7. What are the five types of attacks the Grid Game can provide in their environment? Can you design a sophisticated attack scenario in different stages, comprising of these attacks?

Further Reading

In addition to research efforts and articles presented in Section 13.4, here are two essential readings on the US electric grid anatomy [39] and false data injection attacks taxonomy [40]. To understand how the detection methodology of the false data injection attacks develops in the recent years, readers can refer to the following list:

1. He, Youbiao, Gihan J. Mendis, and Jin Wei. "Real-time detection of false data injection attacks in smart grid: A deep learning-based intelligent mechanism." IEEE Transactions on Smart Grid 8.5 (2017): 2505-2516.
2. Ashrafuzzaman, Mohammad et al. "Detecting stealthy false data injection attacks in power grids using deep learning." 2018 14th International Wireless Communications & Mobile Computing Conference (IWCMC). IEEE, 2018.
3. Wang, Huaizhi, et al. "Deep learning-based interval state estimation of AC smart grids against sparse cyber attacks." IEEE Transactions on Industrial Informatics 14.11 (2018): 4766-4778.
4. Zhang, Ying, Jianhui Wang, and Bo Chen. "Detecting false data injection attacks in smart grids: A semi-supervised deep learning approach." IEEE Transactions on Smart Grid (2020).

References

1 Clarke, G. and Reynders, D. (2003). Practical Modern SCADA Protocols: DNP3, 60870.5 and Related Systems, p. 548.
2 Yun, J.-H., Jeon, S.-H., Kim, K.-H., and Kim, W.-N. (2013). Burst-based anomaly detection on the DNP3 protocol. *International Journal of Control and Automation* 6 (2): 12.
3 Webstore International Electrotechnical Commission (2006). Telecontrol equipment and systems - Part 5-104: Transmission protocols - network access for IEC 60870-5-101 using standard transport profiles.
4 Khaled, O., Marín, A., Almenares, F. et al. (2016). Analysis of secure TCP/IP profile in 61850 based substation automation system for smart grids. *International Journal of Distributed Sensor Networks* 12 (4): 5793183.
5 Sørensen, J.S. and Jaatun, M. (2008). An analysis of the manufacturing messaging specification protocol, pp. 602–615, 01 2008.

6 International Electrical Commission (IEC) (2010). Basic information and communication structure abstract communication service interface (ACSI). *International Standard IEC 61850, Communication Networks and SYSTEMS for Power Utility Automation*, 2nd ed. Geneva, Switzerland: IEC, 9, 2010.

7 Park, D., Summers, J., and Walstrom, M. (2017). Cyberattack on critical infrastructure: Russia and the Ukrainian power grid attacks.

8 United States Government Accountability Office (GAO) (2017). Federal efforts to enhance grid resilience.

9 C&M-News (2017). Watch how hackers took over a Ukrainian power station.

10 Greenberg, W.A. (2017). How an entire nation became Russia's test lab for cyberwar.

11 ESET and Cherepanov, A. (2017). Win32/industroyer a new threat for industrial control systems.

12 Slowik, J. (2019). Crashoverride: Reassessing the 2016 Ukraine electric power event as a protection-focused attack.

13 Weiss, J. (2016). Aurora generator test. In: Robert Radvanovsky, Jacob Brodsky (eds.) *Handbook of SCADA/Control Systems Security* Routledge, 107.

14 Liu, Y., Ning, P., and Reiter, M.K. (2011). False data injection attacks against state estimation in electric power grids. *ACM Transactions on Information and System Security* 14 (1): 13:1–13:33.

15 Tan, R., Krishna, V.B., Yau, D.K.Y., and Kalbarczyk, Z. (2013). Impact of integrity attacks on real-time pricing in smart grids. *Proceedings of the 2013 ACM SIGSAC Conference on Computer & Communications Security*, CCS '13, pp. 439–450. New York, NY, USA: ACM.

16 Bhattacharjee, S., Thakur, A., and Das, S.K. (2018). Towards fast and semi-supervised identification of smart meters launching data falsification attacks. *Proceedings of the 2018 on Asia Conference on Computer and Communications Security*, ASIACCS '18, pp. 173–185, New York, NY, USA: ACM.

17 Asada, E.N., Garcia, A.V., and Romero, R. (2005). Identifying multiple interacting bad data in power system state estimation. *IEEE Power Engineering Society General Meeting, 2005*, pp. 571–577. IEEE.

18 Chen, J. and Abur, A. (2005). Improved bad data processing via strategic placement of PMUs. *IEEE Power Engineering Society General Meeting*, pp. 509–513. IEEE.

19 Handschin, E., Schweppe, F.C., Kohlas, J., and Fiechter, A.A.F.A. (1975). Bad data analysis for power system state estimation. *IEEE Transactions on Power Apparatus and Systems* 94 (2): 329–337.

20 Garcia, A., Monticelli, A., and Abreu, P. (1979). Fast decoupled state estimation and bad data processing. *IEEE Transactions on Power Apparatus and Systems* (5): 1645–1652.

21 Gastoni, S., Granelli, G.P., and Montagna, M. (2003). Multiple bad data processing by genetic algorithms. *2003 IEEE Bologna Power Tech Conference Proceedings*, Volume 1, 6 pp. IEEE.

22 Villez, K., Venkatasubramanian, V., Garcia, H. et al. (2010). Achieving resilience in critical infrastructures: a case study for a nuclear power plant cooling loop. *2010 3rd International Symposium on Resilient Control Systems*, pp. 49–52, August 2010.

23 Shekari, T., Bayens, C., Cohen, M. et al. (2019). RFDIDS: Radio frequency-based distributed intrusion detection system for the power grid. *Proceedings 2019 Network and Distributed System Security Symposium*, San Diego, CA. Internet Society.

24 Formby, D., Srinivasan, P., Leonard, A.M. et al. (2016). Who's in control of your control system? Device fingerprinting for cyber-physical systems. In *NDSS*.

25 Marino, D.L., Wickramasinghe, C.S., Amarasinghe, K. et al. (2019). Cyber and physical anomaly detection in smart-grids.

26 Lin, H., Zhuang, J.-N., and Hu, Y.-C. (2020). DefRec: Establishing physical function virtualization to disrupt reconnaissance of power grids' cyber-physical infrastructures.

27 Lakshminarayana, S., Belmega, E.V., and Poor, H.V. (2019). Moving-target defense for detecting coordinated cyber-physical attacks in power grids. *arXiv preprint arXiv:1908.02392*.

28 Mashima, D. and Cárdenas, A.A. (2012). Evaluating electricity theft detectors in smart grid networks. *International Workshop on Recent Advances in Intrusion Detection*, pp. 210–229. Springer.

29 Barreto, C., Cárdenas, A.A., Quijano, N., and Mojica-Nava, E. (2014). CPS: Market analysis of attacks against demand response in the smart grid. *Proceedings of the 30th Annual Computer Security Applications Conference*, pp. 136–145. ACM.

30 Amini, S., Pasqualetti, F., and Mohsenian-Rad, H. (2018). Dynamic load altering attacks against power system stability: attack models and protection schemes. *IEEE Transactions on Smart Grid* 9 (4): 2862–2872.

31 Giraldo, J., Cárdenas, A., and Quijano, N. (2016). Integrity attacks on real-time pricing in smart grids: impact and countermeasures. *IEEE Transactions on Smart Grid* 8 (5): 2249–2257.

32 Mohsenian-Rad, A.-H. and Leon-Garcia, A. (2011). Distributed Internet-based load altering attacks against smart power grids. *IEEE Transactions on Smart Grid* 2 (4): 667–674.

33 Chen, B., Pattanaik, N., Goulart, A. et al. (2015). Implementing attacks for modbus/TCP protocol in a real-time cyber physical system test bed. *2015 IEEE International Workshop Technical Committee on Communications Quality and Reliability (CQR)*, pp. 1–6. IEEE.

34 Soltan, S., Mittal, P., and Poor, H.V. (2018). BlackIoT: IoT botnet of high wattage devices can disrupt the power grid. *27th USENIX Security Symposium (USENIX Security 18)*, pp. 15–32.

35 Huang, B., Cardenas, A.A., and Baldick, R. (2019). Not everything is dark and gloomy: power grid protections against IoT demand attacks. *28th USENIX Security Symposium (USENIX Security 19)*.

36 Eshghi, K., Johnson, B.K., and Rieger, C.G. (2015). Power system protection and resilient metrics. *2015 Resilience Week (RWS)*, August 2015, pp. 1–8.

37 Rege, A., Parker, E., and McJunkin, T. (2017). Using a critical infrastructure game to provide realistic observation of the human in the loop by criminal justice students. *Proceedings - 2017 Resilience Week, RWS 2017*, pp. 154–160.

38 Rege, A., Biswas, S., Bai, L. et al. (2017). Using simulators to assess knowledge and behavior of "novice" operators of critical infrastructure under cyberattack events. *2017 Resilience Week (RWS)*, pp. 50–56. IEEE.

39 Gilstrap, M., Amin, S., and DeCorla-Souza, K. (2015). *United States Electricity Industry Primer*. US Department of Energy, Office of Electricity Delivery and Energy Reliability.

40 Basumallik, S. (2020). A taxonomy of data attacks in power systems. *arXiv preprint arXiv:2002.11011*.

14

Control Challenges

Quanyan Zhu

New York University, Electrical and Computer Engineering, Brooklyn, NY, USA

Objectives

The objectives of this chapter are to provide an overview of resilient control systems, including cross-layer trade-offs between security at the cyber layer and resiliency at the physical layer, cross-stage resiliency design, including ex-ante planning, interim operation, and ex-post recovery, the games-in-games design paradigm for the multistage and multilayer design of resilient control systems, and control challenges for distributed control systems.

14.1 Introduction

Modern control systems are equipped with information system technologies (ICTs) that can provide situational awareness of the plant and enable a fast response to emergency and security breaches. Although the control systems benefit from the enhanced functionalities and autonomy, the cybersecurity vulnerabilities become prominent issues to resolve. Adversaries can take a sequence of moves and launch multiphase and multistage attacks from the early reconnaissance to the objective of data exfiltration. This structure of attacks is known as the cyber kill chain. The defense against such attacks includes detection of an adversary, disruption of the network system, cyber deception to create uncertainties and costs for the attacker, and many other techniques (see Refs. [1–4]). Despite the effort in developing cyber defense for control systems, the perfect security is not always achievable. Achieving perfect security would either require a cost-prohibitive amount of resources to maintain the security or lead to a degradation of system operability and usability. Hence, it is important to shift the focus from security-centered design to the paradigm of secure and resilient design of control systems. Adding resiliency as an additional dimension to the new paradigm complements the sole reliance on ex-ante perfect security technologies as a solution with interim and ex-post resiliency mechanisms as solutions when the ex-ante security mechanism fails to protect the control system.

Resiliency is a key system concept that focuses on the post-event behaviors of a system. For example, when a cyberattack has successfully reached its target, any security mechanism becomes futile at this point. Reliance should be on resiliency mechanisms that can reduce the impact of this successful attack and enable a fast recovery to restore the operations to their normal state. On the contrary, security mechanisms are often used as tools to prevent successful attacks or attackers

Resilient Control Architectures and Power Systems, First Edition.
Edited by Craig Rieger, Ronald Boring, Brian Johnson, and Timothy McJunkin.

from achieving their objectives. Hence, security and resilience are dual concepts and there are fundamental relationships between them.

First, there exists a trade-off between security and resiliency. When there is perfect security, resiliency becomes unnecessary. The need for resiliency is high when the security is poor. Second, security and resiliency resolutions have to be implemented jointly. One cannot count on either security or resiliency solutions to safeguard the system from adversarial behaviors. The cost for security and resiliency solutions are often different. The level of security and resiliency implemented in the system should be optimized together.

14.2   Resiliency Challenges in Control Systems

Understanding these relationships provides a fundamental understanding toward designing secure and resilient systems. There are two key challenges to design secure and resilient systems. One is the cross-layer challenge and the other is the cross-stage challenge. Cross-layer challenges refer to the fact that the security issues often reside at the cyber layer of the system while the resiliency often deals with the last-mile issue and in the context of control systems, the resiliency issues sit at the physical layer of the systems. In other words, the failure in the protection at the cyber layer can lead to malfunctioning of the physical system performance. The joint design of secure and resilient solutions is naturally cross-layered. It would require understanding the dependency and interdependency among human, cyber, and physical layers of the control system. For example, the failure of cyber defense against an advanced persistent threat would lead to data injection on sensors and the manipulation of the controller of the physical plant. Human negligence in system configurations would lead to cyber vulnerabilities that can be exploited by the attackers to reach targeted physical assets.

The second challenge of resilient control systems arises from the fact that resiliency is a dynamical system concept. Improving resiliency involves multistage planning and design, including ex-ante planning, interim execution, and ex-post recovery. The ex-ante stage refers to the planning stage before the control system starts to run. At this stage, one needs to invest resources and plan contingencies to provide information and physical resources to enable fast recovery at later stages. For example, one can add redundancies such as sensors and power generators to prepare for an attack on sensors and attack-induced loss of power. One can also design secure and switching controllers in advance to prepare for the worst-case operational environment and provide a contingency controller when the control system encounters failures. The offline design of such controller at the ex-ante stage prepares for a set of anticipated attacks in later stages. However, unanticipated attacks can still occur. In addition, the preparation for a large set of events can be expensive. There is a trade-off between what events should be prepared for at the ex-ante stage and what events should be dealt with in later stages. Hence, from the set of anticipated events, the high-impact and high-frequency events should be selected first.

The interim stage refers to the operation stage of the control systems. By running the controller that prepares for a selected set of anticipated events, the control system can run smoothly and intelligently when it encounters these prepared events. The control system can switch to a different control logic or leverage built-in resources to handle events that occurred. In Figure 14.1, at time t_1, an anticipated event has occurred; for example, there is a loss of a sensor. By switching to a redundant sensor, the interim stage operation does not suffer any loss of system performance.

The ex-post stage refers to the stage where an unanticipated event occurs and the recovery process kicks off. The probability of running into an unanticipated event depends on the set of events

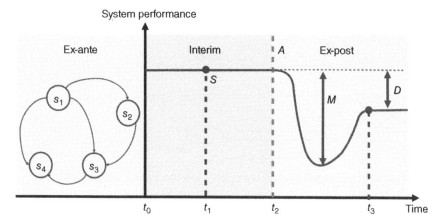

Figure 14.1 Multistage planning and design for resilient control systems: ex-ante stage plans resiliency. Interim stage executes resilient control. Ex-post stage restores system performance after unanticipated attacks.

prepared for by the ex-ante controller. In Figure 14.1, the unanticipated event a occurs at t_2. The expected value of t_2 would depend on the frequency of Event A and whether the ex-ante stage has been taken into account. Since no contingency plan has been made for Event A, the system will suffer a performance degradation. The ex-post resiliency design aims to detect the event quickly and find a self-healing mechanism that can restore the system to its normal operation or an acceptable performance level. Depicted in Figure 14.1, the system performance degrades to its maximum difference M and then gradually recovers to the level D at time t_3. The ex-post resiliency is measured by the total performance loss after the event. The goal of ex-post resiliency is to minimize this loss by responding fast to the event and developing an effective restoration plan. The restoration plan would depend on the configurable resources available to the system and the ex-ante resource investment made for the control system. For example, when an attack has successfully compromised a centrifuge control system, the system can detect and reboot. The cost for reboot depends on whether the control system is well equipped with computational resources and human resources to enable the fast response.

It is clear that the concept of resiliency spans three stages of the control systems. The planning of ex-ante resiliency affects the resource availability of and the need for ex-post resiliency. Hence, the ex-ante resiliency planning and the ex-post resiliency mechanism have to be jointly designed. This chapter will introduce a games-in-games framework to develop a cross-layer cross-stage design framework. We will use examples from unmanned vehicles to illustrate the games-in-games framework.

14.3 Resiliency Design Framework

A natural framework to enable the cross-layer design is to use the games-in-games framework introduced in [5]. Game-theoretic methods have been used to capture different cyberattack scenarios and models including jamming (see Refs. [6–8]), spoofing (see Refs. [9, 10]), and network configurations (see Refs. [11–15]). Interested readers can refer to [16] and [3] for recent surveys. Each game can be used to model a cyberattack scenario. Composing these games together forms a set of

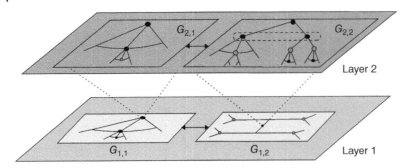

Figure 14.2 Composition of cyber games and physical layer games together to form a cross-layer adversarial model.

anticipated adversarial behaviors at the cyber layer to be considered. Similarly, the attack behaviors at the physical layer can also be modeled by a set of physical layer games. For example, the attacker can choose to inject bad data into sensors while the defender can choose which sensors to use. The defender can determine the switching policy of the controllers while the attacker can determine how to compromise the logic of one of the control laws. Composing the physical layer games together provides a framework to design ex-ante and ex-post resiliency mechanisms. The cyber games and the physical layer games are interdependent. The outcome of one game would lead to a new game. The games-in-games framework is illustrated in Figure 14.2. Cyber games $G_{2,1}$ and $G_{2,2}$ are composed together to form G_2. The physical layer games $G_{1,1}$ and $G_{1,2}$ are composed to form G_1. G_1 and G_2 are interconnected to form a larger game G.

A natural control framework to capture the multistage features of resilient control systems is model-predictive control or moving-horizon techniques. The moving-horizon control looks N steps into the future, prepares for possible events, and finds optimal control strategies to be implemented at the current stage. The ex-ante control design takes into account a credible set of anticipated adversarial models. This design process can be formulated as a game between the control designer and adversaries. The game model can be chosen to capture the cross-layer nature of the control system as illustrated in Figure 14.2. The ex-ante control is implemented immediately after the design. If no unanticipated behaviors occur, this game-based moving-horizon control continues. When unanticipated adversarial behavior happens in the next moving horizon, the control design not only prepares for the anticipated adversarial behaviors but also determines ways to recover from the unanticipated attack. The moving-horizon technique directly incorporates the ex-ante planning by looking into possible future events and the ex-post recovery by immediately reconfiguring the system in the next moving horizon.

14.3.1 Control of Autonomous Systems in Adversarial Environment

To illustrate the moving-horizon resilient control design paradigm, we present a case study of control for mobile autonomous systems in adversarial environment. The objective of the control is to maintain the connectivity of the autonomous system in an environment where an attacker can jam the communications among the autonomous systems. The operator does not know the capability of the attacker (i.e. how many links the attacker can jam, and where and when the attacker will jam the communications). The operator prepares for the anticipated level of attacks and plans the control in a moving-horizon way. At every time k, the operator solves the following problem:

$$Q^k: \quad \max_{x(k+c)} \min_{e \in E} \lambda_2(e, x(k+c)). \tag{14.1}$$

Here $x(k)$ is the configuration of the mobile network at time k (i.e. the position of the mobile agents). Two mobile agents can form a link when they are sufficiently close within a desirable range of communications. Hence, the configuration $x(k)$ includes a network whose connectivity is described by the algebraic connectivity of the network, denoted by λ_2 (i.e. the second-smallest eigenvalue of the associated Laplacian matrix). At each time step k, the operator determines where the agents should move to in the next time step $x(k + c)$, where c is a time interval. The control is constrained by the physical dynamics of autonomous systems. To maintain connectivity, the operator aims to maximize the level of connectivity as much as possible at time $k + c$ by anticipating the worst-case adversarial behaviors given a certain level of attacks, described by the set E. For example, the operator can anticipate one-link removal and determine how agents can move and maintain connected secure to such one-link removal. On one hand, when the capability of the attacker is higher, it becomes more difficult for the operator to find a controller to be secure to the attack level. On the other hand, if the operator underestimates the attacker, his control strategies will not achieve a desirable connectivity. Hence, the operator should anticipate a reasonable level of attacks and design an ex-ante controller that will be implemented at the interim stage. When the attacker's capability does not exceed the attack level anticipated by the ex-ante controller, the network connectivity is maintained as expected. When the attacker's capability exceeds the anticipated level, the network connectivity may not be achieved and the ex-post resiliency mechanism will be designed at time $k + c$. In other words, the problem for the operator at $k + c$, Q^{k+c} includes how to react to the failure of network connectivity and how to design new control laws to heal the broken links while anticipating new attacks at time $k + c$. The anticipation of attack levels can be adjusted from time k to $k + c$.

This moving-horizon framework has been shown to be effective in obtaining the self-adaptability, self-healing, and resilience of the Internet of Battlefield Things (IoBTs) (see Refs. [17–19]). Specifically, the unmanned ground vehicle (UGV) network should coordinate its actions with the unmanned aerial vehicle (UAV) network and the soldier network to achieve a highly connected global network. The designed decentralized algorithm yields an intelligent control of each agent to respond to others to optimize real-time connectivity under adversaries. Figure 14.3 shows an example of a two-layer robotic network that is robust to jamming attack at every step. Figure 14.4 shows the algebraic connectivity over time associated with the two-layer network. Furthermore, the agents can respond to the spoofing attack quickly, which shows the resilience of the control strategy. The developed moving-horizon framework can be further adopted to address the mosaic control design as the framework provides built-in security and resilience for each component in the system, which guarantee the performance of the integrated system.

14.3.2 Cross-Layer Defense for Cloud-Enabled Internet of Controlled Things

To illustrate the games-in-games framework, we consider the Internet of Controlled Things (IoCTs) that integrates computing, control, sensing, and networking. The IoCT relies on local clouds to interface between heterogeneous components. The cloud-enabled IoCT is composed of three interacting layers: a cloud layer, a communication layer, and a physical layer. In the first layer, the cloud-services are threatened by attackers capable of APTs and defended by network administrators (or "defenders"). The interaction at each cloud-service is modeled using the FlipIt game recently proposed by Bowers et al. [20] and Van Dijk et al. [21]. We use one FlipIt game per cloud-service. In the communication layer, the cloud services, which may be controlled by the attacker or defender according to the outcome of the FlipIt game–transmit information to a device that decides whether to trust the cloud-services. This interaction is captured using a signaling

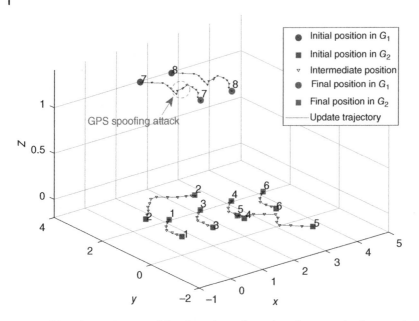

Figure 14.3 Network connectivity: dynamic configuration of secure robotic network. The GPS spoofing attack is introduced at time Step 9, and it lasts for five steps. Source: Chen and Zhu [19].

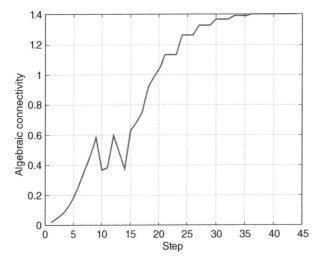

Figure 14.4 The network connectivity over time. Source: Chen and Zhu [19].

game. At the physical layer, the utility parameters for the signaling game are determined using optimal control. The cloud, communication, and physical layers are interdependent. This motivates an overall equilibrium concept called Gestalt Nash equilibrium (GNE). GNE requires each game to be solved optimally given the results of the other games. Because this is a similar idea to the concept of best response in Nash equilibrium, we call the multi-game framework a game-of-games.

A composition of a FlipIt game G_1 (e.g. [20, 21]) and a signaling game G_2 (e.g. [22, 23]), depicted in Figure 14.5, has been used to provide a strategic trust management in Internet of things (IoT) networks vulnerable to advanced persistent threats. The game G_1 describes the strategic interactions

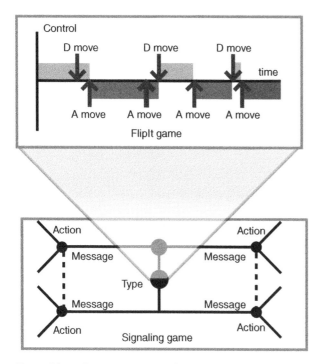

Figure 14.5 Conceptual model of the composed G_1 and G_2. Source: Pawlick and Zhu [58].

between an attacker and a cloud service provider where the attacker aims to gain control of the computing resources and the cloud service provider protects and audits the system. The game G_2 describes the information asymmetry between the sender of the message (i.e. the computational results) and the IoT as the receiver of the message. G_1 and G_2 are composed sequentially as in Figure 14.5. An attacker and a defender play the FlipIt game for control of the cloud. Then, the winner sends a command to the device in the signaling game. The GNE of the meta-game predicts the risk of sequential adversarial interactions. As shown in Figure 14.6, the equilibrium can be computed as the intersection of the dashed curve and the dash-dotted curve in an iterative manner within a finite number of steps.

14.4 Resiliency for Decentralized Control Systems

Resilient design for distributed systems requires engineering agents with flexible interoperability and the capability of self-adaptability, self-healing, and resiliency. It is important that a system can achieve its objective when one node goes away or fails. In addition, a subsystem can respond to other subsystems in a nondeterministic/stochastic way. Such a design increases the composability and modularity of the system design. For example, agents can randomly arrive and respond in a stochastic but structured way to other agents in an uncertain environment. However, the structured randomness leads to emerging system behaviors that manifest desirable properties for the objective of the entire mission.

Systems that have such properties are easily composable and resilient-by-design. Without a preplanned integration among agents, the agents can adapt their responses and reconfigure their own systems based on the type of agents with whom they interact. Agents can be easily composed

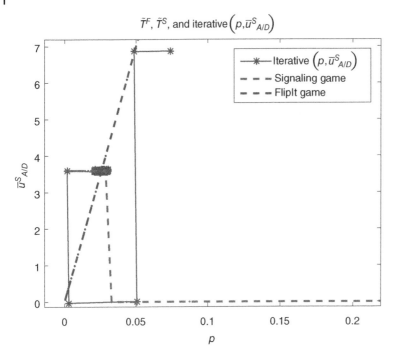

Figure 14.6 Iterations to compute a Gestalt Nash equilibrium. Source: Pawlick and Zhu [58].

to achieve a prescribed objective through an unprescribed path. In adversarial environments, the agents can reconfigure their responses and roles to achieve the global mission in spite of the failures of nodes and links. The system can still operate when one piece is missing. It is the epicenter of the mosaic designs.

We can leverage the games-in-games principle as discussed in [5] and [24] to provide a theoretical underpinning and a guideline for distributed resilient control designs. The games-in-games approach integrates three layers of design for each agent: strategic layer, tactical layer, and mission layer. At the strategic layer, the agents learn and respond to their environment quickly to unanticipated events such as attacks, disruptions, and changes of other agents. At the tactical layer, the agents plan for a more extended period of time by taking into account the long-term interactions with the environment and other agents. The agents can make goal-oriented planning at each stage. At the mission layer, the agents develop stage-by-stage planning of multistage objectives to achieve the mission despite the uncertainties and online changes.

Each layer corresponds to a game of a different scale. For example, at the tactical layer, a game associated with an agent describes its interaction with an adversary (e.g. a jammer, a spoofer, or a sudden loss of a neighboring node). Solutions to this game can prepare nodes for unanticipated attacks and secure the agents. At the strategic layer, an N-person dynamic game describes the long-term interactions among cooperative agents, each seeking control policies to achieve individual stage objectives. The individual control would lead to achieving global objectives (such as connectivity and network formation). At the mission layer, each agent plans their stage objectives for each stage at the tactical layer. This planning is obviously under a lot of uncertainties and needs to be achieved in a moving-horizon way.

The games-in-games framework enables security and resilience by design. From the perspective of security, the framework anticipates the attack behavior and designs a control policy that would prepare to defend against the anticipated attacks. The framework provides a clean-slate design and built-in security for each system component that would lead to the security of the integrated system. From the perspective of resiliency, the framework enables each system to respond to the unanticipated events at each time instant. Each agent can respond to events that inflict damages on the agent and go through a self-healing process that can recover itself from the attacks and failures if possible. If the full recovery is not achievable, the agents will develop control strategies that will allow a graceful performance degradation.

14.5 Summary

This chapter has introduced two key challenges for resilient control systems. One is the cross-layer challenge requiring an integrated cyber–physical perspective. The optimal design pivots on the understanding of the trade-offs between the security at the cyber layer and the resiliency at the physical layer. The other is the cross-stage resiliency mechanism that requires three-stage resiliency design including ex-ante planning, interim operation, and ex-post recovery. The games-in-games framework provides a design paradigm for the multistage and multilayer design of resilient control systems. The framework is particularly useful for distributed control systems by designing modular agents that can work together.

The chapter has presented two applications. One is the control of autonomous vehicles in adversarial environment and the other one is the cross-layer defense for cloud-enabled IoCTs. The methodology presented in these two case studies can be further generalized and made applicable to other systems, including smart grids, transportation systems, and manufacturing systems. The new challenge with these large-scale systems is the scalability of the solutions and the incompleteness of the situational awareness. The resiliency design would need mechanisms that can achieve scalable yet suboptimal resiliency and deal with time delay and nonglobal system state information.

14.6 Thoughtful Questions to Ensure Comprehension

1. Consider a smart electric power system and discuss the cross-layer control challenges of the cyber-enabled power systems. Present methods to improve the resiliency of the power systems. Discuss how to scale the resiliency solutions when the size distribution and transmission network increases. *Hint: The readers can refer to the case studies in* [5].
2. Write down the dynamics of the unmanned aerial vehicles (UAVs) and create a case study of a network of five UAVs. Simulate three steps by solving a sequence of three optimization problems (14.1). Use MATLAB® to present the simulation results. *Hint: The complete description of the algorithm can be found in* [19].
3. In this problem, we will investigate the game-theoretic algorithms for distributed energy systems. We consider an energy system that consists of multiple microgrids that are modeled as self-interested players that can operate, communicate, and interact autonomously to efficiently deliver power and electricity to their consumers. Consider a two-microgrid case where each of them aims to minimize the cost of production and maximize the quality of the power quality. The utility function of this game has been described in [25]. Use GridGame introduced in Chapter 12 to create a simulation platform of two players. Implement the distributed algorithm in (11) in [25] for each microgrid and observe the behavior of the grid. Discuss the resiliency of

the grid under the following two scenarios: (i) there is a false data injection attack on the PMU angle measurement; (ii) One microgrid suddenly fails and cannot generate power to meet the demand. *Hint: The readers can refer to the case studies in Section IV-B of [25].*

Further Reading

Game-theoretic methods have been widely used to model adversarial behaviors in wireless communications (See Refs. [6, 8, 9, 26–29]), network configuration (See Refs. [16, 30–32]), and control systems (See Refs. [5, 10, 33–36]). The applications of security games have addressed the critical infrastructure protection (See Refs. [37–41]), cyber insurance (See Refs. [42–45]), cyber deception (See Refs. [3, 13, 46–49]), adversarial machine learning (See Refs. [50–53]), and network systems (See Refs. [54–58]).

Game-theoretic approaches have also addressed the resilient design of control systems. A games-in-games approach has been proposed. Section 14.3.1 is based on the work of [19, 59] in which two layers of mobile agents are controlled to maintain network connectivity. Section 14.3.2 is based on the work of [60, 61]. The IoCTs extends the concept of IoTs and studies a cloud-enabled sensing and actuation architecture with three layers of interacting systems including the cloud, communication networks, and sensor-actuator networks. The games-in-games principle was first discussed in [5] and has been used for addressing applications of multilayer networks (See [10, 17, 62–65]) and trust management (See [58, 60, 61]). The recent book [66] has provided a comprehensive introduction to game- and control-theoretic techniques for cross-layer cyber–physical systems.

One important application of resilient control systems is critical infrastructures, including power grids, transportation, and manufacturing systems. They are often legacy systems that modern ICT technologies build on to improve their efficiency and expand their functionalities. Due to the large-scale nature of critical infrastructures, their resilience has to be designed in a decentralized manner. It is critical to understand and capture the interconnections between components within the infrastructure as well as the interdependencies among different infrastructures. Network modeling and design approaches (e.g. [67–70]) have been used to provide a holistic framework for understanding cyber–physical interdependencies in infrastructure systems. Readers can refer to [71–75] for further materials on the topic of resilient interdependent infrastructures.

Resiliency for decentralized systems is a fundamental challenge for increasingly complex and connected systems. Recent works [17, 39, 55, 76, 77] have addressed this issue from the perspective of network science. They have provided a design methodology to create networks that are resistant to link removals. In particular, Chen et al. [76] has provided a trade-off analysis between the ex-ante robustness and the ex-post recovery of the network. This framework has also been recently extended to the context of multiagent systems, where multiple mobile agents communicate locally to accomplish a mission (e.g. rendezvous and formation). Interested readers can refer to [25, 62, 78, 78–80] for recent development on this topic.

Chapter 12 of this book introduces Grid Game, a simulation and learning platform for the electric grid and resilient controls. It is a useful tool to simulate and understand the multiagent behaviors described in this chapter in the context of electric power systems. For example, a multiplayer microgrid game has been formulated in [81] and its related extensions to multilayer games have been introduced in [25, 79, 82]. The decentralized architecture of the multiplayer microgrids improves the resiliency of local microgrids when they are subject to attacks or natural failures. Readers can refer to Problem 14.3 for an exercise problem that builds on the Grid Game platform.

References

1 Jajodia, S., Ghosh, A.K., Swarup, V. et al. (2011). *Moving Target Defense: Creating Asymmetric Uncertainty for Cyber Threats*, vol. 54. Springer Science & Business Media.

2 Zhu, Q. and Başar, T. (2013). Game-theoretic approach to feedback-driven multi-stage moving target defense. In: S.K. Das, C. Nita-Rotaru, M. Kantarcioglu (eds.) *International Conference on Decision and Game Theory for Security*, 246–263. Springer: Switzerland.

3 Pawlick, J., Colbert, E., and Zhu, Q. (2019). A game-theoretic taxonomy and survey of defensive deception for cybersecurity and privacy. *ACM Comput. Surv. (CSUR)* 52 (4): 82.

4 Jajodia, S., Subrahmanian, V., Swarup, V., and Wang, C. (2016). *Cyber Deception*. Springer.

5 Zhu, Q. and Başar, T. (2015). Game-theoretic methods for robustness, security, and resilience of cyberphysical control systems: games-in-games principle for optimal cross-layer resilient control systems. *IEEE Control Syst.* 35 (1): 46–65.

6 Zhu, Q., Saad, W., Han, Z. et al. (2011). Eavesdropping and jamming in next-generation wireless networks: a game-theoretic approach. *Military Communications Conference (MILCOM), 2011*, pp. 119–124. IEEE.

7 Song, J.B. and Zhu, Q. (2019). Performance of dynamic secure routing game. In: *Game Theory for Networking Applications*, 37–56. Springer.

8 Zhu, Q., Li, H., Han, Z., and Başar, T. (2010). A stochastic game model for jamming in multi-channel cognitive radio systems. *2010 IEEE International Conference on Communications*, pp. 1–6. IEEE.

9 Zhang, T. and Zhu, Q. (2017). Strategic defense against deceptive civilian GPS spoofing of unmanned aerial vehicles. *International Conference on Decision and Game Theory for Security*, pp. 213–233. Springer.

10 Xu, Z. and Zhu, Q. (2015). A cyber-physical game framework for secure and resilient multi-agent autonomous systems. *2015 54th IEEE Conference on Decision and Control (CDC)*, pp. 5156–5161. IEEE.

11 Zhu, Q. and Başar, T. (2011). Indices of power in optimal IDS default configuration: theory and examples. In: *Decision and Game Theory for Security* (eds. J.S. Baras, J. Katz, E. Altman), 7–21. Springer Berlin: Heidleberg.

12 Zhu, Q., Tembine, H., and Başar, T. (2010). Network security configurations: a nonzero-sum stochastic game approach. *American Control Conference (ACC), 2010*, pp. 1059–1064. IEEE.

13 Huang, L. and Zhu, Q. (2020). A dynamic games approach to proactive defense strategies against advanced persistent threats in cyber-physical systems. *Comput. Secur.* 89: 101660.

14 Huang, L. and Zhu, Q. (2019). Adaptive honeypot engagement through reinforcement learning of semi-Markov decision processes. *International Conference on Decision and Game Theory for Security*, pp. 196–216. Springer.

15 Rass, S. and Zhu, Q. (2016). GADAPT: a sequential game-theoretic framework for designing defense-in-depth strategies against advanced persistent threats. *International Conference on Decision and Game Theory for Security*, pp. 314–326. Springer International Publishing.

16 Manshaei, M.H., Zhu, Q., Alpcan, T. et al. (2013). Game theory meets network security and privacy. *ACM Comput. Surv. (CSUR)* 45 (3): 25.

17 Chen, J., Touati, C., and Zhu, Q. (2019). Optimal secure two-layer IoT network design. *IEEE Trans. Control Netw. Syst.* 7 (1): 398–409.

18 Farooq, M.J. and Zhu, Q. (2018). On the secure and reconfigurable multi-layer network design for critical information dissemination in the internet of battlefield things (IoBT). *IEEE Trans. Wireless Commun.* 17 (4): 2618–2632.

19 Chen, J. and Zhu, Q. (2020). Control of multi-layer mobile autonomous systems in adversarial environments: a games-in-games approach. *IEEE Trans. Control Netw. Syst.* 7 (3): 1056–1068. 10.1109/TCNS.2019.2962316.

20 Bowers, K.D., van Dijk, M., Griffin, R. et al. (2012). Defending against the unknown enemy: applying FlipIt to system security. In: *Decision and Game Theory for Security* (eds. J. Grossklags and J. Walrand), pp. 248–263. Springer Berlin: Heidelberg.

21 Van Dijk, M., Juels, A., Oprea, A., and Rivest, R.L. (2013). FlipIt: The game of "stealthy takeover". *J. Cryptol.* 26 (4): 655–713.

22 Spence, M. (1973). Job market signaling. *Quarterly J. Econ.* 87 (3), 355–374. www.jstor.org/stable/1882010 (accessed 1 September 2021).

23 Banks, J.S. and Sobel, J. (1987). Equilibrium selection in signaling games. *Econometrica* 55 (3): 647–661. www.jstor.org/stable/1913604 (accessed 1 September 2021).

24 Huang, Y., Chen, J., Huang, L., and Zhu, Q. (2020). Dynamic games for secure and resilient control system design. *Natl. Sci. Rev.*, 7 (7): 1125–1141, 10.1093/nsr/nwz218.

25 Chen, J. and Zhu, Q. (2017). A game-theoretic framework for resilient and distributed generation control of renewable energies in microgrids. *IEEE Trans. Smart Grid* 8 (1): 285–295.

26 Zhu, Q., Yuan, Z., Song, J.B. et al. (2012). Interference aware routing game for cognitive radio multi-hop networks. *IEEE J. Sel. Areas Commun.* 30 (10): 2006–2015.

27 Zhu, Q., Song, J.B., and Başar, T. (2011). Dynamic secure routing game in distributed cognitive radio networks. *2011 IEEE Global Telecommunications Conference (GLOBECOM 2011)*, pp. 1–6. IEEE.

28 Farhang, S., Hayel, Y., and Zhu, Q. (2015). PHY-layer location privacy-preserving access point selection mechanism in next-generation wireless networks. *2015 IEEE Conference on Communications and Network Security (CNS)*, pp. 263–271. IEEE.

29 Zhu, Q., Yuan, Z., Song, J.B. et al. (2010). Dynamic interference minimization routing game for on-demand cognitive pilot channel. *2010 IEEE Global Telecommunications Conference (GLOBECOM 2010)*, pp. 1–6. IEEE.

30 Fung, C.J. and Zhu, Q. (2016). FACID: A trust-based collaborative decision framework for intrusion detection networks. *Ad Hoc Networks* 53: 17–31.

31 Zhu, Q., Fung, C., Boutaba, R., and Başar, T. (2012). GUIDEX: A game-theoretic incentive-based mechanism for intrusion detection networks. *IEEE J. Sel. Areas Commun.* 30 (11): 2220–2230.

32 Zhu, Q., Tembine, H., and Başar, T. (2010). Network security configurations: a nonzero-sum stochastic game approach. *Proceedings of the 2010 American Control Conference*, pp. 1059–1064. IEEE.

33 Rieger, C., Ray, I., Zhu, Q., and Haney, M. (2019). *Industrial Control Systems Security and Resiliency: Practice and Theory, Advances in Information Security*. Springer.

34 Zhu, Q. and Başar, T. (2011). Robust and resilient control design for cyber-physical systems with an application to power systems. *2011 50th IEEE Conference on Decision and Control and European Control Conference*, pp. 4066–4071. IEEE.

35 Zhu, Q., Bushnell, L., and Başar, T. (2013). Resilient distributed control of multi-agent cyber-physical systems. In: *Control of Cyber-Physical Systems*, Lecture Notes in Control and Information Sciences, 449 (eds. D. Tarraf). 301–316. Springer: Heidelberg. 10.1007/978-3-319-01159-2_16.

36 Miao, F., Zhu, Q., Pajic, M., and Pappas, G.J. (2018). A hybrid stochastic game for secure control of cyber-physical systems. *Automatica* 93: 55–63.

37 Huang, L. and Zhu, Q. (2019). A dynamic games approach to proactive defense strategies against advanced persistent threats in cyber-physical systems. *CoRR*, abs/1906.09687.

38 Zhu, Q. and Rass, S. (2018). On multi-phase and multi-stage game-theoretic modeling of advanced persistent threats. *IEEE Access* 6: 13958–13971.

39 Chen, J., Touati, C., and Zhu, Q. (2017). A dynamic game analysis and design of infrastructure network protection and recovery: 125. *ACM SIGMETRICS Perform. Eval. Rev.* 45 (2): 128.

40 Chen, J. and Zhu, Q. (2017). Interdependent strategic cyber defense and robust switching control design for wind energy systems. *2017 IEEE Power & Energy Society General Meeting*, pp. 1–5.

41 Rass, S., Schauer, S., König, S., and Zhu, Q. (2020). *Cyber-Security in Critical Infrastructures: A Game-Theoretic Approach, Advanced Sciences and Technologies for Security Applications*. Springer.

42 Zhang, R. and Zhu, Q. (2020). FlipIn: A game-theoretic cyber insurance framework for incentive-compatible cyber risk management of internet of things. *IEEE Trans. Inform. Forensics Secur.*. 15: 2026–2041. 10.1109/TIFS.2019.2955891.

43 Zhang, R., Zhu, Q., and Hayel, Y. (2017). A bi-level game approach to attack-aware cyber insurance of computer networks. *IEEE J. Sel. Areas Commun.* 35 (3): 779–794.

44 Zhang, R. and Zhu, Q. (2016). Attack-aware cyber insurance of interdependent computer networks.

45 Chen, J. and Zhu, Q. (2018). A linear quadratic differential game approach to dynamic contract design for systemic cyber risk management under asymmetric information. *2018 56th Annual Allerton Conference on Communication, Control, and Computing (Allerton)*, pp. 575–582. IEEE.

46 Pawlick, J., Colbert, E., and Zhu, Q. (2018). Modeling and analysis of leaky deception using signaling games with evidence. *IEEE Trans. Inform. Forensics Secur.* 14 (7): 1871–1886.

47 Zheng, J. and Casta nón, D.A. (2012). Dynamic network interdiction games with imperfect information and deception. *2012 IEEE 51st IEEE Conference on Decision and Control (CDC)*, pp. 7758–7763. IEEE.

48 Zhu, Q., Clark, A., Poovendran, R., and Başar, T. (2012). Deceptive routing games. *2012 IEEE 51st IEEE Conference on Decision and Control (CDC)*, pp. 2704–2711. IEEE.

49 Horák, K., Zhu, Q., and Bošanský, B. (2017). Manipulating adversary's belief: a dynamic game approach to deception by design for proactive network security. *International Conference on Decision and Game Theory for Security*, pp. 273–294. Springer.

50 Huang, Y. and Zhu, Q. (2019). Deceptive reinforcement learning under adversarial manipulations on cost signals. *arXiv preprint arXiv:1906.10571*.

51 Pawlick, J. and Zhu, Q. (2017). A mean-field stackelberg game approach for obfuscation adoption in empirical risk minimization. *arXiv preprint arXiv:1706.02693*.

52 Zhang, R. and Zhu, Q. (2015). Secure and resilient distributed machine learning under adversarial environments. *2015 18th International Conference on Information Fusion (Fusion)*, pp. 644–651. IEEE.

53 Zhang, R. and Zhu, Q. (2018). A game-theoretic approach to design secure and resilient distributed support vector machines. *IEEE Trans. Neural Netw. Learn. Syst.* 29 (11): 5512–5527 10.1109/TNNLS.2018.2802721.

54 Chen, J. and Zhu, Q. (2019). Interdependent strategic security risk management with bounded rationality in the Internet of things. *IEEE Trans. Inform. Forensics Secur.* 14 (11): 2958–2971.

55 Chen, J., Touati, C., and Zhu, Q. (2020). Optimal secure two-layer IoT network design. *IEEE Trans. Control Netw. Syst.* 7 (1): 398–409 10.1109/TCNS.2019.2906893.

56 Chen, J. and Zhu, Q. (2017). Security as a service for cloud-enabled internet of controlled things under advanced persistent threats: a contract design approach. *IEEE Trans. Inform. Forensics Secur.* 12 (11): 2736–2750.

57 Yunhan, H. and Quanyan, Z. (2020). A differential game approach to decentralized virus-resistant weight adaptation policy over complex networks. *IEEE Trans. Control Netw. Syst.* 7 (2): 944–955. 10.1109/TCNS.2019.2931862.

58 Pawlick, J. and Zhu, Q. (2017). Strategic trust in cloud-enabled cyber-physical systems with an application to glucose control. *IEEE Trans. Inform. Forensics Secur.* 12 (12): 2906–2919.

59 Chen, J. and Zhu, Q. (2016). Resilient and decentralized control of multi-level cooperative mobile networks to maintain connectivity under adversarial environment. *IEEE Conference on Decision and Control (CDC)*, pp. 5183–5188.

60 Pawlick, J., Farhang, S., and Zhu, Q. (2015). Flip the cloud: cyber-physical signaling games in the presence of advanced persistent threats. *International Conference on Decision and Game Theory for Security*, pp. 289–308. Springer.

61 Pawlick, J., Chen, J., and Zhu, Q. (2018). iSTRICT: An interdependent strategic trust mechanism for the cloud-enabled Internet of controlled things. *IEEE Trans. Inform. Forensics Secur.* 14 (6): 1654–1669.

62 Nugraha, Y., Hayakawa, T., Cetinkaya, A. et al. (2019). Subgame perfect equilibrium analysis for jamming attacks on resilient graphs. *2019 American Control Conference (ACC)*, pp. 2060–2065. IEEE.

63 Xu, Z. and Zhu, Q. (2018). Cross-layer secure and resilient control of delay-sensitive networked robot operating systems. *2018 IEEE Conference on Control Technology and Applications (CCTA)*, pp. 1712–1717. IEEE.

64 Xu, Z. and Zhu, Q. (2016). Cross-layer secure cyber-physical control system design for networked 3D printers. *2016 American Control Conference (ACC)*, pp. 1191–1196. IEEE.

65 Xu, Z. and Zhu, Q. (2017). A game-theoretic approach to secure control of communication-based train control systems under jamming attacks. *Proceedings of the 1st International Workshop on Safe Control of Connected and Autonomous Vehicles*, pp. 27–34. ACM.

66 Zhu, Q. and Xu, Z. (2020). *Cross-Layer Design for Secure and Resilient Cyber-Physical Systems: A Decision and Game Theoretic Approach*. Springer Nature.

67 Zimmerman, R., Zhu, Q., De Leon, F., and Guo, Z. (2017). Conceptual modeling framework to integrate resilient and interdependent infrastructure in extreme weather. *J. Infrastruct. Syst.* 23 (4): 04017034.

68 Zimmerman, R., Zhu, Q., and Dimitri, C. (2018). A network framework for dynamic models of urban food, energy and water systems (FEWS). *Environ. Progr. Sustain. Energy* 37 (1): 122–131.

69 Zimmerman, R., Zhu, Q., and Dimitri, C. (2016). Promoting resilience for food, energy, and water interdependencies. *J. Environ. Stud. Sci.* 6 (1): 50–61.

70 Rinaldi, S.M., Peerenboom, J.P., and Kelly, T.K. (2001). Identifying, understanding, and analyzing critical infrastructure interdependencies. *IEEE Control Syst. Mag.* 21 (6): 11–25.

71 Chen, J. and Zhu, Q. (2019). *A Game-and Decision-Theoretic Approach to Resilient Interdependent Network Analysis and Design*. Springer.

72 Peng, G., Chen, J., and Zhu, Q. (2020). Distributed stabilization of two interdependent Markov jump linear systems with partial information. *IEEE Control Syst. Lett.* 5 (2): 713–718.

73 Huang, L., Chen, J., and Zhu, Q. (2018). Factored Markov game theory for secure interdependent infrastructure networks. *Game Theory for Security and Risk Management*, 99–126. Springer.

74 Huang, L., Chen, J., and Zhu, Q. (2017). A large-scale Markov game approach to dynamic protection of interdependent infrastructure networks. *International Conference on Decision and Game Theory for Security*, pp. 357–376. Springer.

75 Huang, L., Chen, J., and Zhu, Q. (2018). Distributed and optimal resilient planning of large-scale interdependent critical infrastructures. *2018 Winter Simulation Conference (WSC)*, pp. 1096–1107. IEEE.

76 Chen, J., Touati, C., and Zhu, Q. (2019). A dynamic game approach to strategic design of secure and resilient infrastructure network. *IEEE Trans. Inform. Forensics Secur.* 15: 462–474.

77 Chen, J., Touati, C., and Zhu, Q. (2017). Heterogeneous multi-layer adversarial network design for the IoT-enabled infrastructures. *GLOBECOM 2017-2017 IEEE Global Communications Conference*, pp. 1–6. IEEE.

78 Nugraha, Y., Cetinkaya, A., Hayakawa, T. et al. (2021). Dynamic resilient network games with applications to multiagent consensus. *IEEE Trans. Control Netw. Syst.* 8 (1): 246–259.

79 Chen, J. and Zhu, Q. (2017). A stackelberg game approach for two-level distributed energy management in smart grids. *IEEE Trans. Smart Grid* 9 (6): 6554–6565.

80 Chen, J., Zhou, L., and Zhu, Q. (2015). Resilient control design for wind turbines using Markov jump linear system model with Lévy noise. *2015 IEEE International Conference on Smart Grid Communications (SmartGridComm)*, pp. 828–833. IEEE.

81 Maharjan, S., Zhu, Q., Zhang, Y. et al. (2013). Dependable demand response management in the smart grid: a stackelberg game approach. *IEEE Trans. Smart Grid* 4 (1): 120–132.

82 Maharjan, S., Zhu, Q., Zhang, Y. et al. (2015). Demand response management in the smart grid in a large population regime. *IEEE Trans. Smart Grid* 7 (1): 189–199.

15

Human Challenges
Anshul Rege

Criminal Justice, Temple University, Philadelphia, PA, USA

Objectives

The objective of this chapter is to offer insight into one cybersecurity gamification case study utilizing the Grid Game and multidisciplinary students in electrical and computer engineering (ECE) and criminal justice (CJ). Specifically, it discusses the benefits to ECE students in the areas of understanding grid operations, microgrid stability and generation, generator inertia, and energy storage. It also discusses how CJ students benefit from cybersecurity gamification by learning about adversarial and defender decision-making and conducting cyber-field research. The chapter then discusses some challenges and recommendations for future educational direction, and advocates for using gamification tools, like the Grid Game, to enhance collaborative multidisciplinary experiential learning in the area of cyberattacks and cybersecurity.

15.1 Introduction

Rapidly advancing technologies have resulted in highly interconnected information networks and integrated systems, which has made them more accessible and vulnerable to cyberattacks [1]. Advanced persistent threats (APTs) have increasingly targeted U.S. power grids, constantly circumventing traditional reactive security measures, and potentially resulting in large outages and costly damages [1]. APTs cannot easily be detected or defended [2]. Skilled cybersecurity workers on the front lines must be able to detect these highly intelligent and sophisticated threats that can surpass traditional engineered and deployed defenses [2]. Conventional cyberattack management is response-driven, which is an ineffective approach, especially in curbing APTs [1].

Organizations need to develop defenses that can respond rapidly, proactively, and dynamically to more effectively manage APTs [3]. Not surprisingly, cybersecurity experts agree that there is an immediate need for anticipatory defense measures that reflect the adaptive and dynamic nature of APTs [4, 5]. Developing anticipatory cyber strategies require understanding the human aspects of cyberattacks, that is, how APTs organize, strategize, adapt, and function effectively, and how defenders secure grids and make effective decisions in cyber defense and system operation when experiencing cyberattacks. Current education strategies must consider these challenges inherent in the constantly dynamic cyberthreat landscape [2].

Resilient Control Architectures and Power Systems, First Edition.
Edited by Craig Rieger, Ronald Boring, Brian Johnson, and Timothy McJunkin.
© 2022 The Institute of Electrical and Electronics Engineers, Inc. Published 2022 by John Wiley & Sons, Inc.

One popular approach to train the future workforce is by gamifying cybersecurity [3]. Gamification is the "process of enhancing a specific service by implementing game design elements in a non-game context to enhance the user's overall value creation and experience" [4, 5]. Gamification can include progress metrics (points systems), player control (making decisions and performing actions), rewards and penalties, collaborative problem solving, and competition [3].

This chapter discusses how cybersecurity gamification, via the "Grid Game" platform, can be used to enhance the experiential learning of undergraduate students across multiple Science-Technology-Engineering-Math (STEM) fields about cyberattacks and cybersecurity in the context of critical infrastructure, and evaluate and assess this education. The chapter is structured as follows. The next section discusses the Grid Game and the specifics of a joint exercise case study done between the electrical and computer engineering (ECE) and criminal justice (CJ) departments in the Spring 2016 semester. It demonstrates the benefits to ECE students (grid operations; microgrid stability and generation; generator inertia, and energy storage) and CJ students (understanding adversarial decision-making, defender decision-making, and conducting cyber-field research). Next, this chapter discusses the specific challenges emerging from multidisciplinary communication before and during the collaborative exercise. The chapter concludes with recommendations for future applications of cybersecurity gamification that use the Grid Game, such as introducing variations in the exercises, post-exercise multidisciplinary communication, and analyzing technical logs.

15.2   Experiential Learning and the Multidisciplinary Grid Game

Experiential learning describes a process by which knowledge is acquired through praxis, by doing, reflecting, and trying again with improved methods [6]. Students move away from memorization; rather they must take the initiative in guided experiences and receive feedback based on the event [6]. Thus, the learner's subjective experience is at the heart of the experiential learning process.

The Grid Game software program, developed by Idaho National Laboratory (INL), offers a unique and novel critical infrastructure cybersecurity gamification system that can enhance experiential learning for students across multiple STEM fields. The Grid Game is a simulation of a microgrid system, which provides a rich set of possible decisions in both tactical and strategic time frames. The grid can be attacked with a list of attacks that range in severity from small financial harm to a catastrophic impairment of the power grid. The Grid Game has already been used academically in the physical and cognitive sciences [7–9].

15.2.1   Grid Game Case Study

This chapter uses a case study of a joint cybersecurity course project from the Spring 2016 semester between the ECE and CJ departments. The ECE class had 23 students and the CJ class had 18 students [10].

The ECE students had already downloaded the Grid Game software on their laptops from the game website. The ECE students worked in eight groups of three to four members, where they played the role of electric utility administrators responsible for managing consumer loads minute by minute while trading energy with each other, earning profits, and fending off waves of malicious cyberattacks that try to bring the grid down.

Each group was responsible for maintaining its grid while an INL researcher and Temple ECE graduate students played the role of cyberattackers. The exercise lasted for 2.5 hours and was divided into three rounds of 30 minutes each. Each round itself was structured to have 15 minutes of buildup time and 15 minutes to manage an assortment of cyberattacks launched against them. Furthermore, different teams were randomly subjected to different attacks throughout the three different rounds. At the end of each round, students restarted the Grid Game. Thus, Round 1 served as a warmup, while in Rounds 2 and 3, students were more familiar with the game.

Eighteen CJ students observed each of the ECE teams as they progressed through the three rounds. CJ students were also provided with details about the various attacks, their costs (loss of revenue/ points), and what parts of the grid they targeted. CJ students then worked in groups to create (and justify) attack sequences that could be launched against ECE students. The group sequences were then vetted in class and two attack sequences were chosen for two rounds during the joint exercise. CJ students also had the flexibility of launching certain attacks against certain ECE teams. ECE students were not told about the attack sequences or that the CJ students had generated the attack sequences.

15.2.2 Grid Operations and Cybersecurity

The ECE students experienced the impacts of real-time cyberattacks on the power grid, from stability of the grid to power quality evaluation to power markets and learned about various aspects of grid functionality and security.

15.2.2.1 Grid Operations
All teams had specific approaches to generating revenue and effectively maintaining operations. For instance, one team stated that it wanted to balance out what it bought and how many customers it got with regards to storage. Members did not want to overload any one specific area, such as buying so many customers that it crashed their storage. Another team's approach was to store energy and maintain frequency at a functional level. One member stated that the best way to maintain grid operations was to focus most of the team's efforts on keeping the stored energy at mid-range on the grid as this would give them more time and allow them to react to any changes in the grid with greater efficiency. Overall, the team members discussed the best amount of loads and walked through each input, step, and action [9].

15.2.2.2 Microgrid Stability and Generation Control System
The microgrid stability component tested students' understanding of grid stability based on analysis of swing equation under various scenarios due to cyberattacks, such as loss of a generator, loss of the generator controller, and remedial actions of stabilizing the grid. All teams gained a strong understanding of frequency stability of the grid and were able to explain it from energy conservation point of view [9].

15.2.2.3 Generator Inertia
The generator inertia component tested students' understanding of the effects of generator inertia on system stability, and on whether the generator inertia can play any role in maintaining stability in the event of a cyberattack. On effects of generator inertia, one ECE team explained "by using hydro power the inertia of the system increases, which diminishes the effects on the frequency regulation of the power supplied (i.e. it reduces frequency drift)." At the same time, the team was also concerned that "hydro plants cause large environmental impacts around the dam areas." Another

ECE team had the best understanding of effects of inertia on grid stability; it stated: "The more and bigger the machines, the more the inertia. Hydro generators have large spinning machines. More inertia slows down the change in frequency with power imbalance" [9].

15.2.2.4 Energy Storage

The energy storage component evaluated students' understanding of the spinning reserve and its use for maintaining stability and power balance in the events of cyberattacks. One ECE team had a clear view as it explained that "if a supply interruption takes place in the microgrid, the storage energy will provide the power necessary to keep on supplying the critical load." It also understood that having excessive reserve is also a poor business decision "as the generation and the load consumption are not matched, ... production is not very efficient or optimum" (p. 5). Another team observed that adding distributed renewable generation is a good idea for microgrid as the stored energy can be used during outages of conventional generation due to a cyberattack. They also suggested the concept of encouraging customers for shifting demand in case there is not enough storage [9].

15.2.3 Cyber Adversarial Decision-Making

CJ students had to work in groups to generate their own attack schedules. They were provided with a list of different attacks, the type of harm they caused, and the type of remedy available for each attack (Table 15.1).

CJ students then had to design a sequence of attacks and justify this sequence stating why they picked certain types of attacks at certain times. Each group had a different objective. For instance, CJ2's group justified its attack sequence as follows: "This attack is designed to punish teams who do not have total familiarity and expertise in controlling their grid. As the options are to wipe the system or manually control the [grid], groups that have not sufficiently practiced will likely be dealt a lethal or near lethal blow."

Another CJ group's attack sequence is shown in Table 15.2. It decided to launch denial-of-service attacks against two different ECE teams (T6 and T1) because these attacks "require three different security services to defend and it is unlikely the [ECE T6 has] that much defense so early on in the game. It also messes with their contracts while locked out, so it furthers the destruction." The CJ group said they decided to target ECE T1 a little later so that the engineers would "see the other group (T6) being targeted and start to panic, but also not have enough time to buy security measures."

Thus, the CJ students had the opportunity to think offensively. As CJ1 noted: "playing the attackers was fun because we knew what was coming." CJ students reflected about the pros and

Table 15.1 Attack information provided to criminal justice students.

Name	Type	Remedy available
Little Guy	Small persistent financial	Proactive low cost
Big Guy Virus	Large persistent financial	Proactive medium cost
Denial of Service	Communication disruption	Proactive low cost
Gluxnet	Control system gain setting disruption	Proactive high cost
Blue Frog	Automatic control system disruption. Zero Day	Reactive only – requires manual control skillset to survive

Table 15.2 Sample attack schedule from a criminal justice group.

Time	Attack and justification	Teams	Expected student reaction/decisions
2:25	Denial of Service	T6	Immediate frustration
2:30			
2:35	Denial of Service	T1	Panic
2:40			
2:45			
End 1			

Table 15.3 Final attack schedule designed by criminal justice students.

Round 1		
Time (pm)	**Attack type**	**ECE Teams**
2:25	Denial of Service	All teams
2:28	Gluxnet	Teams 1,2,3
2:28	Little Guy	Teams 4,5,6
2:33	Blue Frog	All teams
2:38	Big Guy Virus	All teams

Round 2		
Time (pm)	**Attack**	**ECE Teams**
3:00	Gluxnet	All teams
3:03	Denial of Service	Teams 2,4,6
3:05	Big Guy Virus	Teams 1,3,5
3:09	Denial of Service	Teams 1,3,5
3:10	Gluxnet	Teams 2,4, 6
3:12	Blue Frog	Team 3
3:15	Blue Frog	Teams 1,2,4,5,6
3:19	Blue Frog	Team 3

cons of attack vectors and the relevance of timing these vectors effectively. By engaging in a debate and voting for the best attack sequence, students were able to also tie this in to attack vectors, attacker motivations, and case studies covered in class [10]. The final attack schedule decided by the CJ students is shown in Table 15.3.

The joint exercise gave CJ students a hands-on, tangible example of the cyberattack processes and implications. As one CJ commented: "we got [a glimpse of] everyday [cyberattack] encounters our power grids face." CJ students also learned about the complexities of cyberattacks, the many possible permutations and combinations of attack vectors, and how "thinking offense" was critical to be effective at "thinking defense"; how could attack schedules be used to understand proactive cybersecurity. While they may not possess the technical know-how of engaging in ethical hacking, they did develop the ability to conceptualize, plan/schedule, and justify attack trajectories. As CJ1 and CJ2 noted, they developed an "appreciation of the complex nature of cyberattacks and the challenges of real-time defense" [10].

15.2.4 Cyber Defender Decision-Making

This joint exercise provided the CJ students with several benefits, such as understanding real-time cyber defense: assessing group behavior, dynamics, and decision-making with regards to grid operations and security.

15.2.4.1 Group Dynamics and Divisions of Labor

CJ students had the chance to understand how cybersecurity functioned as a group event when they studied ECE students. They learned about possible group dynamics, divisions of labor, decision-making, and conflict resolution. One CJ student observed an ECE team that had two members with one serving as the grid operator (physically controlled the Grid Game) while the other had an advisory role (providing guidance on buying, storing, and cybersecurity purchases). Another CJ student observed an ECE group that had three members, with one serving in operator capacity and all members contributed equally to the decision-making process and what actions to complete [10].

15.2.4.2 Cybersecurity Preparations

The grid security component assessed students' understanding of the general concept of grid security and decision-making process of grid owners. All teams achieved a strong understanding of grid security and the need for grid protection. The teams used various tools available at their disposal, such as (i) purchasing antivirus, (ii) increasing stored reserve energy, (iii) manually tuning the generator control system, and (iv) increasing generation inertia. However, during cyberattack events, the teams reacted differently based on their own preference. These decisions were more based on human behavior than on engineering principles since defending the grid costs money and affects their decisions on energy trading [9].

One team had not thought out its defense strategy effectively. When asked if it had any preemptive steps to defend against attacks, one member stated: "Good question, I did not think about that. We will probably buy it later, but I am not really sure about this." Influenced by their own response, team members bought the basic antivirus protection measure, which remained a constant approach across all three rounds; the team never progressed to buying any further defenses. During the first round, another team did not focus much on preparing for security. However, after learning from the first round, the team members focused on security. As one member stated: "at this moment we don't need any [security measures], but we should get at least basic security because we never know when an attack will take place. We want to build up. These are investments in our infrastructure so we want to make sure we are buying what we can afford at the time of the purchase." In the first and second rounds, another team's members were novices with regards to their strategy for purchasing and using cybersecurity defenses. For instance, the team had no idea about the number and types of attacks, and as such believed that the best course of action was to buy the advanced antivirus, which the team assumed would protect their grid against most attacks in general [9].

15.2.4.3 Response to Cyberattacks

One team poorly managed cyberattacks against its grid. When the attacks started, the team members did not know what any of the cybersecurity measures did (except for the basic antivirus) or that there were even multiple options available for defense. Occasionally, the team did not even realize that it was under attack. Team 4 essentially ignored the attack aspect of the game in the first round. For the most part, this team's approach was to react: "We implemented some security measures... it's just that we didn't use them properly... So this round we [need more] time to recognize those attacks" ([9], p. 6). One CJ student commented on the reactionary nature of cybersecurity: "I gathered that the defense aspect really is just like playing whack-a-mole with the attacks." Similarly, another CJ student noted that the "thought process that goes into defending against an attack, [involves] a lot of second guessing." CJ students drew parallels in the lack of cybersecurity knowledge, the trade-off between generating revenue and spending on cybersecurity, and confusion over which cybersecurity measures were effective [10].

15.2.5 Cyber-Field Research

This cybersecurity gamification exercise allowed CJ students to do hands-on "cyber field research." Field research involves collecting primary data from an environment without structuring an experiment or survey. In this research method, the social scientist enters a new environment and experiences that world through observations or participations to observe specific behaviors and actions in that environment. The Grid Game joint CJ-ECE exercise provided an ideal field for CJ students to observe and interview defenders (ECE students) in their "environments" (operating a grid facility) as they protected their grids against cyberattacks.

15.2.5.1 Designing and Conducting Research

CJ students were actively involved in designing and implementing stages of the research. They designed interview and observation guides in groups, which were shared and critiqued in class. Figure 15.1 shows a sample interview guide, while Figure 15.2 shows a sample observation guide, both developed by CJ students. After several discussion and iterations, students finalized a set of questions for the interview guide and a set of key components to observe during the live exercise. While many students had taken a basic research methods class, this particular joint course project gave them the opportunity to apply what they had learned in previous classes to develop interview and observations guides for a specific context.

CJ students then engaged in "field research" by observing and speaking with the ECE students, which allowed the CJ students to "learn essential skills of how to interview individuals with a variety of backgrounds, how to use prompts and probes during the actual interview, and how to manage interview contexts" [6]. CJ students had to "think on their feet" as the exercise progressed. CJ students found that the interview and observation guides complemented and supplemented each other. Based on their observations, CJ students would ask new questions (not listed in the interview guides that they had generated) to the ECE students. Thus, CJ students had to change and adapt their observation and interview techniques at various points during the exercise [10].

15.2.5.2 Weaving Multiple Methods in RealTime

The predetermined attack schedule allowed the CJ students to be effective observers as they knew exactly when to observe reactions from their respective ECE teams. As one CJ student noted, "We consistently asked awareness gauging questions in accordance to the attack schedule to gain insight on how they were measuring up to our anticipated results." Another CJ student echoed

Background questions:

Why did you choose to enroll in the EE program?

Why did you choose to enroll in this course?

What is your backgroud in this field?

Which part of our learning will you be able to apply to the Grid Game?

How comfortable are you with applying your learning to the Grid Game?

How many/Which classes have you taken already?

Do you have any family members who are engineers?

First 20 minutes

Why did you choose your teammate?

Do you believe you will work well together?

Have you played this game before?

Figure 15.1 Excerpt of the interview guide developed by a criminal justice student group.

A. Interaction
(Verbal\physical)

B. Structure
(Seating arrangement/roles)

C. Cohesiveness
(Like-gel/trust/support/respect/resilience/compliment)

C. Coflict
(Arguments/failure/accountability/blame)

E. Emotions/other
(Sadness/anger/happiness/confusion/frustration/hope/competitiveness/relief)

F. Decision-making on strategy
(Pre-planning/ adaptability/ reaction-response time/ prioritize/)

Figure 15.2 Excerpt of the observation guide developed by a criminal justice student group.

this point: "knowing what was going to happen … allowed us to focus more on certain aspects of the game" [10].

15.2.5.3 Exposure to All Phases of Research

CJ students were able to understand the various components of the hands-on "cyber-field" research process: designing and refining data collection instruments, doing field research, understanding and appreciating the research environment, respecting research subjects, managing unanticipated events and hurdles, data coding, data analysis, and formal report writing to disseminate findings [10]. Thus, CJ students were exposed to all aspects of the research process and learned how to manage any challenges they experienced.

15.3 Benefits of Gamifying Cybersecurity

15.3.1 Discipline-Specific Benefits

These projects offered several benefits to CJ students. First, they got a small-scale introduction to real-time cyberattack and cyber defense through a controlled simulated classroom exercise. Second, they moved beyond traditional class assignments based on secondary data collection and analysis to doing hands-on research and engaging in primary data collection. Third, CJ students took part in multidisciplinary research where they had to dialog with ECE students. Fourth, CJ students were no longer limited by their non-technical backgrounds to be able to "think offense" effectively. And fifth, they were able to understand and appreciate the complexities and back-and-forth aspects of cyberattacks and cyber defense [10].

ECE students also benefited from the Grid Game. First, while the Grid Game cyberattacks are not representative of actual breaches on the U.S. power grid, it nonetheless allowed ECE students to play the role of simulated microgrid administrators and experience real-time cyberattacks. Second, they had to concisely formulate and justify their decisions with regards to grid functionality and cybersecurity measures. In doing so, the students improved on their analytical ability, verbalized their thought process, and defended their decisions (even if on occasion they made errors) [9].

15.3.2 Challenges

Students from both CJ and ECE domains had to communicate effectively with each other. During the post-exercise debriefing, several CJ students were dissatisfied with their inability to "break the ice" with ECE students and found themselves limited by their understanding of engineering principles. One means of minimizing these issues is to have a "meet and greet" the week before the joint exercise to explain what CJ students would be doing. This would make both CJ and ECE students familiar and comfortable with each other. To address the issue of limited knowledge of ECE principles, CJ students could be exposed to the Grid Game software before and even engage in a hands-on practice session prior to the joint exercise to better understand ECE students' actions.

CJ2 commented on the difficulties he encountered communicating with his ECE team when it experienced cyberattacks: "… in both round[s], ECE students [were frustrated]. If they became frustrated at any moment, they would stop talking about what was happening as well as almost become apathetic on the game itself and give up." While experiencing cyberattacks, it is not surprising that ECE students would be reclusive and unwilling to talk. Some CJ students managed this lack of communication by focusing heavily on observations rather than interviews to get a read on what ECE students might be experiencing. By focusing on body language, facial expressions, and how they moved around in the Grid Game interface, CJ students were still able to get some insight on how ECE students performed when their grids were subjected to cyberattacks.

15.4 Summary

The joint exercise case study detailed in this chapter is highly significant in expanding the cybersecurity education of non-technical students in the area of critical infrastructure protection. To summarize the main benefits for CJ students, they (i) get a small-scale introduction to real-time cyberattack and cyber defense through a controlled simulated classroom exercise, (ii) move beyond traditional class assignments based on secondary data collection and analysis to doing hands-on research and engage in primary data collection, (iii) participate in multidisciplinary research where they must dialog with ECE students, (iv) realize that they are no longer limited by their non-technical backgrounds are able to "think offense" effectively, and (v) understand and appreciate the complexities and back-and-forth aspects of cyberattacks and cyber defense.

As with any educational exercise, there are lessons learned and recommendations made for future iterations:

1. Introduce variations in the exercise by changing the duration of the overall exercise and attack sequences, altering the attack schedules, switching ECE group members to see effects of group dynamics, decision-making, strategy and planning, and approach to cybersecurity.
2. Engage in multidisciplinary dialog after the joint exercise. Once CJ students are done with their analysis and reports, they can meet with their respective ECE teams to share their findings. This serves to not only "close the loop," but also validate the CJ students' findings.
3. Analyze the technical logs from the exercise to get metrics on ECE student performance and reactions to the cyberattacks. This data could then be married to the CJ students' qualitative interview and observation data to get a more holistic understanding of ECE students' performance.

A global shortage of two million cybersecurity professionals is expected by 2022. Cybersecurity gamification, especially through programs like the Grid Game, will make meaningful contributions that can help address this shortage and enlarge and diversify the future U.S. STEM workforce. By training students from multiple STEM domains (psychology, criminology, sociology, anthropology, computer science, engineering, etc.), gamification can widen the scope of cybersecurity education.

To conclude, training a future social science workforce in the area of cybersecurity would ultimately benefit the area of critical infrastructure protection. First, this group could be involved in the design of an assortment of tabletop exercises geared to the training of ECEs or grid operators; the social science personnel could gauge ECEs' ability to manage grid operations in general and during cyberattacks. Second, social science personnel could engage in qualitative research methods of observations, interviews, and focus groups during these exercises, as well as post-exercise debriefings, to understand ECE and grid operators' concerns about cyberattacks and cybersecurity. Finally, social science personnel could combine their analysis of ECE performance and knowledge to develop, implement, and evaluate effective education programs for the ECE workforce. Thus, having a multidisciplinary workforce would offer a more holistic approach to the area of critical infrastructure cybersecurity, which is a vital strategic imperative.

Thoughtful Questions to Ensure Comprehension

1 This chapter offered the many benefits of cybersecurity education via gamification for ECE and CJ. How would other social science disciplines, such as psychology and sociology also leverage cybersecurity gamification to train their students methodologically and in cybersecurity?

2 How could the student learning findings identified in this chapter be generalized? (How) could the proof of concept developed in this chapter be implemented on a larger, national scale, to determine if the implementation, challenges, and findings can be replicated?

3 If you could design a multidisciplinary joint exercises using the Grid Game, what disciplines would you include, how would you structure and implement the exercises, and how would you evaluate/assess the effectiveness of these exercises as they pertain to student learning?

Further Reading

CRS (Congressional Research Service). 2018. Electric grid cybersecurity. https://fas.org/sgp/crs/homesec/R45312.pdf (accessed 2 December 2018).

Fairley, P. (2004). The unruly power grid. *IEEE Spectrum* 41 (8): 22–27. https://doi.org/10.1109/MSPEC.2004.1318179.

Pan, C. and Long, Y. (2015). Evolutionary game analysis of cooperation between microgrid and conventional grid. *Mathematical Problems in Engineering* 2015: 1.

Ulrich, T., Boring, R., McJunkin, T., and Rege, A. (2019). A visualization approach to performing task analysis of time series event log data from a microworld simulation. In: *Proceedings of the Human Factors and Ergonomics Society Annual Meeting*, vol. 63, No. 1, 1867–1871. Los Angeles, CA: SAGE Publications.

References

1 Cloppert, M. (2009). Security intelligence: attacking the cyber kill chain. http://digital-forensics.sans.org/blog/2009/10/14/securityintelligence-attacking-the-kill-chain (accessed 2 February 2014).

2 Assante, M.J. and Tobey, D.H. (2011). Enhancing the cybersecurity workforce. *IT Professional* 13 (1): 12–15.

3 Start-Engineering.com (2018). In cybersecurity, change describes education and threats alike. http://start-engineering.com/ (accessed 28 September).

4 Deterding, S., Dixon, D., Khaled, R., and Nacke, L. (2011). From game design elements to game-fulness: defining gamification. In: *Proceedings of the 15th International Academic MindTrek Conference*, 9–15. New York, NY: Association for Computing Machinery.

5 Huotari, K. and Hamari, J. (2012). Defining gamification: a service marketing perspective. In: *Proceedings of the 16th International Academic MindTrek Conference*, 17–22. Association for Computing Machinery: New York, NY.

6 Kolb, A. and Kolb, D. (2005, 2005). Learning styles and learning spaces: enhancing experiential learning in higher education. *Academy of Management Learning and Education* 4 (2): 193–212.

7 McJunkin, T.R., Rieger, C.G., Rege, A. et al. (2016). Multidisciplinary game-based approach for generating student enthusiasm for addressing critical infrastructure challenges. ASEE Annual Conference and Exposition, New Orleans, LA. doi: https://doi.org/10.18260/p. 25763.

8 McJunkin, T.R., Rieger, C.G., Johnson, B.K. et al. (2015). Interdisciplinary education through edu-tainment. Electric Grid Resilient Control Systems Course Paper presented at 2015 ASEE Annual Conference and Exposition, Seattle, WA. doi: https://doi.org/10.18260/p. 24349.

9 Rege, A., Biswas, S., Bai, L. et al.,(2017). Using simulators to assess knowledge and behavior of "novice" operators of critical infrastructure under cyber attack events. Proceedings of the 10th International Symposium on Resilient Control Systems (ISRCS). Institute of Electrical and Electronics Engineers (IEEE).

10 Rege, A., Parker, E., and McJunkin, T. (2017). Using a critical infrastructure game to provide realistic observation of the human in the loop by criminal justice students. Proceedings of the 10th International Symposium on Resilient Control Systems (ISRCS), Institute of Electrical and Electronics Engineers (IEEE).

Part VI

Additional Design Considerations

Resilient Control Architectures and Power Systems, First Edition.
Edited by Craig Rieger, Ronald Boring, Brian Johnson, and Timothy McJunkin.
© 2022 The Institute of Electrical and Electronics Engineers, Inc. Published 2022 by John Wiley & Sons, Inc.

16

Interdependency Analysis

Ryan Hruska

National and Homeland Security, Idaho National Laboratory, Idaho Falls, ID, USA

Objectives

The objectives of this chapter are to give the reader an overview of infrastructure dependencies and existing techniques used to evaluate them and to provide specific examples of cross-sector dependencies that are directly related to bulk power systems and the potential risk they pose.

16.1 Introduction

The critical infrastructures landscape is a vast, varied pattern of independent, but interconnected systems that are essential to the functioning of our modern society. The United States policy on critical-infrastructure protection has been continually evolving since the Presidential Commission Report on critical infrastructure protection was published in 1997 [1]. This report highlighted the fact that critical infrastructures are complex and interconnected systems and were becoming increasingly vulnerable to physical and cyberattacks that could result in cascading, escalating, and common-cause failure. Presidential Decision Directive No. 63 followed this report in 1998 [2]; it formally established a national program for critical-infrastructure protection, assigning sector-specific lead agencies the responsibility to coordinate and develop sector plans that were to be integrated into a national infrastructure-assurance plan. Following the events of 11 September, numerous executive orders (EOs) and presidential directives (e.g. EO 13228, EO 13231, and Homeland Security Presidential Directive [HSPD]-7) were signed, establishing the Department of Homeland Security and expanding the focus beyond cybersecurity to include physical threats as well. On 12 February 2013, Presidential Policy Directive (PDD)-21 was signed, further broadening the focus to include an all-hazards approach intended to strengthen the nation's security and resiliency posture for critical infrastructure [3]. In response to these policies, federal, state, and local governments and research institutions have invested substantial time and effort into identifying and analyzing critical infrastructures, their functions, dependencies, and interdependencies, in order better understand their vulnerabilities and their risk profiles [4–6].

16.1.1 Dependencies and Interdependencies

The 1996 Critical Infrastructure Working Group (CIWG) Report, created in response to PDD-39, provides one of the earliest known definitions of infrastructure that links it with the concept

Resilient Control Architectures and Power Systems, First Edition.
Edited by Craig Rieger, Ronald Boring, Brian Johnson, and Timothy McJunkin.

of interdependence. The CIWG Report defined critical infrastructure as "a framework of interdependent networks and systems comprising identifiable industries, institutions, and distribution capabilities that provide a continuous flow of goods and services essential to the defense and economic security of the United States, the smooth functioning of the government at all levels, and society as a whole." This definition provided a foundation for EO 13010, which established the President's Commission on Critical Infrastructure Protection. The Commission's report influenced a new body of infrastructure dependency research and has resulted in multiple proposed ontologies and taxonomies intended to contextualize the domain. One of the most widely cited frameworks was proposed by Rinaldi et al. [7] and utilizes the concept of complex adaptive systems (CASs) to provide a methodology to define, characterize, and analyze infrastructure dependencies (physical, cyber, geographic, and logical) to better understand their operational states and risk to potential failure modes (cascading, escalating, and common cause). Other examples of dependency-classification schemes include one by Zimmerman [8], which included functional and spatial types, another by Pederson et al. [9] that included physical, geospatial, policy, and informational types, and Zhang et al. [10], which included functional, physical, budgetary, and market/economic types. Fundamentally, each of these methods has the same goal of achieving a better understanding of the relationships that exist between infrastructure systems and potential risks introduced by specific types of relationships, known as dependencies. A dependency relationship exists when the state of one infrastructure can impact the state of another. Interdependencies are a special type of relationship that is created when feedback loops are established between two or more infrastructure facilities or systems. Figure 16.1 provides a simple example of some common cross-sector interdependencies.

16.1.2 Electric-Grid System Dependencies

An electric grid is typically segmented into three functional areas: generation, transmission, and distribution. As the names imply, the generation segment is responsible for generating electricity, the transmission segment, for transmitting electricity from the generators to the load centers, and the distribution segments for servicing consumers within a load center. Each of the segments

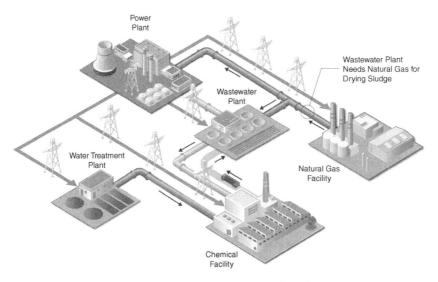

Figure 16.1 Example cross-sector interdependency relationships.

has common and unique intra- and intersystem dependencies. Common dependencies that exist across all segments include electricity and communications. Electricity is required to run the equipment needed to operate and monitor the grid's infrastructure as well as meter the generation, transfer, and consumption of electricity across the grid. The intrasystem dependency on electricity drives significant investment in black-start planning and resources. Like electricity, communications are required to effectively operate and monitor the modern grid. Unlike electricity, the grid's dependence on communications varies depending on the function that a specific system supports. Functional groups include control, protection, operation, dispatching, scheduling, and revenue metering. It is beyond the scope of this chapter to discuss each of these functional groups and the potential impacts of interruption in detail; however, it is important to understand that grid operations are facilitated by both data and voice communications, which are enabled by a range of information and communication technologies (ICTs), such as optical fiber and radio systems. These communication systems are critically dependent on electricity to power the ICT equipment and connectivity to enable all forms of communication. Due to the critical need for communications, regulatory requirements establish a need to maintain multiple redundant forms of communications, often using both private infrastructure and commercial communication providers. The greatest difference between segments exists between the generation segment and the two-circuit base segments. The primary difference is that generation of electricity is critically dependent on fuel supply chains, which in turn are dependent on transportation systems, processing facilities, and key resources. The natural gas and water case studies below provide more-detailed information on these dependencies.

16.2 Approaches to Infrastructure Dependency Analysis

There has been a significant level of effort across government and academia to develop new modeling and simulation capabilities to better characterize and understand the interdependent behavior of infrastructure systems to improve resilience and make better investment decisions. To date, the ability to assess vulnerabilities, conduct risk assessments, and identify priorities for protective and support measures for interdependent critical infrastructure from an all-hazards perspective remains an open and difficult problem. Efforts across the critical-infrastructure community at the federal, state, and local levels often lack an integrated, coordinated, and holistic approach to define and display the current state of security and resilience of critical-infrastructure interconnectedness, interdependencies, and associated all-hazards risks. This is primarily because critical infrastructures display a wide range of spatial, temporal, operational, organizational, and interdependency characteristics that can affect their ability to adapt to changing conditions. The inherent complexity of these systems can introduce subtle interactions and feedback mechanisms that often lead to unintended behavior and consequences during disruption. Understanding of vulnerabilities and risk from an all-hazards perspective is further complicated because, in most cases, the cyber and physical domains have been addressed independently of one another. This separation has led to lack of a shared understanding of the threat and hazard landscape. This section provides a brief overview of some of the approaches used to assess infrastructure dependencies.

16.2.1 Engineering Models

Utilization of industry-vetted high-fidelity engineering models has generally been considered the gold standard for assessing consequences of dependency disruptions on a system of interest.

Example solutions for the bulk-power system that are often put forward include Siemens' PSS/E and PowerWorld's Simulator software platforms. However, these solutions are designed specifically for power-system analysis and have little or no awareness of external dependencies. This deficiency is common across all system- or sector-based engineering models. In order to address this short coming, research groups often employ the concept of federated co-simulation. Co-simulation seeks to leverage system-specific high-fidelity models by encapsulating them in a federated simulation. Early approaches often leveraged the IEEE 1516 high-level architecture (HLA) standard to establish how each infrastructure federate interacts through publish and subscribe mechanisms [11, 12]. Walsh et al. utilized HLA to develop the Critical Infrastructure Protection and Resilience Simulation (CIPRSim) federated model to simulate the effects of natural hazards on both the electrical grid and communication infrastructure while accounting for the interdependencies between the systems. The power federate uses an instance of a real-time digital simulator (RTDS) and the communication federate utilized the QualNet communication-system simulator. More recently, the Hierarchical Engine for Large-Scale Infrastructure Co-Simulation (HELICS) framework was developed by the Department of Energy's (DOE's) Grid Modernization Laboratory Consortium (GMLC) to enable large-scale simulation of regional and interconnection-level power systems. Palmintier et al. [13] utilized HELICS to interface with transmission, distribution, communication, and market simulators to help evaluate the reliability of the bulk-power system.

Federated co-simulation provides tremendous promise for evaluating cross-sector dependencies. However, there are significant challenges that need to be overcome to successfully implement and maintain a co-simulation environment. Challenges include topics like model selection, model resolution, model time synchronization, data availability and maintenance, and result interpretation.

16.2.2 Systems Engineering

NASA defines systems engineering (SE) as a "disciplined approach for the definition, implementation, integration and operations of a system (product or service) with the emphasis on the satisfaction of stakeholder functional, physical and operational performance requirements in the intended use environments over its planned life cycle within cost and schedule constraints"[14]. The SE process is generally composed of the following techniques: requirements analysis, functional analysis, synthesis, system analysis, and controls. Understanding that engineers use this multidisciplinary approach to design systems, it is intuitive to use it to retrospectively model infrastructure dependencies using similar techniques. For example, functional analysis, which is defined as a "systematic process of identifying, describing, and relating the functions a system must perform in order to be successful" has been used by Bobbio et al. [15] to model interdependencies between the power grid and telecommunication networks. Common functional-analysis tools include functional-architecture analysis, functional-flow block diagrams (FFBDs), n-square diagrams, and functional-timeline analysis.

16.2.3 Geospatial Modeling

Geospatial modeling techniques are commonly used to model and assess infrastructure systems due to the ability to provide spatial context of the system of interest and have been widely used to understand how natural hazards can disrupt a system. In addition, geographic information system (GIS) technologies are used to map utility networks for asset-management and response purposes. These capabilities, along with spatial operations such as find nearest, allow for the development of a heuristic to predict functional or dependency relationship. A major challenge

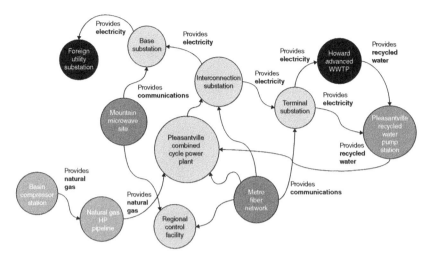

Figure 16.2 Functional dependency model.

of geospatial-dependency modeling is obtaining geospatial datasets for each of the infrastructure systems of interest.

16.2.4 All-Hazards Analysis

The all-hazards analysis (AHA) framework is a dynamic critical-infrastructure dependency-analysis framework that enables knowledge discovery and decision support related to vital and vulnerable infrastructure assets [16]. Developed by Idaho National Laboratory (INL), AHA provides an optimized framework for the collection, storage, analysis, and visualization of critical-infrastructure information and is used to identify dependencies to provide a comprehensive view of interconnected systems. AHA utilizes a dynamic, scalable, graph-style database to develop and populate function-based infrastructure-interdependency profiles from structured (i.e. tabular or database) and unstructured (i.e. textual) data sources as well as through use of geospatial heuristics. The information is presented in the form of nodes (infrastructure assets) and links (dependency relationships) with specific infrastructure properties and attributes (Figure 16.2). This information is maintained through a meta-model that allows for data validation of multiple co-referent data sources, establishment of a confidence metric based on types of data source and their currency, and the ability to tag documents and images to infrastructure nodes and links. AHA allows for visualization of interdependency information in linked views consisting of geospatial-map interfaces and logical graph-based views. AHA can also be used to simulate cascading events, allowing analysts to understand and view nth-order effects associated with infrastructure disruptions. AHA provides a hybrid functional analysis and geospatial modeling approach to cross-sector infrastructure dependency analysis.

16.3 Bulk Power Systems Interdependency Case Studies

16.3.1 Natural Gas Expansion

The demand for natural gas in the continental United States has steadily increased since 2000, primarily due to increases in available reserves and market prices. In order to meet this new

U.S. utility-scale electric generating capacity by initial operating year (as of December 2016)
gigawatts

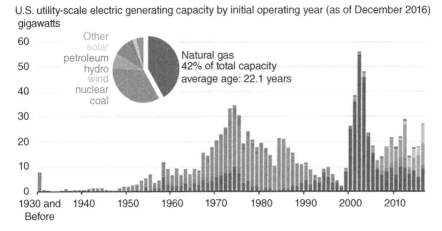

Figure 16.3 Generating capacity by energy source. Source: Data from EIA [17].

demand, natural gas production has risen by more than 66% from 2005 to 2016 [17]. One contributing factor is electricity producers transitioning away from more-costly fuels, such as coal, to natural gas. Figure 16.3 highlights this transition, showing significant new natural gas generation capacity coming online between 2000 and 2005. As of 2016, natural gas fired plants accounted for a record high 43.6% of installed operating capacity and 33.5% of net generation. Further, the Energy Information Administration (EIA) projects that natural gas will continue to expand its market share over the next 30 years, increasing its share of installed capacity to 50%. This increased capacity has occurred and is projected to expand across all generation segments, including baseload, intermediate load, peaking, and distributed generation. It is important to note that the industrial sector is also experiencing a similar significant increase in natural gas consumption, nearly equaling the electricity sector.

As the number of natural gas power plants increases, a system of inter- and intrastate pipelines is being used to transport the natural gas from production wells and processing plants to the point of use or end consumer. Gas deliveries to electric power producers and industrial consumers accounted for 69% of total consumption by volume in 2017. Currently, seven primary production regions have been identified within the United States. These regions account for the majority of the natural gas produced nationally (Figure 16.4). These production regions span from Texas to North Dakota (Anadarko, Bakken, Eagle Ford, Haynesville, Niobrara, and Permian) and within the Marcellus and Utica formations in the Northeast (Appalachia). The Appalachia Region is by far the fastest growing and now largest natural gas producing region in the continental United States, producing 27 562 million cubic feet per day, consistently adding new production year over year since 2009. Pennsylvania is becoming one of the top dry gas producing states in the nation.

Figure 16.4 illustrates which states were net consumers or producers of natural gas in 2016 and demonstrates the national reliance on interstate transportation to meet demand, with 31 states producing less than 10% of the gas required to meet their 2016 demand. Although this growth promises the provision of cheaper, cleaner electricity, an increased reliance on natural gas fired electricity presents potentially myriad regional economic, resilience, and security risks that require consideration.

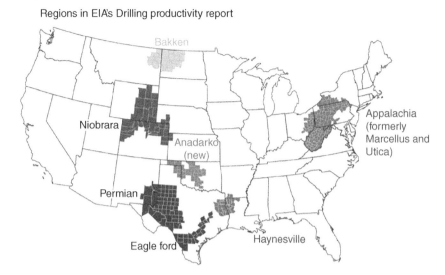

Regions in EIA's Drilling productivity report

Figure 16.4 Primary natural gas production regions. Source: Based on EIA [17].

16.3.1.1 Natural Gas Interdependencies for Electric Generation

The rapid expansion of natural gas fueled electric-power generation has significantly increased the electricity sector's dependency on the natural gas supply systems. At the same time, the natural gas sector has increased its regional dependency on electric power through the installation of electric-powered compressors in upstream gathering fields. Electric compressors are also being installed in urban areas to reduce cost and address noise and air-pollution issues [18]. In order to understand this mutual dependence and its variability, one must consider the operational characteristics of each of the systems in parallel. First, consider the operation of baseload-power generation facilities which require a constant flow of natural gas to produce enough electricity to meet the minimum demand of its consumers. The constant-flow requirement impacts a pipeline's line-pack flexibility and requires operational adjustments to maintain required delivery pressures [19]. Operating a pipeline close to its maximum transport capacity potentially increases its sensitivity to loss of line-pack from production, system, or transfer disruptions. This mutual dependence makes it critical that regional power producers evaluate and determine the appropriate level of installed baseload capacity that should be connected to any one pipeline segment, subsystem, and system. Similar considerations should be taken for intermediate, peak, and distributed power generators utilizing both firm and interruptible agreement contracts.

In natural gas producing regions and urban areas utilizing electric-motor-driven compressors, additional coordination is required to ensure that electrical outages unrelated to the natural gas supply do not cascade into natural gas systems and cause new or escalate the existing outages. In addition, a North American Electric Reliability Corporation (NERC) report, *Special Reliability Assessment: Potential Bulk Power System Impacts Due to Severe Disruptions on the Natural Gas System* [20], identifies unique regional-risk profiles, shown in Table 16.1.

Unlike most other fuels used for electricity generation, natural gas requires a constant flow of fuel to maintain productivity and system integrity. Although the pipeline system includes significant natural gas storage capacity, few natural-gas fired power plants possess enough storage space on-site to keep their units running for any extended period in the event of a pipeline supply disruption. Most natural gas storage facilities are maintained by pipeline operators to ensure that additional firm and interruptible demand during winter months can be met; however, as

Table 16.1 NERC regional risk profiles [20].

2017 NERC Natural gas regional risk profiles	
Area	**Risk description**
Northwest	The northwest does not have significant natural gas storage but also has less reliance on natural gas generation. This area is able to bring in Canadian natural gas supplies as well as domestic supplies in order to meet its natural gas needs
Southern California and Arizona	This area has a high degree of dependence on storage, notably the Aliso Canyon storage facility. Ramping needs, due to an increased penetration of distributed energy resources and utility scale solar photovoltaic, have made storage needs more significant in this area. Limited dual fuel capability adds additional reliability concerns to the reliance on natural gas infrastructure in this area. Natural gas storage may be limited geographically in Arizona due to its proximity to a sole source aquifer for water use
East Texas, Louisiana, and Oklahoma	This area benefits from significant levels of natural gas production and a well-developed system of both interstate and intrastate natural gas pipeline facilities. Additional production area storage facilities provide added deliverability to the area
Southeast	The southeast has significant amounts of storage, production, and pipeline capacity. A sizable amount of electric generation in this area is backed by firm contracts as well as having dual fuel capability
Florida	Florida relies heavily on natural gas generation with close to 70% of its peak requirement relying on natural-gas-fired generation. Firm fuel and dual fuel capabilities provide effective mitigation for this area. Florida has no market area storage and relies on out-of-area supply to meet their demand requirements and out-of-area storage facilities to mitigate supply disruptions or extreme peak conditions
New England	New England has no storage facilities while relying significantly on natural gas and liquefied natural gas supplies. It has limited infrastructure compared to the demand of natural gas in the area for electric generation. Disruption to any of the major trunk lines or deliveries would likely force generation out of service. Under peak conditions demand may not be served; however, under light load conditions some of these outages can be managed by system operators. Lack of firm transportation by electric generators in this area contribute to its risk profile

Source: NERC [20].

natural gas becomes more ubiquitous in electricity generation and other sectors, natural gas storage facilities are being increasingly utilized year round, not just during peak events. While increasing natural gas storage capacity has the ability to make the system more resilient when facing extraction or transmission-pipeline problems, increased reliance on off-site storage units to meet firm electricity-generation demand simply shifts the risk further down the natural gas supply chain. This situation is illustrated by the 2016 temporary closure of the Aliso Canyon underground storage facility due to a major gas leak which resulted in regional gas constraints [21].

16.3.1.2 Seasonal Impacts
As the total share of electricity generated from natural gas has increased, electricity systems have become more reliant on gas-fired generation to meet electricity demand throughout the year, not

just during summer months when interruptible gas pipeline capacity is usually plentiful. While electricity demand is not as high in winter months as it is in summer, demand tends to increase on very cold winter days, which can lead to increased power-sector demand at the same time residential and commercial gas demands are peaking. Such was the case during the winter of 2013–2014 [22].

In early January 2014, the Midwest and parts of the Eastern United States experienced intense periods of cold temperatures, which became known as the Polar Vortex. High natural gas demand during the winter of 2013–2014 contributed to sustained pipeline constraints and price spikes. Natural gas marketers had to contend with demand spikes from both electricity generators and household customers as temperatures dropped and demand spiked. Additionally, the cold temperatures created well-head freeze offs, rendering them inoperable and increasing the level of stress placed on the remaining operational gas-extraction and transportation-system components [22]. As temperatures fell, many electricity generators required more gas than their firm contracts allowed from marketers. Additionally, outages of coal units in PJM's territory on the coldest days of the 2013–2014 winter contributed to an increase in gas demand and prices, as well as concerns about electricity reliability.

Any additional, as-needed gas, known as interruptible supply, is purchased at spot-market prices. However, because pipeline operators design the pipe capacity with firm contract needs in mind, when a majority of electricity generators in a system require the maximum firm supply and additional interruptible supply, the limited capacity of pipeline infrastructure restricts marketers' ability to deliver all of the gas necessary to meet peak demand.

A NERC analysis of grid operations during the event noted that, while natural gas fired power plants provided 40% of electricity-generation capacity in affected areas, they accounted for more than 55% of power outages, due mainly to issues with fuel supply and frozen equipment. Unlike the situation of the electric grid, when portions of the natural gas pipeline network are stressed to capacity or brought offline due to weather or other physical circumstances, it is difficult to reroute gas through other nodes and pipeline segments to ensure that affected gas-fired generators are not brought offline due to localized supply shortages. Although there was enough gas in regional storage facilities due to increased domestic gas production to meet cold weather demand, mostly due to growth in output from the Marcellus Shale, in some cases, there was not enough pipeline space to get gas where it needed to go.

16.3.2   Water Interdependencies

Water plays a fundamental role in the energy sector; for example, water is critical for steam generators and cooling at thermoelectric plants, such as combined-cycle and nuclear facilities, for power production at hydropower plants, and for refining crude oil into fuel. According to the United States Geologic Survey (USGS), thermoelectric plants accounted for 51% of total fresh surface-water and 57% of saline surface-water withdrawals [23]. Figure 16.5 below shows the total estimated water withdrawals made for thermoelectric power production in 2010.

Water systems that supply these plants are regional in nature and can be stressed by regional growth and migration, as well as climate shifts that affect temperature, rain- and snowfall, and evaporation patterns. As these forces increase the demand and, possibly, reduce the available supply, water utilities are constructing recycled-water systems to optimize the supply by transitioning power and industrial plants on to these new systems. This transition adds complexity by introducing two additional systems to the dependency chain: wastewater and recycled-water systems. Now, instead of requiring only electricity for a drinking water system's

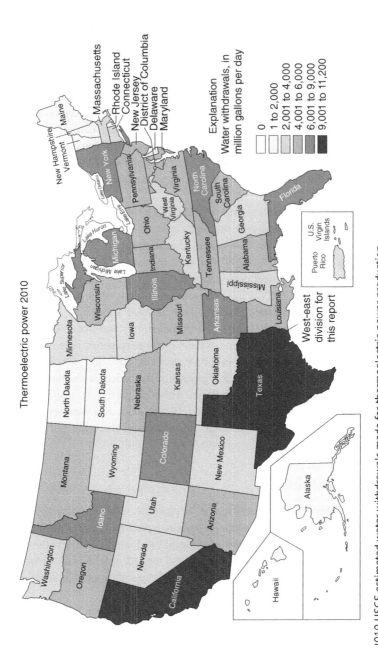

Figure 16.5 2010 USGS estimated water withdrawals made for thermoelectric power production.

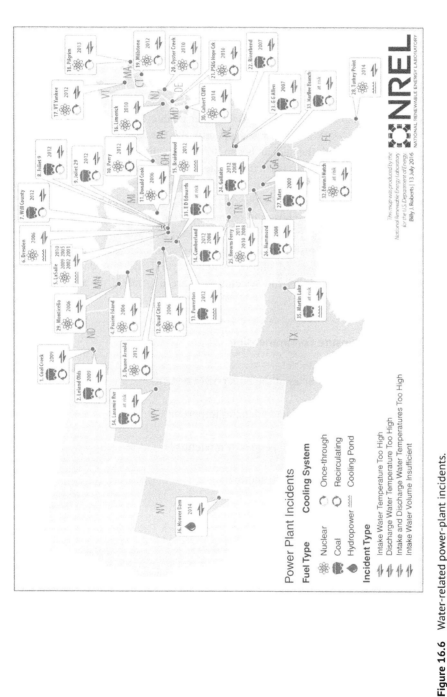

Figure 16.6 Water-related power-plant incidents.

pump stations and water treatment plants, the power systems are required to provide electricity to the wastewater lift station and wastewater treatment plants as well as the recycled-water pump stations to convey the water they require for generation. This added complexity requires careful analysis of load-shed plans and regional drinking-water conservation measures in order to ensure the required water reaches the electric-generation station. Figure 16.6 shows the water-related power plant curtailments from the *National Renewables Energy Laboratory's (NREL) Water-Related Power Plant Curtailments: An Overview of Incidents and Contributing Actors* [24].

16.4 Summary

Critical infrastructure systems provide vital services that enable our modern society and have evolved into a complex network of interacting systems. In the late 1900s, the United States and other governments around the global began to openly discuss and evaluate the interdependent relationships between infrastructure systems and their potential vulnerability to disruptions. Twenty years later governments, utilities, and researchers are still actively developing new methods to identify and assess these critical dependencies and the vulnerabilities they introduce. These new approaches are reimagining traditional methods and leveraging modern computing power to model these relationships. However, to achieve our goal of resilient societies served by resilient infrastructure systems, governments, utilities, and engineers will need to make concerted efforts to look beyond the boundaries of individual systems and consider the resilience of the systems they are dependent on and the ones they enable.

Thoughtful Questions to Ensure Comprehension

1 What are the primary dependencies of the bulk power system?

2 Why is understanding cross-sector dependencies important?

3 What are the benefits of functional-dependency modeling?

4 What is the definition of critical infrastructure?

Further Reading

Laugé, A., Hernantes, J., and Sarriegi, J.M. (2015). Critical infrastructure dependencies: a holistic, dynamic and quantitative approach. *International Journal of Critical Infrastructure Protection* 8: 16–23.

Lewis, T.G. (2019). *Critical Infrastructure Protection in Homeland Security: Defending a Networked Nation*. Wiley.

Setola, R., De Porcellinis, S., and Sforna, M. (2009). Critical infrastructure dependency assessment using the input–output inoperability model. *International Journal of Critical Infrastructure Protection* 2 (4): 170–178.

References

1 President's Commission on Critical Infrastructure Protection (ed.) (1997). *Critical Foundations: Protecting America's Infrastructures*. Washington, DC: The White House.

2 The White House (1998). *Presidential Decision Directive 63*. Washington, DC: The White House.

3 The White House (2013). *Presidential Policy Directive- Critical Infrastructure Security and Resilience*. Washington, DC: The White House.

4 McLean, C., Lee, Y., Jain, S., and Hutchings, C. (2011). *Modeling and Simulation of Critical Infrastructure Systems for Homeland Security Applications*. Gaithersburg, MD, Tech. Rep. NISTIR 7785: US National Institute of Standard and Technology.

5 Ouyang, M. (2014). Review on modeling and simulation of interdependent critical infrastructure systems. *Reliability Engineering & System Safety* 121: 43–60.

6 Yusta, J.M., Correa, G.J., and Lacal-Arántegui, R. (2011). Methodologies and applications for critical infrastructure protection: state-of-the-art. *Energy Policy* 39 (10): 6100–6119.

7 Rinaldi, S.M., Peerenboom, J.P., and Kelly, T.K. (2001). Identifying, understanding, and analyzing critical infrastructure interdependencies. *IEEE Control Systems Magazine* 21 (6): 11–25.

8 Zimmerman, R. (2004). Decision-making and the vulnerability of interdependent critical infrastructure. 2004 IEEE International Conference on Systems, Man and Cybernetics (IEEE Cat. No. 04CH37583). IEEE.

9 Pederson, P., Dudenhoeffer, D., Hartley, S., and Permann, M. (2006). *Critical Infrastructure Interdependency Modeling: A Survey of US and International Research*, vol. 25, 27. Idaho National Laboratory.

10 Zhang, P., Peeta, S., and Friesz, T. (2005). Dynamic game theoretic model of multi-layer infrastructure networks. *Networks and Spatial Economics* 5 (2): 147–178.

11 Eusgeld, I., Nan, C., and Dietz, S. (2011). "System-of-systems" approach for interdependent critical infrastructures. *Reliability Engineering & System Safety* 96 (6): 679–686.

12 Walsh, S., Cherry, S., and Roybal, L. (2009). Critical infrastructure modeling: An approach to characterizing interdependencies of complex networks and control systems. 2009 2nd Conference on Human System Interactions, IEEE.

13 Palmintier, B., Krishnamurthy, D., Top, P. et al. (2017). Design of the HELICS high-performance transmission-distribution-communication-market co-simulation framework. 2017 Workshop on Modeling and Simulation of Cyber-Physical Energy Systems (MSCPES), IEEE.

14 Hirshorn, S.R., Voss, L.D., and Bromley, L.K. (2017). *Nasa Systems Engineering Handbook*. NASA.

15 Bobbio, A., Bonanni, G., Ciancamerla, E. et al. (2010). Unavailability of critical SCADA communication links interconnecting a power grid and a Telco network. *Reliability Engineering & System Safety* 95 (12): 1345–1357.

16 Fisher, R., Norman, M., and Peerenboom, J. (2018). *Resilience History and Focus in the USA, in Urban Disaster Resilience and Security*, 91–109. Springer.

17 EIA (2020). *Today in Energy*. U.S. Energy Information Administration.

18 INGAA Foundation (1992). *Gas Compressor Industry Noise Regulation and Control Review Handbook*. Washington, DC: INGAA Foundation.

19 How Does the Natural Gas Delivery System Work? https://www.aga.org/natural-gas/delivery/how-does-the-natural-gas-delivery-system-work-/ (accessed 01 September 2021).

20 NERC (2017). *Special Reliability Assessment: Potential Bulk Power System Impacts Due to Severe Disruptions on the Natural Gas System*. North American Electric Reliability Corporation.

21 State of California (2018). *Aliso Canyon Mitigation Measures Impact Report (May 2018 Update)*. San Franscisco, CA: Public Utilities Commission.

22 NERC (2014). *Polar Vortex Review*. North American Electric Reliability Corporation.

23 Diehl, T.H., Harris, M., Murphy, J. et al. (2013). Methods for estimating water consumption for thermoelectric power plants in the United States. U.S. Department of the Interior, U.S. Geological Survey.

24 McCall, J., Macknick, J., and Macknick, J. (2016). *Water-Related Power Plant Curtailments: An Overview of Incidents and Contributing Factors*. Golden, Colorado: National Renewable Energy Laboratory (NREL).

17

Multi-agent Control Systems

Craig Rieger

National and Homeland Security, Idaho National Laboratory, Idaho Falls, ID, USA

Objectives

The objective of this chapter is to provide background in the application of agent systems and how they could be applied to the unique dynamical elements of a control system. The foundation of this development is to establish architecture for the future of distributed control that is resilient to disturbances. Thus, this chapter will provide background on the essential features of agents and how multi-agent systems introduce humandecision-making directly into the autonomy but, due to the distributed nature of the system, allow for self-healing and time-sensitive responses to maintain the resilience of the system, which can include loss of elements due to threats.

17.1 Introduction

17.1.1 What Is an Agent?

The suggestion of agents was developed as a mechanism to describe objects, "informing, requesting, offering, accepting, rejecting, competing with, and assisting one another" [1, 2]. Early development of agent identities was directed toward artificial intelligence in computer science applications [3], as compared to the dynamics of control. The definition of an agent is varied, depending upon its application [4]. For the purposes of this paper, the characteristics of an intelligent agent are the most applicable. In general, an intelligent agent can be described as a semi-autonomous entity that evaluates the state of its environment and applies control actions toward achieving goals within this environment, such as is depicted in Figure 17.1. By semi-autonomous, the implication is that the agent has a predefined sphere of influence that establishes the boundaries of its autonomy. Considering the definition for a resilient control system, further refinement can be made. It follows that a "resilient" intelligent agent is one that maintains a state awareness of its environment and responds to disturbances to maintain operational normalcy within this environment.

17.1.2 Intelligent Agent

An intelligent agent can generally be described as a semi-autonomous entity that evaluates the state of its environment and applies control actions toward achieving goals within this environment [6]. While an intelligent agent could be fully autonomous, for the purposes of our discussion on control

Resilient Control Architectures and Power Systems, First Edition.
Edited by Craig Rieger, Ronald Boring, Brian Johnson, and Timothy McJunkin.
© 2022 The Institute of Electrical and Electronics Engineers, Inc. Published 2022 by John Wiley & Sons, Inc.

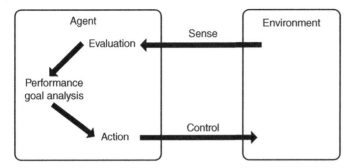

Figure 17.1 Agent attributes. Source: Rieger et al. [5].

system, it is assumed that some sphere of influence and goals are established by a separate entity. These corrections could continue throughout the life of the agent in a distributed control system.

These intelligent reasoning in these agents can be based upon artificial methods such as fuzzy logic, neural networks, etc. or a combination of reasoning methodologies. It could also include other data-driven or physics elements, but it would still require the inclusion of a reasoning aspect that represents the ability to emulate human behaviors. The application of these agents can include both event- and time-based, the former defined by a processing activity executing based upon the triggering of some event and the latter the execution occurring at a predetermined real-time interval [7].

17.1.3 Resilient Agent

A resilient agent is an intelligent agent that maintains a state awareness of its environment and responds to disturbances to preserve operational normalcy within this environment. The concept is to align the definition of resilience, the ability to achieve goals are now multi-faceted and considered from the cognitive, cyber–physical standpoint.

While this distinction can seem minimal from an intelligent agent, the importance of the concept is the establishment of a resilience threshold or manifold that defines the operation that is characterized by many types of disturbances. As part of the inter-agent dynamic, individual agents may negotiate for resources to achieve overarching goals. For example, if their agents were associated with adjacent microgrids and the main transmission source was lost, individual agents could negotiate priorities based upon predetermined management layer criteria to maintain a local area or regional power supply to critical loads.

17.1.4 Multi-agents and Benefit to Resilience

The next generation of control systems should have a threat-based approach to develop control systems that are resilient by nature. As such, the ability to not only detect but also correlate the impact on the ability to achieve minimum normalcy is a necessary attribute. Unlike fault tolerance approaches, what follows are ill-defined interdependencies that distinguish resilience from reliability [6].

- Unexpected condition adaptation
  - Achievable hierarchy with semi-autonomous echelons – The ability to have large-scale, integrated supervisory control methodologies that implement graceful degradation.

○ Complex interdependencies and latency – Widely distributed, dynamic control system elements organized to prevent destabilization of controlled system.
- Human interaction challenges
  ○ Human performance prediction – Humans possess great capability based upon knowledge and skill but are not always operating at the same performance level.
  ○ Cyber awareness and intelligent adversary – The ability to mitigate cyberattacks is necessary to ensure the integrity of the control system.
- Goal conflicts
  ○ Multiple performance goals – Besides stability, security, efficiency, and other factors influence the overall criteria for performance of the control system.
  ○ Lack of state awareness – Raw data must be translated to information on the condition of the process and the control system components.

With the provided context, this chapter develops a hierarchical, multi-agent dynamic system (HMADS) architecture that specifically addresses unexpected condition adaptation. However, this framework also establishes the foundation to simplify the human interactions, in that it integrates the traditional human dynamics of defining goals into the HMADS itself, as part of real-time operation that will be formalized in the remaining sections. As compared to the organic development of control systems, the integration with the physical plant, and different control systems and domains, this framework looks to reduce the complexity, both the dependency and ability to adapt to degradation.

17.2 Control System Design

17.2.1 Tiers of Control

While multiple layers can be imagined for a HMADS, for the purposes of illustration three layers are suitable to identify distinct and separate functionality (defined as follows and illustrated in Figure 17.2) [8]:

- Upper layer–Management – This layer provides the overall philosophical goals and priorities for operation. The sources for this design range from management, regulators, physical constraints of the system, etc.
- Coordination layer–Coordination – This layer provides potential realignment of resources that best enables meeting the dictated philosophy. This layer drives the execution layer.
- Lowest layer–Execution – This layer provides direct monitoring of sensors and control of field devices.

17.2.2 Decomposition of Operational Philosophy into Management and Coordination Layers

Several factors influence the philosophies that govern how performance goals are set. These factors require consideration when establishing the policy of the management layer. Nevertheless, it is important to understand what these factors are and to develop a method to decompose them down to constraints of operation. In some cases, the constraints are cut and dried. However, in other cases, an interpretation must be made by those in authority. Below is a list of some factors that should be considered with control system decomposition:

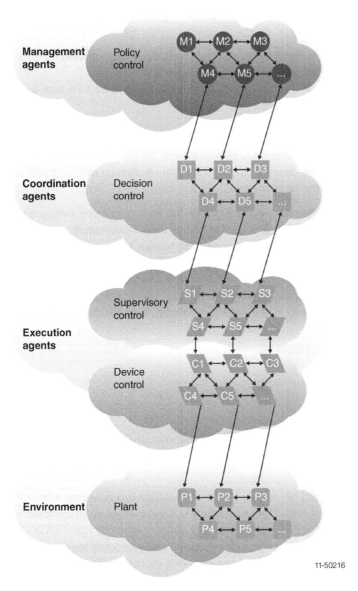

Figure 17.2 HMADS hierarchy. Source: Rieger et al. [5].

- Regulatory and ownership requirements – Considering primarily governmental agencies that regulate the operation or its products in some fashion is a key responsibility at this layer.
- Desired performance – Whether a production rate or an efficiency objective, this aspect comes from a desire to maximize profit for the organization using the control system, which includes fairness criterion.
- Physics-based limitations – The physics of the design affects the limits of the operation. While this might seem obvious, one must collectively include this when considering the trade-offs of performance.

Establishing a coordination layer that could be considered the distribution coordination organization provides a mechanism to connect management policy to execute dynamics, which establishes an adaptive capacity to maintain operational normalcy. In its simplest form, this connection can be considered set points or even a mathematical relationship that allows flexibility in operation but also constraints execution to a given set of dynamics. Unlike traditional concepts of set points established to prevent violation of operational limits, here the discussion is the development of overall resilience buffers by the nature of the design to achieve optimal performance and prevent loss of critical operation. Below are some of the aspects that can be considered in the coordination layer decomposition:

- The dynamics of the system require tracking of the optimum path or trajectory to achieve optimum stability, in addition to achieving local stability. Stated differently, the system performance remains within its constraints for operation. This implies a performance goal that considers path and endpoints.
- The ability to share, and ultimately negotiate generation, reactive power, and storage resources is limited by the scale of assets coordinated. In this case, the level of uniformity is defined as the ability to provide a necessary functionality in the fulfillment of a load. Decisions for shifting of resources can occur at different layers of the hierarchy, with control action taken at both the middle and lower layers. However, the goal of consensus is the same regardless of the level, which is to adjust resources to reach optimum performance. The difference lies in the sphere of influence of the HMADS aggregate. That is, the coordination layer has responsibility to orchestrate shifts in operation to accommodate the performance goals of the management layer.

17.2.3 Decomposition of Operational Philosophy into Execution Layer

The responsibility to operate based upon direction lies within the execution layer. Whereas in a traditional control system, these directions come from procedures, orders, and judgment, this hierarchical framework directly ties policy to execution. Some of the interactions at the execution layer include:

- Decomposition to minimize, and as a result, simplify resource interactions and complex dynamics.
- Optimize control interactions with consensus theory to achieve a common objective.
- Optimize control interactions with applicable control engineering, soft and hard computing, defined by most relevant to situation. The ultimate goal is to stabilize the shared manipulated and controlled variables.

Considering the execution layer of a three-tiered HMADS, a decomposition principle is already implied by current control system designs. That is, execution layer elements (supervisory and device control) are associated with unit operations, substations, or some optimally stabilizable entity. This can be seen from looking at power generation plants where a collection of separate unit operations make up an integral plant. In this case, the unit operation defines an area of local optimization. Within the operation, many physical variables may exist.

In a plant made up of many operations, the process of determining the optimally stabilizable entities normally results in a minimization of the interactions between individual operations. That is, normally only a few physical variables will make up the interactions between operation. For example, the flow and thermodynamic characteristics of steam from the boiler to the generator turbine must remain within a specified range, as the downstream operation is designed to be stabilized for operation within that range.

The process of breaking down operations suggests an HMADS approach, at least at the execution layer. As a transfer function would provide a conventional approach to controlling individual operations, extrapolating this concept would seem reasonable. That is, a transfer function should exist that describes the interaction and provides the dynamics of operation. From a hierarchical context, if a similar concept can be extrapolated to the coordination and management layers, existing control engineering can be extended to stabilizing an HMADS. However, while the execution layer interactions are control system centric (i.e. directly associated with supervisory and device control) the coordination and management layers are not.

17.2.4 Data-driven Methodology for Application of Tiered Control

Bayesian Belief Networks (BBN) provide one method to integrate the consideration of hierarchical control. As shown in Figure 17.3, the considerations of a management layer can introduce shifts to the operation of both the coordination and execution [8]. However, the shifts at the management layer are specific to the overall philosophy of operation. The coordination layer provides transition of the generation and transmission/distribution to achieve these goals, and the execution the application of the time-based control strategy to affect the response in the power assets. Interaction would be specific to larger impact issues, and it complements data analysis occurring at the execution layer itself that reacts to loss of sensory capacity. Summarizing, the coordination layer will need to take the high-level operational philosophy from the various contributors to provide corresponding direction to the supervisory control and device agents in the execution layer.

As per Figure 17.3, shifts in generation sources and any feeder changes consider the management layer policy changes, which considers efficiency, economics, and cost. The decision that would be influenced in the coordination layer is based upon a probabilistic table that suggests the historic likelihood to achieve the desired outcome (i.e. reduction in cost). For example, some generators cost more to operate than others. A determination of the final generator mixes for the "Negotiate shift" agent then could be based upon a straight-forward cost optimization problem. However, shifts in generation will also consider efficiency (including security) and stability agents to allow a shift. The resulting formulation will consider the priority to achieve the desired resilience. In the case of the power system, this is stability first, then economics, and finally efficiency. In addition, the state estimation (conditional probability) is now based upon data from the execution layer agents that characterizes estimations in the ability of the generator to support shifts in demand.

17.2.5 Cyber–Physical Degradation Assessment

Unlike traditional considerations of fault tolerance, which have been based upon high-confidence, reproducible physical failures, recognition of all types of failures that include cyberattack must also be considered. The basis for reliability has been in the assumption that sensing of the failure exists, can be confirmed reliable, and inherent in high-reliability industrial control system (ICS). While no hardware–software design is perfect, the reliability or availability of ICS has been greater that 99.9999% for many of these designs [9]. However, the ability to corrupt such systems through cyberattack has become a game changer. The ability to generate failures in system operation at the hardware to application layer is a reality, and when considering HMADS design, no autonomous system will be viable if such cyberattacks are not recognized and mitigated at the agent level. Such mitigation may require isolation of a specific agent, say at the execution layer, for the purposes of protecting the whole and the control functions of that agent shifted or negotiated by a different agent.

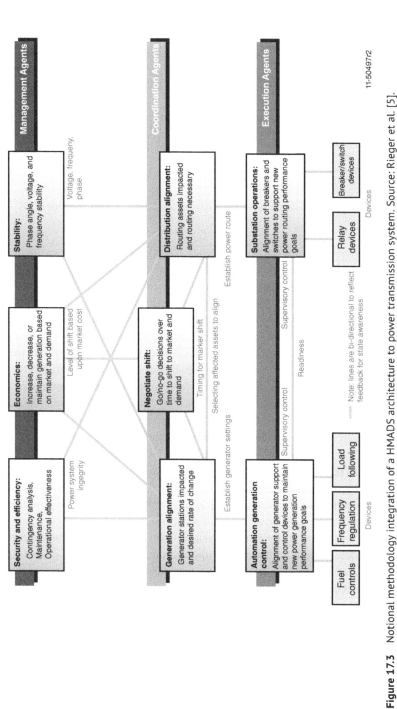

Figure 17.3 Notional methodology integration of a HMADS architecture to power transmission system. Source: Rieger et al. [5].

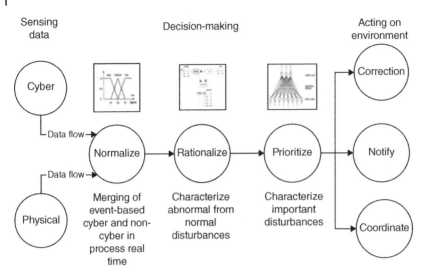

Figure 17.4 Cyber–physical degradation assessment framework. Source: Rieger and Villez [10].

A framework for correlating and integrating cyber–physical degradation assessment is shown in Figure 17.4 [10]. The sensing includes both cyber and physical data. Cyber data includes data that would be gained through network packet analysis, the hosts that are communicating, packet rate, ports used, etc. The physical data includes power-specific information, such as voltages, current, etc. This normalization aspect will be addressed in the combination of fuzzy membership functions based upon the two types of data (physical and cyber), correlated to a common sensor. While the degradation impacts to the sensor can be indicated by anomaly analysis of the physical data provided by the sensor, it may also be characterized by cyber-detection anomalies that indicate modification of the sensor data. The task of the "Normalize" aspect will be to associate the incoming data with an individual sensor or type of information. The "Rationalize" aspect will characterize anomalies. Through the comparisons of both the cyber and physical aspects, a judgment will be characterized and a statistic provided that will be proportional to the belief that the sensor has not been impacted. The "Prioritize" aspect will determine the level of importance in mixed initiative (human + automation) response to correlate to the different anomalies detected. The "Correction," "Notify," and "Coordinate" aspects will be provided by a mixed initiative response that provides corrective responses that can be at a local control loop level as well as at a supervisory level, and notify, which implies a human in the loop for appropriate response.

Figure 17.4 aligns most specifically to the execution layer, where the trustworthiness of data must be understood before using for decisions. However, one might suggest a related framework for the management and coordination layers to ensure malicious intent is not present. This will be discussed briefly in a following section. This design has been applied using all data-driven methodologies, as well as the hybrid integration of data-driven methodologies for cyber behaviors and first principle models for physical behaviors [11]. There are benefits and downsides with each. The primary benefit of the former is the ability to automate or partially automate the model development through clustering methodologies and training, but it may be susceptible to maligned datasets. Toward the latter, the exactness of the physics removes the dependency of maligned data, assuming the degradation is also recognized in the physical, but is time-consuming as models must be developed for each application.

17.3 Control System Application

17.3.1 Human Decision Integration into Management and Execution Layers

While the benefit of autonomy has historically been categorized as one that is beneficial due to the principal of reproducible action, as compared to the human, the HMADS architecture is dependent upon engendering the human reasoning into its framework. In addition, it is not exclusive of the continued interactions for humans to incorporate changing considerations for variables such as markets. The challenges associated with formulating optimal correlations of humans, such as the leadership who formulates the management layer contributions and the power system subject matter experts who formulate the coordination layer, will be ongoing. The challenges lie not in the methodologies required to codify, such as a BBN, but in the methods of establishing a comprehensive framework and weighting the responses to achieve anything that might be considered optimal in the mathematical realm.

While different methodologies of integrating decisions are available, Figure 17.5 presents a methodology where decisions are weighted higher for those that share the decision, but nonlinear for various reasons, which could be based upon experience, confidence, or other criteria. $Q(r)$ is a regular increasing monotone quantifier that establishes the relationship weighting. Given that $Q(r)$ is a quadratic function r^2, the relationship correlates a nonlinear preference to shared decision. Therefore, the implication of this method of aggregating is that the more individuals satisfied ultimately determine the decision selected. In so doing, the assumption is that more confirmation of a decision can also vet out malicious intent in any particular decision. However, if this type of technique were directly applied to a data-driven methodology, such as the BBN discussed above, further protections would need to be introduced.

A simple formulation for the human studies considered is provided in Figure 17.6. In this study, the appropriate decision makers for either the management layer or power system subject matter experts for the coordination layer would be aggregated. Normal and abnormal scenarios, not unlike is already the basis for decision making that affects the goals and operation of the power grid, are evaluated by the individuals. Some decisions will be exact, as they have a first principles or regulatory basis, and others will not, as they are more reflective of profit intent or less exactness in how to achieve goals, for example. The resulting decisions can be weighed based upon the prior methodology or some other reasoning. The final outcome is to use this as a basis for integrating the decision into the data-driven methodology, allowing for automating negotiations.

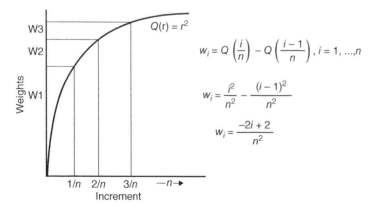

$$w_i = Q\left(\frac{i}{n}\right) - Q\left(\frac{i-1}{n}\right), i = 1, \ldots, n$$

$$w_i = \frac{i^2}{n^2} - \frac{(i-1)^2}{n^2}$$

$$w_i = \frac{-2i + 2}{n^2}$$

Figure 17.5 Decision weighting from expert opinion. Source: Rieger et al. [5].

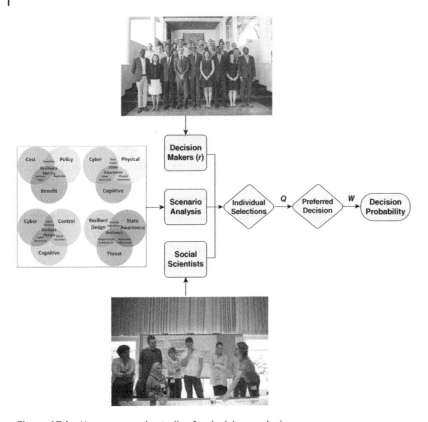

Figure 17.6 Human scenario studies for decision analysis.

17.3.2 Distributed Control and the Execution Layer Formulation

The development of distributed control systems occurred several decades ago. However, instead of the focus on the ability to adapt and be threat resilient, it allowed for centralized monitoring and decentralized action to chemical process, power generation, and like infrastructure facilities. Within the context of a HMADS, the ability to include semi-autonomous echelons allows for the ability to recognize and respond, without necessarily having to depend on ongoing centralized communications. It does this through the distribution of spheres of influence and a set of allowable operations. So, while an overarching set of goals allows for orchestration, if communications or another threat is severed, the control system and operations would continue. Distribution of the controls into agents also addresses certain challenges in control theory, such as the computation of complex physics relationships. The more complex the overall modeled process, the more complex the set of equations must be developed and solved for a control response. When a large system is decentralized, so is the sets of physics equations. As an example, consider the design of an execution layer agents, depicted in Figure 17.7, for a notional power system [12].

As inputs from the coordination layer provide the basis for supervisory control actions transmitted to the device layer, the supervisory layer will need to translate these into device actions. Considering this as a consensus problem [13, 14], the ultimate goal is to stabilize the inter-agent dynamics shown in the figure. Consensus theory provides an optimal method of achieving collective consensus, and demonstrates how agents can cooperate to achieve agreement. Considering the

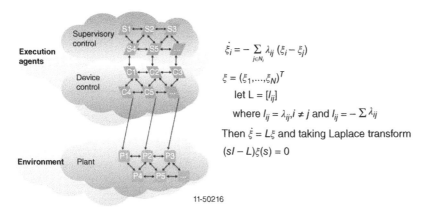

Figure 17.7   Execution layer agents and consensus equations. Source: Rieger et al. [5].

inter-agent dynamics in the form of an equation:

$$\dot{\xi} = L\xi$$

Formulation of ξ_i "belief" about global or collective consensus and λ_{ij} gains for a system-digraph relationship [8]. Given agents are "connected," Laplacian approach causes all ξ_i converge to a "consensus" belief ξ. And L is called the graph Laplacian. If the underlying graph implied by the matrix L is connected, meaning that there is a least one agent from which there is a path to every other agent, then all the variables ξ_i will converge to the same value.

Taking the Laplace transform of the result provides a resulting relationship provides basis to then apply to broader execution agent layers. The resulting shifts in operation would occur based upon this consensus, also consuming the degradation assessment information to formulate judgments in trustworthiness of belief.

In terms of a physical system, the weights between supervisory and device control layers will need to codify this interaction in terms of establishing control actions or set points within the device layer in one direction, and feedback in the other that capture the desired behavior. In terms of a real and reactive power system controls, the decision to adjust the generation to maintain the real power flow or capacitor adjustment to affect the power factor is a control, and sensor information of voltage, frequency, and phase is a feedback to ensure desired results. Where renewable generation is present, control actions can include adjustment of offsets between load requirements, storage recharge, and bulk power availability. More regional corrective actions, including bulk power system generation, could be imagined based on the inability to achieve results while recognizing a particular substation has a critical load.

17.3.3   Domain Application

A HMADS, as depicted in Figure 17.8, is composed of the three tiers: management, coordination, and execution. The top tiers are both event-based, informed, and developed from human-based decisions with the management philosophy or goals as the top tier, the optimal configuration and transition from power system subject matter experts as the second tier. The final tier is that of execution, in alignment with traditional time-based control and based upon the set points and shifts dictated by the coordination layer, which establishes supervisory control of the individual power assets, including generators, substations, real/reactive power controls, etc.

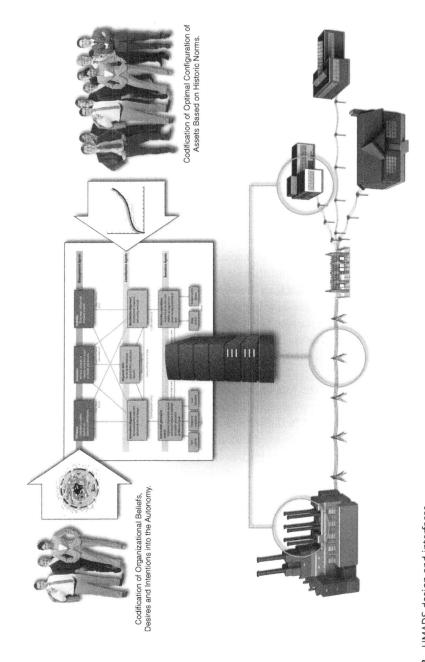

Codification of Optimal Configuration of Assets Based on Historic Norms.

Codification of Organizational Beliefs, Desires and Intentions into the Autonomy.

Figure 17.8 HMADS design and interfaces.

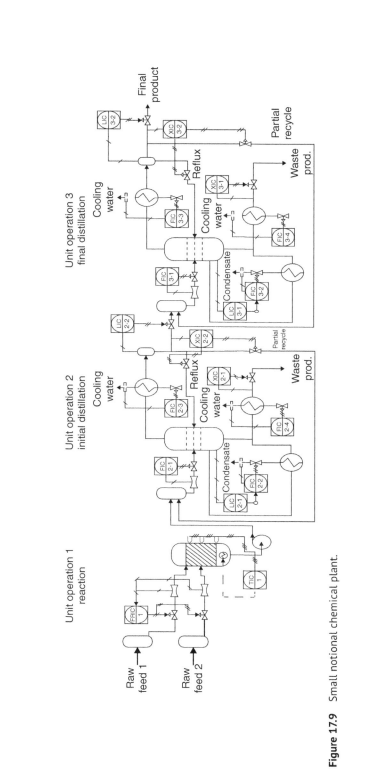

Figure 17.9 Small notional chemical plant.

The precepts of operation vary from domain to domain or even within the same domain, such as transmission and generation in the power system. The goal of these two power domains is to generate a certain level of power and transmit the power of several generation plans to wide-area loads, respectively. However, the generation goal is normally affected by transmission operation, meaning that the power output of the generator is adjusted to meet the needs of the loads, but also considering other factors, such as economics. The HMADS of the generator has a set of management goals designed to fulfill this mission. The HMADS formulation of the layers for the transmission system has higher level goals to achieve stability of operation, economics of trade-offs between generators in service, and efficiencies of operation (considering Figure 17.3). Therefore, HMADS can be designed at different levels of granularity when considering the overall goals and physics of operation and the future of the power grid decentralization – the transformation from a bulk grid to one that is built upon individual microgrids. In this form, the economics, control strategies, and interfaces would change, but the goal would be to enhance resilience as the fast control actions in response to disturbances, such as we have considered in the degradation assessment, can be affected more rapidly.

While the domain focus of this chapter and the text is the power system, the intent of the HMADS is to be broadly applicable. This consideration indicates the architecture and methodologies applied at each layer are extensible to the particular domain application of interest. For instance, a chemical plant could be selected for this application (as shown in Figure 17.9) [15]. As chemical plants are normally made up of a series of unit operations, with complex control dynamics in each but primarily only flow rate as the variable concern between each unit operation, there is a natural breakdown of agents. Each unit operation could have its own HMADS three-layers, but as the intent is to maintain flow in a certain band, so as not to overflow tanks if it is too high or if it affects equilibrium stability if too low, the management goals will be based upon the overarching goals established through the chemical plant HMADS. This would parallel our power example as the transmission system provides the overarching HMADS and the generator is one element or a unit operation.

17.4 Summary

The aspiration to achieve distributed and hybrid (time and event based) has been ongoing within control engineering circles for some time. The development of a multi-agent hierarchy naturally frames these aspirations into a foundation for future research. In achieving resilience, however, the interdisciplinary and competing goals are considered that suggest a much broader team and theory than just control engineering. This chapter provides a basis for future designs on the pathway to greater autonomy through this appropriate foundation.

Thoughtful Questions to Ensure Comprehension

1   What is the general architecture of a HMADS and the purpose of each layer?

2   What is the suggested resilience benefit of using HMADS architecture?

3   How might negotiation be achieved between agents and how are the methodologies to achieve this different between layers?

4   Why is cyber–physical degradation assessment important for any autonomous control architecture?

5 How is consensus achievement different for the management and coordination layer, as compared to the execution layer?

6 What types of modeling methodologies or approaches might be used to formalize a multi-agent architecture for practical application?

7 How might the HMADS architecture be applied to different domains?

Further Reading

An overview of a HMADS for power system applications:

Badiru, A.B. (2006). Ergonomics of Design. *Handbook of Industrial and Systems Engineering*, Chapter 24. Boca Raton, FL: Taylor and Francis.

Rehtanz, C. (2003). *Autonomous Systems and Intelligent Agents in Power System Control and Operation*. Berlin: Springer-Verlag.

A comprehensive survey of consensus theory for agents:

Olfati-Saber, R., Fax, J., and Murray, R. (2007). Consensus and cooperation in networked multi-agent systems. *Proceedings of the IEEE* 95 (1): 215–233.

References

1 Shoham, Y. (1991). AGENT-0: a simple agent language and its interpreter. In: *Proceedings of the Ninth National Conference on Artificial Intelligence*, vol. II, 704–709. Anaheim, California: MIT Press.

2 Shoham, Y. (1993). Agent oriented programming. Artificial Intelligence, vol. 60, No. 1, North-Holland, 51–92.

3 Huhns, M.N. and Singh, M.P. (1998). *Readings in Agents*. San Francisco: Morgan Kaufmann.

4 Luckand, M. and d'Inverno, M. (2001). A Conceptual framework for agent definition and development. *Computer Journal* 44 (1): 1–20.

5 Rieger, C.G., Moore, K.L., and Baldwin, T.L. (2013). Resilient control systems: A multi-agent dynamic systems perspective. IEEE International Conference on Electro-Information Technology (1–16, May 2013), EIT.

6 Petrie, C.I. (1996). Agent-based engineering, the web, and intelligence. *IEEE Expert* 11 (6): 24–29.

7 Scheler, F. and Schroeder-Preikschat, W. (2006). Time-triggered vs. event-triggered: a matter of configuration? ITG FA 6.2 Workshop on Model-Based Testing, GI/ITG Workshop on Non-Functional Properties of Embedded Systems, 13th GI/ITG Conference Measuring, Modelling, and Evaluation of Computer and Communications, Nürnberg, Germany, 1–6.

8 Rieger, C.G., Moore, K.L., and Baldwin, T.L. (2013). Resilient control systems: a multi-agent dynamic systems perspective. *IEEE International Conference on Electro-Information Technology, EIT* 2013: 1–16.

9 Pietrzyk, A. (2011). *Achieving High Availability in Process Applications*. Rockwell Automation Publication.

10 Rieger, C.G. and Villez, K. (2012). Resilient control system execution agent (ReCoSEA). 2012 5th International Symposium on Resilient Control Systems, 143–148.

11 Amarasinghe, K., Wickramasinghe, C., Marino, D. et al. (2018). Framework for data driven health monitoring of cyber-physical systems. 2018 Resilience Week (RWS), 25–30.

12 Rieger, C. and Zhu, Q. (2013). A hierarchical multi-agent dynamical system architecture for resilient control systems. 2013 6th International Symposium on Resilient Control Systems (ISRCS), San Francisco, CA, 6–12.

13 Olfati-Saber, R., Fax, J., and Murray, R. (2007). Consensus and cooperation in networked multi-agent systems. *Proceedings of theIEEE* 95 (1): 215–233.

14 Hara, S., Shimizu, H., and Kim, T.H. (2009). Consensus in hierarchical multiagent dynamical systems with low-rank interconnections: analysis of stability and convergence rates. 2009 American Control Conference. 5192–5197.

15 Rieger, C.G. (2014). Resilient control systems Practical metrics basis for defining mission impact. 7th International Symposium on Resilient Control Systems. 1–10.

18

Other Examples of Resilience Application

Meghan G. Sahakian and Eric D. Vugrin

National Security Programs, Sandia National Laboratories, Albuquerque, NM, USA

Objectives

The objectives of this chapter are to introduce and define four resilient design capacities that commonly apply across electrical power and other critical infrastructure systems, to provide examples of tangible resilience enhancement features that contribute to each of the four capacities, and to describe additional considerations, issues, constraints, and tradeoffs that commonly arise during the resilient infrastructure design process.

18.1 Introduction

The previous chapters focus on resilient architectures, evaluation, control, and associated challenges for the electrical power grid[1]. The emphasis on this "uniquely critical infrastructure"[2] is deserved because the strong dependence of all other critical infrastructure and, more generally, modern society on electrical power. The loss of power for extended and sometimes even brief periods of time can have significant economic, health, and security impacts to communities, regions, and industries.

When considering "how" to design resilient infrastructures, the electrical power grid provides many options and examples for doing so; however, many examples of good (and sometimes bad) resilient design features can be found in other infrastructure systems. With some experience and study across many infrastructure systems, one can start to identify resilient design commonalities that exist, at a conceptual level, across electrical power, and across other critical infrastructure systems. Furthermore, regardless of the infrastructure system under consideration, designers must decide which designs and architectures best achieve resilience objectives while satisfying a myriad of other infrastructure goals and constraints.

---

1 This chapter describes objective technical results and analysis. Any subjective views or opinions that might be expressed in the paper do not necessarily represent the views of the US Department of Energy or the United States Government. Sandia National Laboratories is a multimission laboratory managed and operated by National Technology and Engineering Solutions of Sandia, LLC, a wholly owned subsidiary of Honeywell International Inc., for the US Department of Energy's National Nuclear Security Administration under contract DE-NA0003525.
2 In 2013 Presidential Policy Directive 21 (PPD-21) Critical Infrastructure Security and Resilience designated energy and communications infrastructure as *uniquely critical* "due to the enabling functions they provide across all critical infrastructure sectors."

Resilient Control Architectures and Power Systems, First Edition.
Edited by Craig Rieger, Ronald Boring, Brian Johnson, and Timothy McJunkin.
© 2022 The Institute of Electrical and Electronics Engineers, Inc. Published 2022 by John Wiley & Sons, Inc.

This chapter discusses issues that infrastructure owners and operators commonly face when attempting to design resilient infrastructure systems. The discussion herein will apply to both control systems for electrical power systems and other various infrastructure systems. The chapter begins with an introduction to and definition of four resilient design capacities. At an abstract level, these capacities describe fundamental system properties that can determine the resilience of a system. The chapter further provides examples of resilience enhancement features that contribute to one or more capacities. These examples describe tangible technologies, designs, and procedures that can be implemented to contribute to the overall resilience of the system. The chapter concludes with a discussion of common trade-offs that designers must consider when planning for resilience. Rarely, if ever, is resilience the primary and sole consideration of designers. Rather, designers must find an acceptable balance between resilience and cost, regulatory, environmental, and other priorities.

Please note that this chapter does not provide a detailed, step-by-step process on how to optimally design a resilient system in a constrained decision space; the body of resilient design research is not so mature that doing so is possible yet. Still, by providing a discussion on resilient design capacities and trade-offs, this chapter will familiarize the reader with many of the issues and challenges that designers commonly face.

18.2 Resilient Design Capacities

Vugrin et al. [1] first defined resilient design capacities to be fundamental system properties that collectively contribute to or detract from a system's overall resilience. At an abstract level, these four capacities are a part of all infrastructure systems. Vugrin et al. [1] further defined resilience enhancement features to be the tangible, infrastructure design features that are implemented and put in place with the intent of improving the resilience of infrastructure systems to a variety of threats.

Vugrin et al. [1] first introduced three resilient design capacities: absorptive, adaptive, and restorative. Since the capacities were first introduced, cyber–physical systems, including the electrical power grid and other infrastructure, have increasingly become the targets of stealthy cyberattacks (e.g. Dragos Inc. [2, 3]). These instances have highlighted the need for a fourth capacity that describes a system's ability to detect threats and monitor operations to foster increased resilience, which from hereon is referred to as the anticipative capacity. This chapter describes and expands upon Vugrin et al.'s initial specification of resilient design capacities.[3]

18.3 Anticipative Capacity

A system's anticipative capacity is its ability to identify, categorize, predict, and provide advanced warning of threats to enable a rapid, proactive response. Alone, the anticipative capacity has no direct impact on mitigating threats. However, features contributing to the anticipative capacity can

---

3 Several resilience frameworks include concepts analogous to these four resilience capacities. For example, PPD-21 mentions *preparing, withstanding, adapting, and recovering* in its definition of resilience. NIST's cyber resilience systems engineering standard *NIST SP 800-160* similarly defines anticipate, withstand, recover, and adapt as resilience goals [4]. The four concepts in these two frameworks, as well as those mentioned in others, map reasonably well to the absorptive, adaptive, restorative, and anticipative capacities and can be used almost interchangeably.

act as a catalyst for other capacities to make absorptive, adaptive, and restorative features more effective. Anticipative features are typically most effective when implemented before the onset of threats and when they can recognize threats before negative consequences are realized.

Examples of resilience enhancement features that contribute to a system's restorative capacity include:

- *Intrusion detection systems (IDSs)* – IDSs monitor operations to detect possible threats and alert operators. Commonly used in cyber applications, IDSs may monitor network traffic and other behaviors to detect threats. In physical systems, IDSs may include infrared sensors, vibration detectors, cameras, and other monitors to detect physical threats.
- *State-of-health monitoring* – Whereas IDSs detect and report the presence of a threat, state-of-health monitoring reports when system functions are failing and not operating properly. Such alerting can notify operators to investigate and remediate the cause.
- *Stress testing* – During stress testing, system operators voluntarily expose their systems to threats (in a controlled manner) so that the operators can identify potential risks and how to effectively address them before the threats occur in an uncontrolled environment. These tests can take on many forms; they can be simple thought exercises or drills. Sometimes, red teams are engaged to probe defenses and find vulnerabilities. Netflix has gone so far as to use the Chaos Monkey tool to randomly terminate instances on operational systems, which ensures high reliability of its streaming platforms [5].
- *Threat intelligence* – Greater awareness of the threat environment and information sharing can notify system operators of potential threats, giving operators time to prepare for threats and possibility eliminate vulnerabilities.

18.4 Absorptive Capacity

The absorptive capacity is the extent to which a system can automatically absorb or withstand the impacts of a threat and minimize negative consequences with relatively low levels of effort. Ideally, system features contributing to the absorptive capacity are installed prior to the realization of a threat and take little effort during and after the threat realization to provide benefit to system operations. Hence, system features contributing to absorptive capacity can sometimes be sufficient for mitigating the effects of a lesser threat.

Examples of resilience enhancement features that contribute to a system's absorptive capacity include:

- Compartmentalization (segregation) – If a portion of the control systems are compromised, compartmentalization can limit the extent to which adversaries can negatively affect system operations in other portions of the system.
- Decentralization – Spreading system functions and operations to different locations can limit damage from natural disasters. Against human attackers, decentralization can increase resources and time required to cause detrimental effects.
- Excess capacity – Excess capacity can facilitate continued operations if some portion of the infrastructure system is rendered nonfunctional or is stressed. Excess capacity in electric power generation permits ramping up when demand is high. Excess bandwidth is a common design feature in communication networks to handle spikes in demand or denial-of-service attacks.
- Redundancy – Inclusion of multiple devices and components that can execute the same operation can mitigate consequences if a device fails or is attacked. Triple modular redundancy is a common design aspect in safety and reliability systems.

- Diversity – Redundancy may not be effective against a cyberattack. (If one device contains a vulnerability, another identical device likely includes that same vulnerability.) Diversity in ecosystems can lead to increased sustainability; researchers have postulated that diversity in design and components can potentially enhance resilience.
- Storage – Energy storage is becoming a more-commonly used technology to protect against disruptions of power systems. This approach is also used in manufacturing and supply chains, agriculture, and other infrastructure systems.

18.5   Adaptive Capacity

The adaptive capacity is the extent to which a system can adapt and change to nonstandard operating modes in an attempt to overcome the effects of a threat. Activation of adaptive capacity features typically requires an active, dynamic effort and may incur greater costs (measured in terms of money, manpower, time, and other resources) and be less efficient to implement than proceeding according to normal operating procedures. The benefit of their use decreases negative consequences and increases loss avoidance. Because of the greater costs, system operators will generally focus on first activating absorptive capacities features; features contributing to the adaptive capacity will only be activated if the absorptive capacity alone is insufficient to mitigate threat effects or if the perceived consequences of not activating them are deemed to exceed the cost of implementing nonstandard operating modes. Adaptive measures are generally implemented temporarily, with systems returning to normal operations when the threat is overcome.

Examples of resilience enhancement features that contribute to a system's adaptive capacity include:

- Rerouting – Using alternative transport pathways is a common approach when normal pathways are congested or unavailable. Rerouting is used in the transportation sector (e.g. railroads), communications, and other sectors, in addition to the power systems.
- Substitution – Replacing a component or operating procedure with an equivalent or comparable one can continue operations if shortages or attacks occur. Power production with alternative fuel sources (e.g. gas versus coal) is a common example in power systems; switching from wireless to wired communication may be an option to cope with a radio frequency (RF) jamming attack.
- Islanding – Islanding is a form of dynamic compartmentalization that utilities may use in response to a disruption and to prevent larger cascading failures.
- Moving target defense (MTD) and frequency hopping: MTD technologies attempt to change fundamental system attributes (e.g. IP addresses) in a coordinated manner that appears to be random to potential adversaries. The frequent changes are intended to confuse adversaries, ultimately preventing or delaying compromise of systems. MTDs are being researched to protect against physical and cyber threats. Frequency hopping is a related concept for changing RF wireless communication channels in a seemingly random manner to protect against jamming attacks.
- Conservation and rationing – When systems are stressed or under attack, cutting off nonessential functions may enable continuation of essential functions. For example, when hospitals lose power, they will frequently cancel or delay nonessential procedures to ensure the power from backup generators can be used for critical, life-saving procedures.
- Deception networks: This technology is a relatively new concept for defending against cyberattacks. Deception networks emulate real networks so well that attackers will explore and attack these decoys instead of exploiting the real targets. Furthermore, network defenders can observe attacker techniques and build defenses against them.

- Ingenuity: Though difficult to characterize and plan for human ingenuity can sometimes significantly affect the resilience of infrastructures.

18.6 Restorative Capacity

When a system is damaged or compromised, its restorative capacity is the extent to which the system can be repaired rapidly and efficiently. The restorative capacity can be considered the last line of defense because it may not be necessary if the other capacities are sufficiently effective; furthermore, activation of restorative features is generally more costly than activation of anticipative, absorptive, and adaptive features. The effects of restorative features are generally intended to be permanent and longer lasting than those from adaptive features.

Examples of resilience enhancement features that contribute to a system's restorative capacity include:

- Graceful degradation and fail-safe modes – When system operators have advanced notice of threats, they may switch to fail-safe operating modes and elect to gracefully degrade to prevent significant losses from an unplanned shut down. For example, petrochemical refineries in the Gulf Coast region often proactively shut down operations 48–72 hours before a hurricane makes landfall in the region. Doing so protects equipment from further damage from an unplanned shutdown; furthermore, the cost of not operating for one week is far less than the cost of additional equipment repairs and not operating for weeks or months while repairs are made.
- Intrusion protection system (IPS) – In cyber networks, IPSs can automatically implement defensive measures, such as closing access points and reconfiguring firewalls. Significant research is ongoing to develop autonomous methods that learn to detect and mitigate automatically cyber threats. IPS typically operates only when an IDS is used in combination with it.
- Fault detection and forensics – Fault detection technologies can notify operators about issues and enable a rapid response to repair them. Forensics, especially in cyber applications, are needed to identify the cause of failure or attack so that appropriate steps can be taken to repair the systems and prevent against future attacks.
- Reciprocal aid agreements – Utilities and industries will often participate in reciprocal aid agreements so that if one member is under duress, other members will share staff and/or equipment to limit damages and benefit the overall group. Electric power utilities, telecommunications providers, and emergency services (e.g. fire departments) often exercise these agreements.

18.7 Considerations for Resilient Design

The resilience capacities and enhancement features can be used in a variety of manners. They can be incorporated into the design of new systems or they can be used to improve and address resilience deficiencies in existing systems. However, no singular combination of resilience enhancement features is optimal across all systems. Rather, operators and designers need to determine which features best achieve system operating goals, address threats of concern, and meet budget, regulatory, and other goals. This section discusses the many issues that must often be considered when making resilient design decisions.

18.8 System of Interest

Though saying so may seem trivial, the first factor that ought to be considered in the resilient design process is the infrastructure system being designed. Infrastructure systems can be large, complex systems with many functions and outputs. Trying to ensure that every element of an infrastructure system is protected and resilient can seem, if not be, impossible. Hence, when making design decisions, understanding of the critical infrastructure elements that enable completion of the most important missions is key. Designers need to address the following mission questions:

- What is the infrastructure system's mission(s)? Oftentimes, infrastructure mission can be described in terms of goods and services that the system provides.
- Of these missions, which are most critical and of highest priority? These missions should be the primary beneficiaries of resilience enhancements, and lower priority missions may need to be excluded from analyses if resources do not permit addressing every mission element. Prioritization requires analysts' preferences be included. For electrical power systems, priority may be given to loads supporting hospitals, police stations, and other emergency services. For the gas and oil industry, customers will sometimes opt for "interruptible" contracts that specify in times of shortages and crises, these customers are of the lowest priority and may be cut off, if necessary. In some commercial industries, the highest priority may be given to the goods that bring the company the highest revenue.
- How does the infrastructure system achieve its mission? Understanding the manner in which mission is achieved will identify critical components, devices, and processes that should be the focus of design improvements.

Another consideration is to determine the various time scales that exist within the system.

- How soon will negative consequences be realized if the infrastructure mission is compromised? Power failures are realized almost immediately, possibly resulting in rapid economic, security, and health impacts if the failures are not quickly remediated (within minutes to hours). Financial systems often move at rapid timescales; fractions of a second can make large monetary differences for stock trading companies. Petroleum transmission systems operate at a slower timescale. Reserves in storage facilities (and in the transmission pipes themselves) permit for disruption recoveries to occur at a slower pace, with lesser impacts to downstream customers. As a thought experiment, try to recall the last time a black out occurred and affected your day. How much did your day differ from how you expected it would go? Now, try to remember the last time when a petroleum outage or shortage affected your day. Was your day significantly affected, aside from paying a few extra dollars at the gas pump?
- How long will the recovery take, and how do operations vary throughout the recovery duration? Extended recoveries may require rationing of resources to make sure they last for the entirety of the recovery. Additionally, activities may vary from one stage of the recovery to the next. For example, when a hurricane affects transportation networks, response and recovery begin with rerouting of shipments away from the affected area. After weather conditions stabilize, rerouting will likely continue while debris removal operations initiate. Transportation will only commence in the affected region after debris removal is sufficiently advanced and fuel is available.

Other timescale examples include:

- During a natural disaster, a refinery may be required to provide fuel throughout the long-time scale of recovering from the disaster. Although anticipative and absorptive features can provide immediate benefits, features contributing to the adaptive and restorative capacities will likely be needed to continue meeting goals throughout the extended recovery period.

- A nuclear reactor has hard, real-time communications requirements for safety controls. Rerouting communications in response to a cyberattack in a manner that adds too much latency to this time critical system would not be an acceptable option. An IDS that passively monitors the controls and alerts to anomalies might be acceptable if it does not add latency to the critical communication paths.

A last consideration in identifying the system is to consider the degree of automation within the system. Degree of automation is closely related to timescales within the system. Highly automated systems may enable a faster response; however, they may introduce additional vulnerabilities. Attackers may know or learn the effects of automated responses and use them to their benefit by creating false positives. Conversely, automation may be required to attain necessary response speeds to address coordinated cyberattacks. An IDS may be effective for alerting to the presence of a threat. However, if threat mitigation is not automated and relies on a human response, then the response and recovery may not be on a fast-enough timescale to prevent substantial damage. Cyberattacks are a common example of this temporal asymmetry between attackers and defenders. Attacks may be automated, but responses to cyberattacks in the electrical power sector and other industries typically rely on predetermined playbooks in which decisions must be made by humans that move at "human speed." A sophisticated attacker with a highly automated approach can often accomplish his/her goal well before the human response takes effect.

18.9 Threat Space

Resilience is a contextual concept and must be considered in the context of a specific disruptions and threats. A system may be highly resilient to one set of threats but vulnerable to other threats, so resilient design activities should identify the threats of greatest concern before proposing resilience enhancing features. Oftentimes, the number of scenarios postulated can seem endless, and constraints (time, resources, etc.) prevent consideration of all scenarios. In these cases, key stakeholders ought to be consulted to identify and prioritize which threat scenarios should be included in design efforts. With the selection of threat scenarios, designers can then match the mitigations that best address those scenarios.

Development of the scenarios should include specification of the disruption, the effect that the disruption has on the system, timing of effects, and system response mechanisms – planned or implemented. Common disruptions considered in resilience analyses consist of natural disasters, accidents, and malevolent events. Examples of threat space considerations and how they affect the choice of resilience enhancing features include:

- Buildings concerned about power loss during storms and flooding should not house backup generators and related equipment in basements or floors that could experience flooding.
- HAZMAT suits and protective equipment necessary to respond to dangerous chemical explosions should not be located in the path of expected chemical plumes.
- Information technology networks concerned about insider threats should not rely solely on firewalls preventing unauthorized, external access. These networks likely require additional network analytics that can recognize anomalous behaviors by users with credentialed access.

18.10   Operational Constraints

Resilient design activities often focus on the effect that proposed mitigations have upon reducing negative consequences from threats. However, in practice, designers need to consider the effects of the mitigations on other operational considerations. Common considerations include:

- Budget and cost – What is the cost for the proposed features? Cost estimates should include both upfront investment and the cost to operate, maintain, and implement when threats are realized. These costs need to be weighed against available budgets, the potential benefits from the mitigation (which may include loss avoidance), decreased insurance premiums, competitive advantage gained from decreased down time, etc. For example, building a fully redundant system may increase resilience; however, doing so is generally cost prohibitive.
- Reliability – Electrical power utilities (and other industries) are regulated according to rigid reliability standards. Design modifications intended to enhance resilience that would negatively affect reliability of power systems would likely be rejected immediately.
- Safety – Safety is another common requirement (and regulatory basis) that must be satisfied. Safety is the primary requirement of nuclear power and chemical manufacturing facilities, so resilient design features must be proven to ensure continued compliance with safety requirements.
- Other regulations – Beyond reliability and safety, many industries must comply with additional regulations. Banking and healthcare industries have privacy standards that they must maintain, so the proposed cyber-resilience measures must not expose client information. Food preparation companies and supply chains must meet health standards; therefore, rerouting efforts that keep foods in transit for periods of time that violate health standards would not be effective.
- Size, weight, and power (SWaP) – Cars, planes, satellites, and other vehicles have stringent SWaP restrictions that resilient design efforts must consider. Inclusion of sophisticated, computationally intense IDS in satellites with limited battery capacities would likely be rejected due to SWaP constraints.

18.11   Summary

Four resilience capacities (anticipative, absorptive, adaptive, and restorative) contribute to an infrastructure's resilience. Resilience enhancement features are the tangible technologies, designs, and procedures that are implemented, contribute to one or more of the capacities, and determine the overall resilience of the system.

Design of resilient systems does not include a "one-size-fits-all" approach. Selection of the right resilience enhancement features requires consideration of several factors. Designers need to consider the infrastructure system's core missions, the potential threat space, and a variety of operational constraints. Development of formal, resilient design processes is a subject of continuing research, and this section detailed resilient design considerations.

Thoughtful Questions to Ensure Comprehension

1   Give an example of a resilience enhancement feature that can be viewed as contributing to more than one resilience capacity.

2   Resilience enhancing features may be contradictory (i.e. they may enhance one capacity while degrading another). Give an example of such a feature.

3 It may not be financially possible to invest in all resilience enhancing features. Describe a method you would use to choose a subset of features that you would propose to an infrastructure manager.

4 Cascading failures between system components is an important resilience consideration when defining the system of interest. Give an example of an infrastructure with the potential for cascading failures. What resilience enhancing features could benefit your example?

5 Give an example of two contrasting infrastructure systems: one that needs to obtain a certain level of performance at all costs and one that can tolerate degraded performance in order to conserve resources.

6 Give an example of a system where humans should be considered part of the infrastructure system itself. How does this affect which resilience enhancing features may be considered?

7 Systems can be studied at a variety of different scales. Suppose the threats and proposed resilience enhancing features to the system have already been chosen, what are some questions one could ask to determine the proper level of system granularity to consider for the analysis?

8 An important consideration when considering resilience enhancing features is identifying when the response/recover is deemed to be complete. Give an example identifying these criteria for a system under a specific disruption.

9 Many experiments and analyses include sources of uncertainty. Give some examples of uncertainty as they relate to resilience enhancing features discussed in this chapter.

Further Reading

Biringer, B., Vugrin, E., and Warren, D. (2013). Discussion and application of resilience capacities: Chapter 10. In: *Critical Infrastructure System Security and Resiliency*. Boca Raton, Florida: CRC Press.

Ross, R., Pillitteri, V., Graubart, R. et al. (2019). *Developing Cyber Resilient Systems: A Systems Security Engineering Approach. Cyber Resilience Framework*: NIST SP 800-160, vol. 2.

Rieger, C.G., Gertman, D.I., McQueen, M.A. (2009). Resilient control systems: next generation design research. 2nd IEEE Conference on Human System Interaction, Catania, Italy (May 2009).

The White House (2013). *Presidential Policy Directive 21 (PPD-21) Critical Infrastructure Security and Resilience*. The White House.

References

1 Vugrin, E.D., Warren, D.E., and Ehlen, M.A. (2011). A resilience assessment framework for infrastructure and economic systems: quantitative and qualitative resilience analysis of petrochemical supply chains to a hurricane. *Process Safety Progress Journal* 30: 280–290. https://doi.org/10.1002/prs.10437.

2 Dragos, Inc. (2017). CRASHOVERRIDE Analysis of the Threat to Electric Grid Operations. Technical Report Version 2.20170613, Dragos Inc., Hanover, MD.

3 Dragos, Inc. (2017). TRISIS Malware Analysis of Safety System Targeted Malware. Technical Report version 1.20171213. Dragos Inc., Hanover, Maryland.

4 Ross, R., Pillitteri, V., Graubart, R. et al. (2019). *Developing Cyber Resilient Systems: A Systems Security Engineering Approach*, vol. 2. Gaithersburg, Maryland: National Institute of Standards and Technologies, NIST SP 800-160.

5 Netflix (2019). Github – Netflix/chaosmonkey: Chaos Monkey is a resiliency tool. https://github .com/Netflix/chaosmonkey (accessed 9 December 2019).

Part VII

Conclusions

Part VII, "Conclusions," summarizes the book and challenges the students to consider the future and the new science of resilience.

    Chapter 19: The previous chapters of this book take an interdisciplinary approach to discussing resilience in control systems, designed to encourage students from diverse disciplines to consider this critical concept. This chapter concludes that resilience is not a design layer. Instead, it is a philosophy. This chapter summarizes the challenges of designing resilient control systems and the relationship between humans and automation. Autonomy is not the final goal, but one tool to achieve a resilient system.

Resilient Control Architectures and Power Systems, First Edition.
Edited by Craig Rieger, Ronald Boring, Brian Johnson, and Timothy McJunkin.
© 2022 The Institute of Electrical and Electronics Engineers, Inc. Published 2022 by John Wiley & Sons, Inc.

19

Summary and Challenge for the Future

Craig Rieger

National and Homeland Security, Idaho National Laboratory, Idaho Falls, ID, USA

19.1 Introduction

The basis for this text is to provide an interdisciplinary perspective for students on the challenges of resilient control systems. The power grid provides a relevant domain for application of control system concepts, so students were provided an operational understanding for consideration of how resilience principles might be applied. It is intended that this text will provide a perspective that inspires the student to dig into areas that are outside of their discipline to develop and respect the contributions from others. The result is to understand that resilience is not just about an understanding of design considerations, but instead a philosophy of teaming and optimized results. For the future of control systems, this is the road to autonomy.

19.2 Resilience Is Not a Design Layer, It Is a Philosophy

How often have developers stated that they will design a new advanced energy or manufacturing system technology, which will also be secure and resilient? The latter terms are added to fulfill some desire to appear that everything is considered in the design or to use the latest modern term. But by the very structuring of the sentence, it is altogether unlikely that these considerations are at the core of the developed technologies. With this in mind, one might ask how secure and resilient will the technology be? The answer will depend upon the design. For security, it will likely inherit existing, well-recognized security methods that include border defenses, password protections, selective encryption, etc. While that might provide some level of comfort, when it comes to resilience you will more likely receive very selective or no resilience benefit. Resilience technologies, to meet the definition [1], have state awareness of degradation and adapt or transform to meet desired system performance.

 As we consider the resilience of our infrastructure to cyberattack, we must directly consider the control systems that provide a human-in-the-loop mechanism to monitor and control the power, water, and other utilities we depend upon. As the desire to automate and achieve efficiencies of labor and operation has grown, so has the investment in control systems that allow for integrating different operations, facilities, utilities, and infrastructures. However, the evolutionary integration of control systems has led to complexities of failure, human interaction, and security vulnerability. As control systems evolve toward greater autonomy, reducing/changing the role of the human,

Resilient Control Architectures and Power Systems, First Edition.
Edited by Craig Rieger, Ronald Boring, Brian Johnson, and Timothy McJunkin.

the need to consider resilience becomes more profound. Autonomous systems can react quickly to anomalous conditions, ensuring we have power even if a transformer fails. However, it can also cause a quick escalation to a cascading fault if the autonomy has been corrupted by cyberattack or unrecognized failure. Enabling the human in the loop will be necessary throughout, ensuring their ability to adapt to anomalous conditions the control system cannot.

The next generation of control systems should have a threat-based approach to develop the systems that are resilient by nature. As such, the ability to not only detect, but correlate the impact on the ability to achieve minimum normalcy is a necessary attribute. What follows is a synopsis of the current complexity challenges that will need to be addressed in future control systems designs. First, current automation environments are the result of an organic interconnection of control systems and the inability to recognize and prevent resulting, unrecognized faults. Second, benign human error as the result of data overload and lack of information is an ongoing issue, and for the malicious human, current perimeter protections are insufficient and not designed to adapt rapidly to attacks in order to prevent compromise. Finally, current control systems have multiple performance goals, but without the necessary identification and prioritization can lead to undesirable response from both the human operation and the automation design [2].

19.3   Resilience and the Road to Autonomous Systems

Many companies and individuals discuss the advancement of autonomy for autonomy's sake. The benefits seem obvious, from greater energy efficiencies to greater safety – the things we need are all done automatically. And speaking of that, why not combine all functions into one black box? Great, the all-in-one internet of things device. While some level of consideration is taken for ensuring reliability, unwittingly the lack of resilience has opened Pandora's box for less-obvious black swan events.

The cyber hacker rubs his hands together in deep anticipation to find out what the automation will now do with one small tweak. No longer needing to work on understanding the big picture, the increased complexity of the autonomy does the work for him. Even without the threat of a hacker launching a cyberattack, we have the failures from component degradation and software bugs – how do they interact and where might the holes in the Swiss cheese holes line up to bypass system protections? Enter now the human operator who has even less understanding of how this black box contraption works.

The obvious predecessor to autonomy is a consideration of resilience. Not only is an understanding of degradation necessary to be distributed in the architecture and evaluated across the architecture, trade-offs between aspects such as efficiency are necessary to ensure resilience – the assurance that the contraption will work in spite of the unexpected. Perhaps the designers will recognize this limitation at some point and understand that it is not autonomy they are after but resilience, which builds a platform where autonomy can be logically placed.

References

1 Rieger, C.G. (2014). Resilient control systems practical metrics basis for defining mission impact. 7th International Symposium on Resilient Control Systems (August 2014).

2 Rieger, C.G. (2010). Notional examples and benchmark aspects of a resilient control system. 3rd International Symposium on Resilient Control Systems (August 2010).

Index

Resilient Control Architectures and Power Systems, First Edition.
Edited by Craig Rieger, Ronald Boring, Brian Johnson, and Timothy McJunkin.
© 2022 The Institute of Electrical and Electronics Engineers, Inc. Published 2022 by John Wiley & Sons, Inc.

**IEEE Press Series
on Power and Energy Systems**

Series Editor: Ganesh Kumar Venayagamoorthy, Clemson University, Clemson, South Carolina, USA.

The mission of the IEEE Press Series on Power and Energy Systems is to publish leading-edge books that cover a broad spectrum of current and forward-looking technologies in the fast-moving area of power and energy systems including smart grid, renewable energy systems, electric vehicles and related areas. Our target audience includes power and energy systems professionals from academia, industry and government who are interested in enhancing their knowledge and perspectives in their areas of interest.

Printed and bound by CPI Group (UK) Ltd, Croydon, CR0 4YY

16/04/2025

14658427-0001